# Phytomanagement of Fly Ash

# Phytomanagement of Fly Ash

**VIMAL CHANDRA PANDEY**

Department of Environmental Science,
Babasaheb Bhimrao Ambedkar University,
Lucknow, India

ELSEVIER

Elsevier
Radarweg 29, PO Box 211, 1000 AE Amsterdam, Netherlands
The Boulevard, Langford Lane, Kidlington, Oxford OX5 1GB, United Kingdom
50 Hampshire Street, 5th Floor, Cambridge, MA 02139, United States

**Notices**

Knowledge and best practice in this field are constantly changing. As new research and experience
broaden our understanding, changes in research methods, professional practices, or medical
treatment may become necessary.

Practitioners and researchers must always rely on their own experience and knowledge in
evaluating and using any information, methods, compounds, or experiments described herein. In
using such information or methods they should be mindful of their own safety and the safety of
others, including parties for whom they have a professional responsibility.

To the fullest extent of the law, neither the Publisher nor the authors, contributors, or editors,
assume any liability for any injury and/or damage to persons or property as a matter of products
liability, negligence or otherwise, or from any use or operation of any methods, products,
instructions, or ideas contained in the material herein.

**British Library Cataloguing-in-Publication Data**
A catalogue record for this book is available from the British Library

**Library of Congress Cataloging-in-Publication Data**
A catalog record for this book is available from the Library of Congress

ISBN: 978-0-12-818544-5

For Information on all Elsevier publications
visit our website at https://www.elsevier.com/books-and-journals

Publisher: Joe Hayton
Acquisitions Editor: Marisa LaFleur
Editorial Project Manager: Vincent Gabrielle
Production Project Manager: Surya Narayanan Jayachandran
Cover Designer: Christian J. Bilbow

Typeset by MPS Limited, Chennai, India

Working together
to grow libraries in
developing countries

www.elsevier.com • www.bookaid.org

# Dedication

Dedicated to my beloved wife—Deepti Pandey

# Contents

# About the author

## Dr. Vimal Chandra Pandey

Dr. Vimal Chandra Pandey is currently a CSIR-Pool Scientist at the Department of Environmental Science, Babasaheb Bhimrao Ambedkar University, Lucknow, India. He also worked as a Consultant at CSTUP and a DST-Young Scientist at the Plant Ecology and Environmental Science Division, CSIR-National Botanical Research Institute, Lucknow, India. He is well recognized internationally in the field of phytomanagement of fly ash/polluted sites. His research interests include phytoremediation, revegetation, and restoration of fly ash dumpsites, red mud dumpsites, heavy metal-polluted sites, degraded lands through ecologically and socioeconomically valuable plants with minimum inputs, least risk, and low maintenance toward nature sustainability in terms of raising rural livelihoods and maintaining ecosystem services. He is a recipient of a number of awards/honors/fellowships such as the CSTUP-Young Scientist award, the DST SERB-Young scientist award, the UGC-Dr. DS Kothari Postdoctoral Fellowship, and the CSIR-SRA award and a member (MNASc) of National Academy of Sciences, India (NASI), a Commission Member of IUCN-CEM Ecosystem Restoration, and a member of the BECT's Editorial Board. He has published more than 50 peer-reviewed articles including review and research papers in reputed international journals with high impact factor and 11 book chapters. He is the author and editor of three books published by Elsevier with several more forthcoming. His high-quality research papers have good citation and have been cited worldwide by researchers within short time. He is also serving as a potential reviewer for several journals from Elsevier, Springer, Wiley, Taylor & Francis, etc., and received Elsevier Reviewer Recognition Awards from the editors of many journals. Email address: vimalcpandey@gmail.com; ORCID iD: https://orcid.org/0000-0003-2250-6726; Google Scholar: https://scholar.google.co.in/citations?user = B-5sDCoAAAAJ&hl.

# Foreword

Disposal of heaps of fly ash generated during the production of electricity is a global problem. As it contains heavy metals and other chemicals that are injurious to health, it is not safe for the environment. Various uses of fly ash such as in building and pipes for the disposal of wastes and others have been suggested but are not viable due to heavy transportation costs. In recent years, efforts are being made for use of fly ash as resource for the production of biomass and plant species of economic value. In this context, the *Phytomanagement of Fly Ash* provides advanced knowledge of all aspects of fly ash management through green technologies in the form of chapters that reveal applied and viable ways in managing and remediating fly ash through adaptable plants with economic returns. The book provides ecologically, friendly, and cost-effective solutions to decontaminate fly ash–polluted sites that can also yield biodiesel, aromatic oils, fibers, biofortified products, and pulp paper biomass.

The book also focuses on novel topics such as afforestation on fly ash catena, potential sink for carbon sequestration, and fly ash ecosystem services. The adaptable and valuable plant-based phytoremediation of fly ash dumpsites can be a fantastic strategy toward environmental and economic sustainabilities that have a productive future across the nations. It can help in meeting the sustainable development goals and contribute to the bioeconomy of the nation.

I appreciate and compliment the efforts of Dr. Vimal Chandra Pandey, who has made significant contributions in this area, for bringing out this valuable edition through a leading global publisher, Elsevier Publishing. The book is easy to read with 10 chapters covering various aspects of the fly ash management through ecologically and socioeconomically important plants. The book has global importance, because most of the countries are dependent on coal-based thermal power plants for their energy. I am sure that the book will be an asset for students, researchers, environmental scientists, coal-based industries, practitioners, policy makers, stakeholders, and entrepreneurs alike.

**Prof. P.K. Seth**
NASI Senior Scientist Platinum Jubilee Fellow
Former CEO, Biotech Park
Former Director, Indian Institute of Toxicology Research
Lucknow, India

# Preface

*Phytomanagement of Fly Ash* is a new book that provides all aspects of green technologies that allow sustainable phytoremediation of fly ash deposits, revitalization of these polluted sites, and new opportunities and challenges in fly ash utilization, primarily in agriculture. This book provides current state of research of phytoremediation of fly ash catena with advances in knowledge on fly ash—soil-plant system, application of fly ash with different amendments, bioremediation by microbes, carbon sequestration, and commercial crop-based phytoremediation. Currently, this book presents sustainable phytomanagement of fly ash, offering a great prospect in developing policy frameworks to promote cleanup of polluted sites worldwide. Therefore eco-efficient and cost-effective phytomanagement of fly ash are essential for the selection of remediation technology, because they connect the three pillars of sustainability, such as environment, economy, and society. This valuable book will support researchers, students, environmental scientists, regulatory agencies, practitioners, policy makers, and stakeholders.

The first chapter provides general information to help readers better understand fly ash properties, threats, management, and its multiple uses. The second chapter describes the main limitation of fly ash for plant growth and establishment as well as the limitation of microbial remediation of fly ash deposits, highlighting holistic approach. The third chapter provides a thorough overview of the impact of fly ash incorporation in the soil-plant system, soil amelioration, and coapplication of fly ash with different amendments, whereas fourth chapter is focused on the beneficial role of fly ash in paddy fields. The detailed review about microbial processes and enzymatic activities in fly ash—aided soil is presented in the fifth chapter. The sixth chapter focuses on fly ash application in the reclamation of degraded soils. Ecological engineering for effective afforestation of fly ash catena with detailed information about flora, fauna, and microbes is presented in the seventh chapter, whereas the eighth chapter is focused on fly ash deposits as potential sink for carbon sequestration. Finally, fly ash ecosystem services, science, and policy lessons for the fly ash catena, risk assessment, and utilization of commercial crops in phytomanagement of fly ash were described in the 9th and 10th chapters.

**Vimal Chandra Pandey**

# Acknowledgments

I sincerely wish to thank Candice Janco (Acquisitions Editor), Vincent Gabrielle and Billie Jean Fernandez (Editorial Project Manager), and Indhumathi Mani (Copyrights Coordinator) from Elsevier for their excellent support, guidance, and coordination of this fascinating project. I would like to thank all the reviewers for their time and expertise in reviewing the chapters of this book. Finally, I must thank my respective families for their unending support, interest, and encouragement, and apologize for the many missed dinners!

# CHAPTER 1

# Fly ash properties, multiple uses, threats, and management: an introduction

## Contents

## 1.1 Introduction

Most of the nations are largely dependent on the coal-based thermal power stations to meet its energy demands. Global coal production is approximately $3.5 \times 10^9$ tons/year (Australian Uranium Association, 2007). It is estimated that world coal consumption will increase by 49% from 2006 to 2030 (International Energy Outlook, 2009). In India, the coal fulfills 70% of the total energy demand that is generally used in thermal power stations. This exercise in India will continue in the near future due to huge coal assets (Lior, 2010), which is projected to be $287.0 \times 10^9$ tons (Ram and Masto, 2014). Countrywise coal consumption for electricity generation is presented in Table 1.1. The coal combustion generates electricity in the thermal power stations, and thus produces solid wastes that are known as fly ash (FA). Global FA production is vast and projected to go beyond 750 million tons/year, but merely less than 50% of its generation is used (Izquierdo and Querol, 2012).

*Phytomanagement of Fly Ash*
ISBN: 978-0-12-818544-5
DOI: https://doi.org/10.1016/B978-0-12-818544-5.00001-8
1

**Table 1.1** Countrywise coal consumption (million tons) by thermal power stations in 2009 (International Energy Agency, 2013).

| S. no. | Country | Lignite | Bituminous coal | Subbituminous coal |
|---|---|---|---|---|
| 1 | United States | 58.04 | 332.09 | 438.74 |
| 2 | India | 28.14 | 393.79 | – |
| 3 | China | – | 1439.5 | – |
| 4 | EU 27 | 289.07 | 115.27 | 2.02 |
| 5 | Australia | 644.6 | 27.40 | 28.38 |
| 6 | Turkey | 62.64 | 5.781 | 0.19 |
| 7 | Japan | – | 86.36 | – |
| 8 | Canada | 10.08 | 3.58 | 28.75 |
| 9 | South Africa | – | 121.39 | – |
| 10 | Indonesia | – | – | 33.52 |

Countrywise FA generation (million tons) and utilization (%) are presented in Fig. 1.1A and B. In India, the FA production by the coal-based thermal power stations is estimated to increase to 300 million tons/year by the year 2016–17 (Skousen et al., 2012). After the fast depletion of higher quality coal, the use of coal with lower calorific value and high ash content for burning in the thermal power stations to meet energy demands would result to manifold increase in FA production. FA has been recognized as the most abundant by-product of coal combustion that poses several environmental and human health problems due to the presence of pollutants (Pandey et al., 2009a; Pandey et al., 2011). Besides, considerable research effort has been focused to utilize FA in construction, as a low-cost adsorbent for the removal of flue gas (generally includes particulates, mercury, CO, $CO_2$, $NO_x$, and $SO_x$), metal(loid)s, and organic compounds; lightweight aggregation; backfilling of coal mine voids; road embankment; and zeolite formation. Moreover, FA is explored as potential soil amendments to improve the degraded soil quality and fertility (Pandey et al., 2009a; Pandey and Singh, 2010). Converting FA into zeolites not only alleviates the disposal problem but also converts a waste material into a marketable commodity. Besides multiple uses of FA, many millions of tons of FA are still disposed of in landfills/lagoons/ponds at nearby thermal power stations all over the world. Therefore the FA can be considered as the world's fifth-largest raw material resource (Mukherjee et al., 2008). The present chapter provides information on the FA in terms of its properties, pollutants, pollution, threats, other problems, multiple uses, and management.

(A)

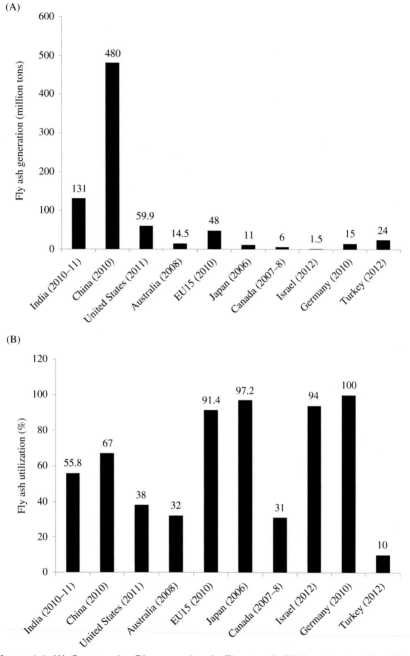

(B)

**Figure 1.1** (A) Countrywise FA generation (million tons). (B) Countrywise FA utilization (%) (Central Electricity Authority, 2011; Tang et al., 2013; American Coal Ash Association, 2011; ADAA, 2009; Caldas-Vieira and Feuerborn, 2013; Japan Coal Energy Centre, 2007; CIRCA (Association of Canadian Industries Recycling Coal Ash), 2008; NCAB, 2012; VGB Power Tech, 2010; Uyanik and Topeli, 2012).

## 1.2 Fly ash

There are two major types of coal ash, such as FA and bottom ash (BA), which are generated in thermal power stations. Mainly, the FA constitutes 80% of the total ash produced in coal-based thermal power stations. It exits from a combustion chamber in the flue gas and is captured by the air pollution control equipment such as electrostatic precipitators and fabric filters; it mostly consists of very small particles with lighter fraction and high surface area and a higher proportion of metal(loid)s on the surface. Usually, the range of FA color is tan, gray, and black, which depends on the unburned carbon amount in the ash (Kassim and Williams, 2005). The FA fraction is more reactive chemically and has a finer texture (0.01−100 μm) than the BA fraction. The BA fraction is heavier and coarse (>100 μm), which falls down in the furnace bottom. BA consists of agglomerated ash particles that are too large to be carried in the flue gases and instead adhere to the boiler walls or fall through open grates into an ash hopper at the bottom of the boiler; they generally consist of larger chunks of relatively inert material. The BA particles are mostly unequally shaped, sand-to-gravel sized, and angular (Meawad et al., 2010). Both the ashes are mixed with water to form slurry, which are normally transported to ash ponds through iron alloy pipelines. Hereafter, both the disposed ashes in the ash pond will be referred to as FA in this chapter and in the following chapters. The FA has chemical and physical properties that make it useful products in a variety of applications. FA is classified as class C and class F. Class C of FA has higher amount of CaO and is found in lignite or subbituminous coal, while class F of FA has lower amount CaO and is present in bituminous coal. On the basis of ASTM standards, bituminous coal and subbituminous coal produce class F ash, and lignite produces class C ash with a high degree of self-hardening capacity (ASTMC618). The classifications of FA to determine its logical uses are mostly based on the contents of Si, Fe, Ca, and Mg oxides as well as on the amorphous and reactive water-soluble phases in FA (Dewey et al., 1996). In this context, Vassilev and Vassileva (2006) also presented an innovative method that is based on the origin, chemical composition, phase-mineral composition, FA behaviors, and other characteristics.

The physicochemical and mineralogical properties of FA are reviewed by several researchers (Page et al., 1979; Adriano et al., 1980; Carlson and Adriano, 1993; Mohapatra and Rao, 2001; Bhattacharjee and Kandpal, 2002). The comparison of FA characteristics produced from lignite, bituminous coal, and subbituminous coal is given in Table 1.2. In general, the FA properties can vary and depend on the combustion method, coal source, coal quality, type of emission control devices, particle size, storage and handling methods, and age and extent of weathered ash (Singh and Kolay, 2002; Jala and Goyal, 2006; Pandey and Singh, 2010; Ram and Masto, 2014). In general, FA consists of glasslike fine particles that are mostly sphere-shaped, either hallow or solid, and normally amorphous (glossy) in nature. Sphere-shaped particles make the most portion of the FA, particularly in the finer fractions. FA particles are hollow, empty spheres (cenospheres) filled with smaller amorphous particles and crystals (plerospheres). Chemically, FA predominantly consists of $SiO_2$ (silica), $Al_2O_3$ (alumina), $Fe_2O_3$ (iron oxide), $CaO$ (calcium oxide), $MgO$ (magnesium oxide), $K_2O$ (potassium oxide), $Na_2O$ (sodium oxide), $SO_4^{2-}$ (sulfate), and unburned carbon (Pandey and Singh, 2010; Singh et al., 2011a). Lignite FA has high $CaO$, $SiO_2$, $Al_2O_3$, $SO_3$, and $MgO$ compared with bituminous/subbituminous ash (Table 1.2). The FA is a diverse combination of crystalline and amorphous phases, which has lower bulk density, higher electrical conductivity, higher specific surface area, lower cation exchange capacity, and higher moisture retention capacity than normal soil (Pandey and Singh, 2010; Ram and Masto, 2014). In general, the specific surface area of FA ranges from 170 to 1000 $m^2 \, kg^{-1}$, while its specific gravity may vary from 2.1 to 3.0 (Roy et al., 1981; Tolle et al., 1982; Mattigod et al., 1990).

**Table 1.2** Comparison of FA generated from thermal power stations through combustion of lignite (pH = 11.0) and coal (pH = 6.0−11.0).

| S. no. | | Lignite ash[a] | Bituminous/subbituminous coal ash[b] |
|---|---|---|---|
| 1 | $SiO_2$ | 48.4 ± 0.99 | 38.0−63.0 |
| 2 | $Al_2O_3$ | 29.8 ± 0.81 | 27.0−44.0 |
| 3 | $Fe_2O_3$ | 5.4 ± 0.68 | 3.3−6.4 |
| 4 | $CaO$ | 7.9 ± 0.53 | 0.2−0.8 |
| 5 | $MgO$ | 2.6 ± 0.30 | 0.01−0.5 |
| 6 | $K_2O$ | 0.2 ± 0.02 | 0.04−0.9 |
| 7 | $Na_2O$ | 0.40 ± 0.03 | 0.07−0.43 |
| 8 | $SO_3$ | 2.8 ± 0.22 | − |
| 9 | $TiO_2$ | 1.4 ± 0.09 | 0.4−1.8 |
| 10 | $P_2O_5$ | 0.4 ± 0.03 | − |
| 11 | LOI | 5.7 ± 0.22 | 0.2−3.4 |

[a]Singh et al. (2011a).
[b]Ram and Masto (2010).

The Si, Al, and Fe are the key elements of FA along with significant fractions of Ca, K, Na, and Ti. The FA is also noticeably rich in metal (loid)s such as Pb, As, Hg, Co, Ni, Zn, Cd, and Cr, which are concentrated in the minor ash particles (Davison et al., 1974; Adriano et al., 1980). The dominant cation of FA is Ca, followed by Mg, Na, and K (Mattigod et al., 1990). The Al element is frequently bound in insoluble aluminosilicate structures, which significantly inhibits its biological toxicity. Thus from an agricultural point of view, the FA has micronutrients such as Mn, Fe, Zn, Co, Cu, Mo, and B and macronutrients including K, P, Ca, S, and Mg; in which most of them are the essential plants nutrients. The FA pH varies from 4.5 to 12.0, and is mainly dependent on the sulfur content of coal. Its pH is linearly related to the $CaO/SO_4$ ratio or the CaO content (Mattigod et al., 1990). If the Ca/S ratio in FA is less than 2.5, then it will be acidic FA, while when the Ca/S ratio is more than 2.5, then it will be alkaline FA (Anisworth and Rai, 1987). The majority of the FAs are alkaline all over the world including India. The FA has largely amorphous aluminosilicate spheres that are similar to soil particles (El-Mogazi et al., 1988). The FA has been utilized as a potential soil ameliorator (Pandey et al., 2009a; Pandey and Singh, 2010). The Ca-rich, alkaline FA has been explored to neutralize agriculture acidic soils (Mishra and Shukla, 1986; Taylor and Schuman, 1988; Pandey and Singh, 2010) as well as to facilitate revegetation of degraded lands. In general, the humus and N are absent in FA and can be added with the help of organic amendments. Besides, the FAs also contain some metal(loid)s such as Hg, As, Cd, Pb, Se, Cr, Ni, Mo, Cu, Be, and Co, radionuclides ($^{238}$U, $^{226}$Ra, $^{232}$Th, and $^{40}$K), and organic pollutants including polycyclic aromatic hydrocarbons (PAHs), polychlorinated dibenzo-$p$-dioxins (PCDDs), and polychlorinated dibenzofurans (PCDFs). The content of these elements in FAs mostly depends on the coal type, coal source, FA particle size, combustion conditions, volatilization−condensation mechanisms, occurrence of important elements, and their association with the organic and inorganic parts of the coal (Vassilev et al., 2005; Ward and French, 2006; Jiménez et al., 2008; Frandsen, 2009; Pandey and Singh, 2010; Singh et al., 2011a).

## 1.3 Fly ash pollutants

Coal-based electricity production comes with substantial costs to our environment and human health. In general, FA contains some toxic pollutants such as metal(loid)s, radioactive elements, organic pollutants, and atmospheric emissions that contaminate air, soil, and water reservoirs, and thus affecting the flora and fauna nearby coal thermal power stations. FA and its pollutants are also damaging the agricultural activities surrounding the coal thermal power station. These FA pollutants will be explored separately in the following subsections.

*Metal(loid)s*—Almost every plant essential element exists in coal and thus in its by-products such as FA. On the basis of their concentration in FA, they are grouped in three categories. These are major, minor, and trace elements as $>1\%$, $0.1\%-1\%$, and $<0.1\%$, respectively (Vassilev et al., 2005). Particularly, Cd, As, Zn, B, Cu, Pb, Co, and Ni are predominantly found in the finer FA particles (Clarke, 1993), whereas the extremely volatile Hg, Br, and F are almost totally released through the flue gas stack (Clarke, 1993). Certain elements have a tendency to accumulate at the surfaces of the FA particles by coatings or single-spot coatings, while other elements tend to bind inside the inner matrices of the FA particles (USGS, 2002). Certainly, environmental concerns of FA are strictly associated with the existence of these elements, as surface bound elements will be more easily susceptible to leaching than the inner matrix bound elements. For example, Ca, As, Mg, Cr, Sr, and B are likely to be concentrated on the surfaces of FA particle, while Si, Al, Pb, and K do not exhibit a particular enrichment in any phase (Zandi and Russell, 2007). Flue gas desulfurization (FGD) plant removes approximately 90% of sulfur dioxides ($SO_2$) from the flue gas. FGD also sequesters particulate matter and volatile elements. Usually, FGD residue has lower concentrations of trace elements than FA. The volatile elements as Se and Hg are removed and collected in FGD residue (Cheng et al., 2009). Whereas high Hg elimination ability will considerably contribute in reducing Hg emission to the environment, the positive use of such FGD residue may be conceded. Major chemical species of some metal(loid)s evolved during coal combustion are As (elemental form), $As_2O_3$, $As_2S_3$, Cd (elemental form), CdO, CdS, $PbCL_2$, PbO, PbS, and Pb.

*Radioactive elements*—The feed coal used in thermal power stations generally contains natural radionuclides such as $^{40}K$, $^{232}Th$, $^{238}U$, $^{226}Ra$, $^{228}Ra$, $^{210}Pb$, $^{222}Ru$, and $^{220}Ru$ (Tadmore, 1986; Sharma, 1989; Mandal and Sengupta, 2003; Papastefanou, 2008; Amin et al., 2013), each with several physiochemical features with respect to the decay modes, decay products, decay energies, leachability, their associated half-lives, etc. During the process of coal combustion, these radionuclides are favorably accumulated in smaller FA particles, with the more volatile species, for instance, $^{210}$Pb and $^{210}$Po, showing the highest enrichment (IAEA, 2003). The activity concentrations of $^{238}$U, $^{226}$Ra, $^{210}$Pb, $^{228}$Ra, and $^{40}$K radionuclides are determined in Greece FAs, and ranged from 263 to 950, 142 to 605, 133 to 428, 27 to 68, and 204 to 382 Bq kg$^{-1}$, respectively (Papastefanou, 2008). The results revealed that there is an enrichment of the radionuclides in FA comparative to the feed coal. The enrichment factors of $^{238}$U, $^{226}$Ra, $^{210}$Pb, $^{228}$Ra, and $^{40}$K ranged from 0.60 to 0.76, 0.69 to 1.07, 0.57 to 0.75, 0.86 to 1.11, and 0.95 to 1.10, respectively (Papastefanou, 2008). Balkans FAs are found to be among the most radioactive, with a total activity of up to 12,000 Bq kg$^{-1}$ (IAEA, 2003). The FGD installations also decrease the discharge of the extremely volatile radionuclides into the environment. These radionuclides are discharged into the environment, and thus may pose health impacts. The assessment of public radiation exposure from FA in the surrounding area of a coal-based thermal power station critically depends on the amount of radioactive elements in the FA, the nature of intake, the duration of exposure, as well as the modes of dispersal to the atmosphere. The concentration of natural radionuclides discharged into the environment through FA is also dependent on the FA content of the coal, partitioning between FA and BA, the effectiveness of the emission control systems, and the combustion temperature. Most of the FAs are not significantly enriched in natural radionuclides than the common type of rocks or soils (Zielinski and Finkelman, 1997). On the basis of the study by Mittra et al. (2005), the radioactivity analysis (Bq kg$^{-1}$) of FA and soil amended with FA at 40 t ha$^{-1}$ showed that higher radioactivity of $^{226}$Ra, $^{228}$Ac, and $^{40}$K was noted in the latter than the former, while the activity of $^{137}$Cs was contrary. The radioactivity due to FA incorporation was subjected to dilution effect in soil though these marginal variations remained with the safe limit. In addition, Goyal et al. (2002) reported that the activity levels of gamma emitting radionuclides $^{40}$K, $^{226}$Ra, and $^{228}$Ac were within the acceptable limits, and mixing of FA with soil at 24% (v/v) was of no significance. Thus the quality of groundwater due to FA disposal remained unaffected with regard to radionuclide pollution (Cothern and Smith, 1987; Zielinski and Finkelman, 1997). Fig. 1.2 represents the world average activity concentration (Bq kg$^{-1}$) of the radionuclides in FA.

*Organic compounds*—Some organic pollutants such as PAHs, PCDDs, and PCDFs are associated with coal combustion processes (Dellantonio et al., 2010). Coal-based power stations accounted for 40% of the atmospheric PCDD and PCDF emissions occurring in the United Kingdom in the 1990s (Luijk et al., 1993). PCDFs and PCDDs are mostly formed in the electrostatic precipitator of coal-based power stations by strong catalytic influence of Cu and Fe at temperatures between 200°C and 400°C (Luijk et al., 1993). Coal-based thermal power stations are the well-known and utmost important anthropogenic source of PAHs in the atmosphere (Mastral and Callén, 2000; Liu et al., 2002; Arditsoglou et al., 2004; Sahu et al., 2009). The organic matter in coal contains a molecular soluble phase, a macromolecular insoluble fraction, PAHs, and aliphatic hydrocarbons (Liu et al., 2000; Mastral and Callén, 2000). At the time of coal combustion, the coal structure undergoes physical and chemical changes, resulting in the release of organic fractions such as PAHs and others organic compounds through reactions (Liu et al., 2000). The formation and release of PAHs during the coal combustion depends mostly on the coal characteristics and on the operating conditions (Liu et al., 2001, 2002; Sahu et al., 2009): the quantity of PAHs emissions reduces with the increase in combustion temperature and in coal rank. Usually, PAHs can be released in the gaseous stage at high combustion temperatures, while adsorbed and/or deposited on FA particles when the temperature decreases (Mastral and Callén, 2000). The PAHs with less than four aromatic rings exist in gas phase, whereas PAHs with six or more rings are present in solid phase. Moreover, PAHs with the four and five rings may exist in both forms (Liu et al., 2000; Mastral and Callén, 2000). It is also supported by Readman et al. (2002) that the high combustion temperature yields pyrolytic PAHs with four to six rings.

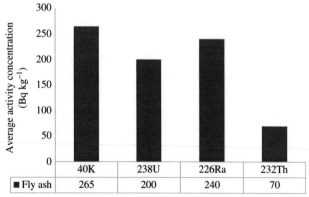

**Figure 1.2** World average activity concentrations (Bq kg$^{-1}$) of $^{40}$K, $^{238}$U, $^{226}$Ra, and $^{232}$Th in FA (UNSCEAR, 1982).

Therefore the FA lagoons can be an anthropogenic source of environmental pollution for the release of organic pollutants, thus posing a threat to water, soil, and biodiversity (Achten and Hofmann, 2009). These organic compounds are specially bounded to the particulate matter but some, such as PAHs, can be identified in water stages (Schwarzbauer, 2006). The fate of organic pollutants in the environment mostly depends on their physicochemical characteristics such as photooxidation, volatilization, chemical oxidation, microbial degradation, adsorption onto soil particles, and leaching (Wild and Jones, 1995). Gohda et al. (1993) reported the concentrations of PAH (183 mg kg$^{-1}$) in FA, which is a result of low-temperature combustion or incomplete combustion, while the PAH levels in various Greece FAs normally did not go beyond the levels (1.1 mg kg$^{-1}$) reported in urban soils (Arditsoglou et al., 2004; Blume, 2004). Furthermore, it is a pressing need to evaluate the potential impacts of organic compounds from FA disposal on the environment, human health, and ecosystems.

*Atmospheric emissions*—Coal combustion is one of the major anthropogenic atmospheric emission sources around the world for sulfur dioxide ($SO_2$), nitrogen oxides ($NO_x$), particulate matter ($PM_{2.5}$ and $PM_{10}$), carbon monoxide (CO), carbon dioxide ($CO_2$), volatile organic compounds (VOCs), and various trace elements like F, As, Se, Hg, and Sb (Querol et al., 1992; Schroeder and Munthe, 1998; Finkelman, 2007; Ito et al., 2006; Sekine et al., 2008; Nelson et al., 2010; Chen et al., 2013; Guttikunda and Jawahar, 2014). Most importantly, for human health, coal combustion releases atmospheric emissions through flue gas stacks that can scatter this pollution in the atmosphere over large areas. The 2010 global burden of disease study listed the outdoor air pollution (ozone and PM pollution) among the top 10 health threats in India, with as-projected 695,000 annual premature deaths from compromised immune systems, cardiovascular conditions, and respiratory illnesses (IHME, 2013). Of the projected yearly anthropogenic emissions in India, the coal-based thermal power stations account for $\sim$50% of $SO_2$, $\sim$30% for $NO_x$, and $\sim$15% for $PM_{2.5}$ (GAINS, 2012). The elements As, F, Se, Sb, and Hg are easily volatilized trace elements during coal combustion (Table 1.3). The majority of F releases directly into the atmosphere after coal burning (Luo et al., 2002). The loss rates of As during coal combustion ranged from 6.4% to 64%, with an average of 31.25% at 815°C (Zhou, 1998). The As could be volatilized and released with small particulates into atmosphere during coal combustion (Sun et al., 2004). More than 78% of F and 97% of Se in

**Table 1.3** Atmospheric emission factors (%) of arsenic, fluorine, selenium, antimony, and mercury from coal combustion.

| Elements | Emission factors (%) | Pollution control equipment | References |
|---|---|---|---|
| Arsenic | 1.2 | ESP | Otero-Rey et al. (2003) |
| | 48.2 | ESP | Guo et al. (2004) |
| | 1.6 | ESP | Reddy et al. (2005) |
| | 3.9 | ESP | Helble (2000) |
| | 94 | ESP | Han et al. (2002) |
| | 0–4 | FGD | Alvarez-Ayuso et al. (2006) |
| | 13.8, 19.6 | ESP, FGD | Tian et al. (2011) |
| | 1.6 | ESP + FGD | Ito et al. (2006) |
| | 1.7, 25, 0.4 | ESP, FGD, ESP + FGD | Meij and te Winkel (2007) |
| | 7.7, 2.8 | High-temperature combustion, low-temperature combustion | Luo et al. (2004) |
| Fluorine | 0–4 | FGD | Alvarez-Ayuso et al. (2006) |
| | 13.1 | ESP + FGD | Ito et al. (2006) |
| | 80.3, 5.1, 4.1 | ESP, FGD, ESP + FGD | Meij and te Winkel (2007) |
| | 95.9, 77.6 | High-temperature combustion, low-temperature combustion | Luo et al. (2002) |
| Selenium | 3.7 | ESP | Otero-Rey et al. (2003) |
| | 0–4 | FGD | Alvarez-Ayuso et al. (2006) |
| | 5.04 | ESP | Goodarzi (2006) |
| | 67.1 | ESP | Reddy et al. (2005) |
| | 76.7 | ESP | Wang et al. (1995) |
| | 50.9 | ESP | Helble (2000) |
| | 26.2, 25.1 | ESP, FGD | Tian et al. (2011) |
| | 17.6, 34.4, 6.1 | ESP, FGD, ESP + FGD | Meij and te Winkel (2007) |

*(Continued)*

**Table 1.3** (Continued)

| Elements | Emission factors (%) | Pollution control equipment | References |
|---|---|---|---|
| Antimony | 0.78–1.1 | ESP | Ondov et al. (1979) |
| | 1.5 | ESP | Helble (2000) |
| | 27.9 | ESP | Wang et al. (1995) |
| | 0.06 | ESP + FGD | Ito et al. (2006) |
| | 0 | Pulverized coal combustion | Demir et al. (2001) |
| | 1.1, 17.9, 0.2 | ESP, FGD, ESP + FGD | Meij and te Winkel (2007) |
| Mercury | 40 | FGD | Alvarez-Ayuso et al. (2006) |
| | 50–70 | FGD | Pacyna et al. (2006) |
| | 56.5 | ESP | Goodarzi (2006) |
| | 69.4 | ESP | Streets et al. (2005) |
| | 71.1 | ESP | Helble (2000) |
| | 76.5 | ESP | Reddy et al. (2005) |
| | 83 | ESP | Guo et al. (2007) |
| | 86.6 | ESP | Zhou et al. (2008) |
| | 66.8, 42.8 | ESP, FGD | Tian et al. (2011) |
| | 27.6 | ESP + FGD | Ito et al. (2006) |
| | 29.1 | ESP + FGD | Yokoyama et al. (2000) |
| | 88.5, 86.3 | ESP, ESP + FGD | Zhang et al. (2008) |
| | 50.4, 20, 10.1 | ESP, FGD, ESP + FGD | Meij and te Winkel (2007) |

ESP, Electrostatic precipitator; FGD, flue gas desulfurization.
Source: Modified from Chen,J., Liu, G., Kang, Y., Wu, B., Sun, R., Zhou, C., Wu,D., 2013. Atmospheric emissions of F, As, Se, Hg, and Sb from coal-fired power and heat generation in China. Chemosphere 90, 1925–1932.

coal will be volatilized at 800°C during coal combustion (Han et al., 2009). Approximately 90% of Hg in coal is emitted during combustion (Guo et al., 2003) and starts to release at 100°C−150°C (Wagner and Hlatshwayo, 2005). Almost all the F and Hg in coal were volatilized during combustion process, while only 10% of As was released (Jiao, 1998). About 75% (1900 tons) of the global Hg emission in 1995 were attributed to particular coal combustion for electricity and heat generation in China, India, and South and North Korea (Pacyna et al., 2003), and the power generation facilities, as the largest source of atmospheric Hg, released 498 tons (26% of the total anthropogenic emission) of Hg in 2005 (Pirrone et al., 2009; UNEP, 2010).

## 1.4 Fly ash pollution

Huge amount of FA remains unutilized and dumped in lagoons, which poses risks to local environment owing to its fine particles and pollutants, that is, metal(loid)s, radioactive elements, organic contaminants, and atmospheric emissions. The main adverse effects of FA are (1) respiratory problems to human beings; (2) increased cycling of these toxic elements through food chain (Carlson and Adriano, 1993); (3) changes in plant elemental composition; and (4) leaching of potentially toxic substances from FA into surface water, groundwater, and soils. Fig. 1.3 describes the main pathways of FA pollution from FA deposited sites.

*Air pollution*—FA, a major source of air pollution, becomes dry in summer season as temperature increases and gets airborne for a long time that poses health threats to the local people. It causes irritation to skin, eyes, nose, respiratory tract, and throat. Recurrent inhalation of FA dust having crystalline silica can cause lung cancer, silicosis (scarring of the lung), and bronchitis. The high levels of inhaled FA dust may result in asthma−like symptoms. FA may also increase the risk of scleroderma. In addition, FA degrades environment and corrodes exposed metallic structure in its vicinity (Davison et al., 1974; Finkelman et al., 2000). Some plant species with self-regenerating potential and high dust capturing efficiency are *Abutilon indicum, Artocarpus heterophyllus, Callicarpa tomentosa, Calotropis gigantea, Cocculus hirsutus, Cordia dichotoma, Ficus benghalensis, Ficus glomerata, Ficus hispida, Guazuma ulmifolia, Heterophragma quadriloculare, Lantana camara, Mallotus philippinensis, Nyctanthes arbor-tristis, Ocimum basilicum, Spathodea campanulata, Tectona grandis, Thespesia populnea, Trema orientalis,* and *Ziziphus mauritiana* (Chaphekar and Madav, 1999). The potential

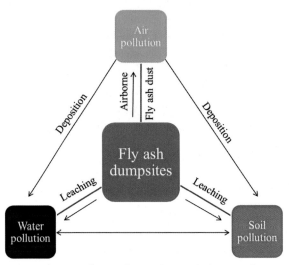

**Figure 1.3** The key pathways of FA pollution from ash deposited sites.

approaches must be implemented to mitigate air pollution nearby coal-based power stations, including an electrostatic precipitator with high dust removal capacity (95%–96%), controlling the combustion temperature that decreases the volatilization of elements, and planting efficient tree species with high dust capturing capacity. Moreover, atmospheric FA pollution up to a certain level can be minimized by burning high-quality coal in thermal power stations.

*Water pollution*—The impact of FA on the surface and groundwater is dependent on the FA physicochemical properties, and climatic and hydro-geologic conditions of the dumping site. The leaching of Cu, Pb, and Zn was noticed in higher level in tube well water situated near the ash pond in Orissa, whereas Pb, Cu, Zn, and Mn were the main pollutants in groundwater (Praharaj et al., 2002). The leaching of toxic elements in an aqueous media mostly depends on the ash source, pH of the ash–aqueous medium, leaching time, and bonding of these elements on the ash parti-cles. Usually, the degree and amount of leaching are greater in acidic con-ditions. Weathered FA dumpsites result in more groundwater pollution due to the occurrence of higher concentrations of soluble salts. Quality of groundwater due to FA application stayed unaffected relating to radionu-clide pollution (I.I.T. Kharagpur, 1999). Proper approaches must be implemented to minimize water pollution in the vicinity of coal-fired power stations. In this regard, there are many strategies such as the lining of ash pond, stabilization of ash pond through vegetation, covering the upper surface of ash pond by organic wastes or soil, planting saplings (tolerant against hash conditions) on ash pond, and the disposal practice of FA in the form of "ash mound," which not only saves the water and land but also limits the risks of water pollution, by lessening leaching.

## 1.5 Fly ash threats

Disposal of FA is a major global problem, which will increase manyfolds in the next few decades (Pandey and Singh, 2010; Pandey et al., 2011). FA disposal occupies thousands of hectares of land and causes threat to environment. Production of 1 MW of electricity requires about 1 acre of land for the disposal of FA (Sahu, 1994). Therefore disposal of FA is a burning problem today and it could aggravate in future course of time. There are two types of disposal methods of FA: the wet disposal and the dry disposal. In general, the wet disposal method is followed in India and abroad. In this method, FA is mixed with water in the ratio of 1:8 and the formed slurry is pumped to the ash catena through iron pipelines. The environmental impact of FA depends on its physicochemical, elemental composition, and the nature of coal used. Usually, FA is dumped in unmanaged open ash ponds that serve as a major source of environmental pollutant in that area through FA erosion, leachate toxicity, thereby raising the suspended particulate matter (SPM) in air and atmospheric fallouts (Fig. 1.1A). Leachable FA contaminants pollute lakes, rivers, and other aquatic systems located around the thermal power plants (Ruhl et al., 2010). Being fine in particle size, it can readily escape into the atmosphere along with flue gases and can become a source of air pollutants. FA particles remain suspended in the air for a long period of time and cause severe difficulty to animal and human respiration (Costa and Dreher, 1997; Chakraborty and Mukherjee, 2009; Celik et al., 2007; Becker et al., 2002; Ruhl et al., 2009). Due to FA pollution, a number of problems related to human health have been noticed among the residents living near coal-based thermal power plants (Borm, 1997; Ruhl et al., 2009). Moreover, some researchers also reported the threat to aquatic fauna adjacent to thermal power plants (Hopkins et al., 1999, 2000; Rowe, 1998). They disturb the physiological performance of higher plants. FA atmospheric fallouts also corrode the structure of historic monuments and other buildings situated near the thermal power plants. The following subsections will explore several FA-related diseases in aquatic and terrestrial fauna including human being.

*FA and cancer disease*—FA is also a reason for cancer in human beings living near the thermal power plants. According to US Environmental Protection Agency (EPA), cancer risk for the residents living nearby the unlined FA basins was as high as 1 in 50 residents, which was caused due to the arsenic from FA contaminated drinking water. It is 2000-fold higher than the EPA's goal of decreasing cancer threat to 1 in 100,000 inhabitants (US EPA, 2007). Lead and mercury in FA are also the causes of concern for cancer or neurological damage. Ruhl et al. (2009) reported on the basis of survey that the surface release of FA with high concentrations of toxic pollutants (As = 75 mg/kg; Hg = 150 µg/kg) and radioactivity ($Ra^{226} + Ra^{228}$ = 8 pCi/g) in the atmosphere can pose a health risk to local communities residing in the nearby areas of power plants.

*FA and genetic problems*—Human cells that were exposed to FA showed genetic mutations and damage (Chakraborty and Mukherjee, 2009). Workforce engaged in the task of cleaning at the coal-based thermal power plants and also those who are regularly exposed to FA and other combustion by-products are more susceptible to chromosomal aberrations indicating genetic damage (Celik et al., 2007). Recently, Markad et al. (2012) demonstrated that FA-induced oxidative stress, destabilization of lysosomal membranes, DNA damage, and DNA−protein crosslink in earthworm coelomocytes through comet assay. Dwivedi et al. (2012) also showed that ROS induced by FA nanoparticles, oxidative DNA damage, chromosomal breakage, and also mitochondrial membrane cause damage in the peripheral blood mononuclear cells through comet and cytokinesis-blocked micronucleus assays. FA-induced genetic problem is also observed in the plants, in which Love et al. (2009) reported higher levels of DNA damage in the leaf tissues of the *Cassia occidentalis* plant growing wild on FA catena than *C. occidentalis* growing on soil. They found that the foliar concentrations of As, Ni, and Cr were two to eight times higher in plants growing naturally on FA catena than the plants growing on soil.

−*FA linked respiratory and cardiovascular diseases*—It has been observed that workers with long exposure to FA were more susceptible to decreased lung function in comparison with workers with less exposure (Schilling et al., 1988). Native people with chronic obstructive pulmonary disease and asthma may be more vulnerable to airway irritation from FA (Becker et al., 2002). Repeated inhalation of fine dust particles of FA can cause bronchitis and silicosis (scarring of lung) due to the presence of crystalline silica in FA. The studies on animals have also confirmed that inhaling SPM from FA is associated with airway reactivity and inflammation (Costa and Dreher, 1997). Repeated exposure to FA can cause irritation in eyes (watering and redness), skin, nose, throat, and respiratory tract (coughing and sneezing). Two isotopes of radium ($Ra^{226}$ and $Ra^{228}$) are the key sources of ionizing radiation in FA and may remain in the human lung for several months after their inhalation. These two isotopes regularly enter the circulatory system of the human body and finally get deposited on the bones and teeth for the lifetime (Ruhl et al., 2009). Similar health problems of $Ra^{226}$ and $Ra^{228}$ have been studied among the cleaning workers of the nuclear accident that took place at the Chernobyl nuclear power plant. The workers, who inhaled large quantities of these two isotopes, later suffered from bronchial mucosa lesions (in some cases preneoplastic) with an increased susceptibility to the invasion of microorganisms in bronchial mucosa (Poliakova et al., 2001; Chizhikov et al., 2002). FA constitutes different size of particles that are inhaled in the lungs during respiration and finally go to the circulatory system. Mainly three size of particles, namely the coarse particle ($2.5-10\ \mu m$ in diameter), the fine particle (less than $2.5\ \mu m$ in diameter), and the ultrafine particle (less than $0.5\ \mu m$ in diameter), have been identified in coal FA (Linak et al., 2007; US EPA, 2008). Inhaling of SPM of FA has been attributed to the factors such as decreased lung function, aggravated asthma, irregular heartbeat, heart attacks, development of chronic bronchitis, and premature death in people with lung or heart disease (US EPA, 2008). Local population with heart or lung diseases is the most susceptible to the health hazards of exposure to SPM pollution of FA.

*FA and aquatic fauna*—FA has considerable influence on the aquatic ecosystems of the surrounding area of coal-based thermal power plants. Decantation from FA catena into the adjacent river or aquatic body through seepage or leaching increases turbidity, decreases primary productivity, and affects fishes and other aquatic biota (Pandey et al., 2011). Ruhl et al. (2009) have documented that the release of FA effluents from Kingston plant in Tennessee has threatened the nearby aquatic biota. The leaching of pollutants from the FA catena had caused contamination of surface water, but only trace levels were found in the downstream of Emory and Clinch rivers due to river dilution (Ruhl et al., 2009). Apart from it, the accumulation of Hg- and As-rich FA in river sediments affects fish-poisoning and methyl mercury formation in anaerobic river sediments (Ruhl et al., 2009). Initial studies on the releases of FA into Belews Lake and Hydro Reservoir (both in North Carolina) and Martin Creek Reservoir (Texas) documented the uptake of pollutants by aquatic biota, mostly in the inhabiting fish groups (Lemly, 1996, 2002). Several other studies reported the adverse effects of FA contaminants on amphibians, fish, and reptiles inhabiting in the settling impoundments, which receive slurried ash from the coal-fired power plant located on the Savannah River Site in United States (Rowe et al., 1996; Hopkins et al., 1998, 2003; Nagle et al., 2001). Few studies have reported the impacts on amphibians that are contaminated with coal FA. Both *Bufo terrestris* (adult southern toads) and *Palaemonetes paludosus* (freshwater grass shrimp) have been shown to accumulate trace elements including arsenic and cadmium from FA polluted areas. Similarly, elevated trace element concentrations were also reported in banded water snakes (*Nerodia fasciata*; Hopkins et al., 1999; Rowe, 1998). *B. terrestris* (larval southern toads) and *Rana catesbeiana* (larval bullfrogs) were also found struggling for their survival due to the threatening physiological impacts (Hopkins et al., 2000). Reash et al. (2006) studied the ecological hazards of selenium (Se) from coal FA discharge. They collected Bluegill sunfish (*Lepomis macrochirus*) and caddis flies (*Trichoptera: Hydropsychidae*) from a stream receiving FA discharge and the nearby reference streams to determine the tissue levels of Se and other metals. They concluded that Bluegill sunfish—inhabiting a stream that received coal FA—had a raised Se concentration in the whole body and gonad tissue (9- to 10-fold higher than reference fish). A long-term persistence of the bluegill population may be linked to the antagonistic metal interactions, which were also elevated in the fish. The effluent from FA basins affects metabolism and reproduction negatively in a number of fish species. However, the mosquito fish (*Gambusia holbrooki*) is known to maintain viable population S in FA contaminated areas. Staub et al. (2004) reported the effects of FA on the reproduction and standard metabolic rate (SMR) of mosquito fish inhabiting an FA contaminated settling basin. They observed that SMR of mosquito fish from an FA contaminated site and a reference site was not significantly different. Despite enhanced pollutant levels in the FA basin, females, and their offspring, their brood sizes and offspring viability did not differ from clutches collected from the FA basin and the reference site. The results showed high degree of tolerance of mosquito fish against the exposure to aquatic FA disposal (Staub et al., 2004).

*FA and terrestrial fauna*—FA contains multiple pollutants including toxic metal(loid)s (Cd, As, Cr, Cu, Pb, Ni, Se, etc.), which are harmful for the biota of terrestrial ecosystems and which cause ecological damage (Pandey et al., 2009a, 2011). Few studies reported the adverse effects of FA releases on avian species. However, Bryan et al. (2012) demonstrated the maternal transfer of toxic metal(loid)s and provided evidence of reproductive impacts. They analyzed the diet and tissues of Common Grackle nestlings associated with FA basins for the determination of trace elements. The concentrations of As, Cd, and Se elements in ash basin diets were five times higher than those of reference diets. Concentrations of these elements were elevated in feather, liver, and carcass. About 15% of the total body burden of Se, As, and Cd was sequestered in the feathers of old nestlings. At the Martin Creek Reservoir, King (1988) and King et al. (1994) also noted maternal transfer of Se to eggs by Red-winged Blackbirds (*Agelaius phoeniceus*) and Barn Swallows (*Hirundo rustica*), both insectivorous species, along with suppressed ( > 50%) egg hatchability by blackbirds. Bryan et al. (2003) also reported the maternal transfer of Se to eggs in Common Grackle (*Quiscalus quiscala*) nestling associated with aquatic ash disposal sites. It was further noted that there were significantly higher concentrations of Se in eggs from the ash basins than a reference site. It was also reported that the elevated Se concentrations were found to be greater than background, but did not exceed the suggested levels of concern that was expected to influence hatchability in most birds (Heinz, 1996; Fairbrother et al., 1999; Janz et al., 2010). Peles and Barrett (1997), first time, examined in situ condition to know the effects of FA toxins on species of small mammals in FA ecosystems. No evidence of genetic damage was noticed as a result of exposure to FA toxicants in both small mammals. However, enhanced metal levels were observed in tissues of both species collected from FA deposit compared with the reference site. The concentrations of As, Cu, and Ni in liver tissue of *Sigmodon hispidus* collected from the FA habitat were significantly more than the reference site. Similarly, concentrations of Cd, Cr, Cu, and Zn were significantly more in kidneys of *S. hispidus* from the FA habitat. In addition, the concentrations of As, Cu, Mn, Pb, Se, and Zn were significantly greater in the liver tissue of *Oryzomys palustris* associated with the FA habitat than the reference site. No significant differences were reported between sites regarding metal(loid)s concentrations in kidney tissue of *O. palustris*. Thus these results have important suggestions about potential bioaccumulation of metals in the food chain, because it is well known that small mammals are an important link for the transfer of toxic metals to higher trophic levels (Peles and Barrett, 1997).

*Atmospheric emissions and human health impacts*—Atmospheric emissions of the coal-fired power stations are accountable for both environmental and human health issues. In 2010 and 11 the 111 coal-fired power stations with a fixed capacity of 121 GW used 503,000,000 tons of coal and generated a projected 580,000 tons of $PM_{2.5}$, 1,200,000 tons of $PM_{10}$, 2,100,000 tons of $SO_2$, 2,000,000 tons of $NO_x$, 1,100,000 tons of CO, 100,000 tons of VOC, and 665,400,000 tons of $CO_2$, thus resulting in an estimated 80,000−115,000 premature deaths and 20,900,000 asthma cases due to $PM_{2.5}$ pollution, which cost the public and the government a projected USD 3.2−4.6 billion (160,000,000,000−230,000,000,000). Besides, the other estimated human health impacts owing to atmospheric emissions from the coal-fired power stations in India are child mortality under 5 years of age (10,000), chest discomforts (8,400,000), chronic bronchitis (170,000), and respiratory symptoms (625,000,000). In India, the electricity production from the coal-based thermal power stations will be continually high due to energy demands and huge coal reserves at least over 2030 (Prayas, 2011, 2013). The study indicates that forceful pollution control guidelines, for example, the reduction of emission standards for all pollutants, updating techniques for environment impact assessments, and installing FGD, are necessary for cleaning air and minimizing human health issues.

## 1.6 Other problems

A number of problems based on an extensive survey have been noted due to FA in the surroundings of coal-based thermal power stations. Here, some observations related to corrosion, deposition, skin problems, asthma, bronchitis, affected agriculture and horticulture, and contamination of water and soil during field visits of coal-based thermal power stations across India are:

- corrosion of fencing and tin sheets within thermal power station complex;
- corrosion of iron pillar, iron window, and iron door of the main entrance;
- corroded dish antennas, agriculture implements, vehicle chassis, tin roofs, well pulleys, mesh cover to wells, etc. in nearby houses;
- black dust deposition on vegetation (sacred grove, etc.);
- salt deposition evident from saline taste of mist (deposited on foliage);

- chlorosis, necrosis, and leaf and flower burn (chilly, wheat, banana, mango, etc.);
- reduced population of avifauna (dwindling peafowl population) indicator of enhanced pollution levels in the environment as birds are bioindicators;
- skin rashes, lesions, and nail deformation in humans (children as well as adults);
- higher coughing among local people (respiratory ailments such as asthma);
- bronchitis, impact of minute FA dust in mucus membrane;
- skin disease in livestock;
- affected agriculture and horticulture fields;
- contamination of groundwater and surface water sources;
- reduced fish yield in the aquatic reservoirs nearby FA ponds; and
- emigration of local people due to the prevailing adverse environmental conditions.

## 1.7 Multiple uses of fly ash

Globally, the multiple uses of FA have been recognized well in wide-ranging areas such as adsorbents for cleaning of flue gas (sulfur compounds, adsorption of $NO_x$, removal of mercury, and adsorption of gaseous organics), removal of toxic metal(loid)s from wastewater, removal of other inorganic components from wastewater (removal of phosphate, fluoride, and boron), removal of organic compounds (i.e., phenolic compounds, pesticides, and other organic compounds) from wastewater, removal of dyes from wastewater (azo dyes, thiazine dyes, xanthene dyes, arylmethane dyes, and other dyes), synthesis of zeolite, construction work/industry, lightweight aggregation, roadway and embankment, reclamation of low lying areas, and coal mine backfill (Ahmaruzzaman, 2010). Moreover, In India, FA is generally used in cement, bricks, concrete, tiles, blocks, roads, embankments, mine backfilling, and a soil conditioner as source of plant nutrients in agriculture and forestry (Central Electricity Authority, 2011). Thus it appears that FA has a lot of potential for many applications as above discussed. However, the global average FA utilization is approximately 25% (Wang, 2008). Consequently, it is clear that a major amount of the global FA is unutilized and wants urgent attention to recognize potential opportunities.

Besides, the FA can be utilized in soil improvement. Several reviews revealed that FA has potential as a soil ameliorator (Iyer and Scott, 2001; Gupta et al., 2002; Yunusa et al., 2006; Jala and Goyal, 2006; Pandey et al., 2009a; Basu et al., 2009; Pandey and Singh, 2010; Ram and Masto, 2014; Shaheen et al., 2014). Thus the application of FA as a soil additive will solve some problems such as its disposal and environmental concerns, and will also be helpful in tapping economic potentials. Several studies reveal to the wider potential of FAs to enhance soil quality and fertility as well as to ameliorate degraded land for agriculture (Pandey et al., 2009b, 2010; Ram et al., 2006; Srivastava and Ram, 2010; Singh et al., 2011a,b, 2016; Singh and Pandey, 2013). The use of FA in agriculture has been advocated for three decades to ameliorate nutrient poor soils due to its favorable physicochemical characteristics including significant amounts of plant macro- and micronutrients (Pandey and Singh, 2010). The FA is deficient in N and available in P, because N is oxidized into gaseous form during coal combustion as well as excessive Fe and Al change soluble P to insoluble P compounds (Pandey et al., 2009a; Pandey and Singh, 2010). The deficiency of N and P could be avoided in FA amended soil by using waste organic materials such as farmyard manure, press mud, poultry, and cattle manure (Ram and Masto, 2014).

## 1.8 Fly ash management

There are several uses of huge amount of FA in wide-ranging sectors. However, maximum quantities of FA are unused and dumped as waste material in ash landfills/lagoons, which cause many environmental problems and human health impacts. The FA dumpsites have been regarded as hazardous sites on the earth due to the presence of its fine particles and pollutants. These fine particles of FA are easily blown and remain airborne for a long period and cause health hazards to the local residents of coal-based thermal power station. Thus the threats of FA dust particles are more pronounced, and have been well recognized in the vicinity of ash disposal sites due to inhalable form of FA particles. Besides, the metal(loid)s of FA can contaminate soil and water systems by leaching process. To avoid all these problems, there are several methods to bind FA particles such as sprinkler system, soil layer, organic wastes, phytoremediation/ revegetation, and handling of wastewaters. The sprinkler system is generally used to bind FA particles so that these particles cannot blow nearby areas of coal power station. The soil layering on the FA dumpsites is also an effective approach to bind and mitigate FA dust pollution, and offers revegetation on the harsh conditions of FA dumps. In this context, mostly topsoil is used, which contains the highest concentration of microorganisms and organic matter. It is that place where most of the Earth's biological soil activity occurs. However, most of the researchers discourage for using topsoil in this type of practices because of being most precious resource to humans, which is formed over the course of many decades. In general, the FA has soil-like properties, and has been used as a topsoil substitute in mineland reclamation to secure the amount of topsoil (Gorman et al., 2000; Matsumoto et al., 2019). FA is also used as a rooting media (Pandey and Kumar, 2012). Furthermore, organic waste amended FA landfill generally provides a friendly substrate for the plant establishment and growth. The incorporation of animal composts, biosolids, poultry manures, press mud, farmyard manure, etc. into the upper surface layers of FA landfill could enhance the success of revegetation/rehabilitation efforts (Haynes, 2009; Pandey et al., 2009a; Pandey and Singh, 2012; Belyaeva and Haynes, 2012). The environmental issues of FA could be effectively managed by the rehabilitation of FA dumps via the selection of appropriate plant species, amendments, and ecological engineering (Pandey and Singh, 2012). Some promising plant species such as *Typha latifolia, Azolla caroliniana, Ricinus communis, Thelypteris dentata, Saccharum munja, Saccharum spontaneum, Ipomoea carnea,* and *Z. mauritiana* have been assessed for phytoremediation/revegetation/restoration of FA deposited sites with low inputs and minimum care, suggested as potential species for the remediation of recently deposited FA landfills (Pandey, 2012a,b, 2013; Kumari et al., 2013; Pandey et al., 2014, 2015a,b; Pandey and Mishra, 2018). Thus those plants having some desirable characters such as perennial, native, economically valuable, extensive root system, vegetative propagation, unpalatable, and tolerance against high FA stress conditions should be suggested for revegetation/stabilization of FA dumpsites (Pandey and Singh, 2011). In addition, assisted phytoremediation of FA dumps by spontaneously colonized native plants through ecological engineering can be employed for the rehabilitation programs with multiple benefits (Pandey and Singh, 2012; Pandey, 2015). Aromatic and energy crops are more valuable than a promising crop in phytoremediation (Pandey et al., 2012a, 2016, 2019; Verma et al., 2014; Pandey and Singh, 2015; Pandey, 2017) and provide market opportunities in phytoremediation (Pandey and Souza-Alonso, 2019). Jamil et al. (2009) analyzed FA trapping and metal accumulating efficiency of trees for the implication of green belt around coal-fired power stations. They found several plants with dust trapping capacity in the order of *Mussaenda frondosa* > *Haldina cordifolia* > *Pedilanthus tithymaloides* cv. Variegates > *P. tithymaloides* > *Duranta erecta* > *Delonix regia* > *Anthocephalus cadamba* > *Mangifera indica* > *Polyalthia longifolia* > *Mimusops elengi*, while metal accumulation index was highest in *H. cordifolia* followed by *M. frondosa*. The emphasis should be on the treatment/handling of wastewater from coal-based thermal power stations and its reuse within the power plant.

## 1.9 Conclusions

FA, a coal combustion residue, can be used as a resource material in huge amount in various novel applications. However, it is not fully utilized in most of the countries. But few countries have utilized 100% of the produced FA. Several studies are going on to identify promising applications that can significantly contribute in maximizing the use of FA. In addition, to avoid FA pollution from the ash dumping sites, phytomanagement is the best approach for both remediation and revenue generation, and can be achieved by using ecologically and socioeconomically valuable native plant species as well as by applying the waste organic amendments with minimum inputs and low maintenance.

## References

Achten, C., Hofmann, T., 2009. Native polycyclic aromatic hydrocarbons (PAH) in coals — a hardly recognized source of environmental contamination. Sci. Total. Environ. 407, 2461–2473.
ADAA, 2009. Ash Development Association of Australia. <http://www.adaa.asn.au/documents/ADAA_membership_infov2009.pdf>.
Adriano, D.C., Page, A.L., Elseewi, A.A., Chang, A., Straughan, I.A., 1980. Utilization and disposal of fly ash and other coal residues in terrestrial ecosystem: a review. J. Environ. Qual. 9, 333–344.
Ahmaruzzaman, M., 2010. A review on the utilization of fly ash. Prog. Energy Combust. Sci. 36 (2010), 327–363.
Alvarez-Ayuso, E., Querol, X., Tomas, A., 2006. Environmental impact of a coal combustion–desulphurisation plant: abatement capacity of desulphurisation process and environmental characterisation of combustion by-products. Chemosphere 65, 2009–2017.
American Coal Ash Association, 2011. Production and use survey 2011. <http://www.acaa-usa.org/associations/8003/files/News_Release-coalAshProductionandUse_2011.PDF>.
Amin, Y.M., Khandaker, M.U., Shyen, A.K.S., Mahat, R.H., Nor, R.M., Bradley, D.A., 2013. Radionuclide emissions from a coal-fired power plant. Appl. Radiat. Isotopes 80, 109–116.

Anisworth, C.C., Rai, D., 1987. Chemical Characterization of Fossil Fuel Wastes. EPRI EA-5321. Electric Power Research Institute, Palo Alto, CA.

Arditsoglou, A., Petaloti, C., Terzi, E., Sofoniou, M., Samara, C., 2004. Size distribution of trace elements and polycyclic aromatic hydrocarbons in fly ashes generated in Greek lignite-fi red power plants. Sci. Total. Environ. 323, 153–167.

Australian Uranium Association, 2007. Energy for the world—why uranium? <http://www.uic.com.au/whyu.htm>.

Basu, M., Pande, M., Bhadoria, P.B.S., Mahapatra, S.C., 2009. Potential fly-ash utilization in agriculture: a global review. Prog. Nat. Sci. 19, 1173–1186.

Becker, S., Soukup, J.M., Gallagher, J.E., 2002. Differential particulate air pollution induced oxidant stress in human granulocytes, monocytes, and alveolar macrophages. Toxicol. In Vitro 16, 209–218.

Belyaeva, O.N., Haynes, R.J., 2012. Comparison of the effects of conventional organic amendments and biochar on the chemical, physical and microbial properties of coal fly ash as a plant growth medium. Environ. Earth Sci. 66, 1987–1997.

Bhattacharjee, U., Kandpal, T.C., 2002. Potential of fly ash utilisation in India. Energy 27, 151–166.

Blume, H.-P. (Ed.), 2004. Handbook of Soil Protection (in German). Ecomed, Landsberg am Lech, Germany.

Borm, P.J., 1997. Toxicity and occupational health hazards of coal fly-ash (CFA). A review of data and comparison to coal mine dust. Ann. Occup. Hyg. 41, 659–676.

Bryan Jr., A.L., Hopkins, W.A., Baionno, J.E., Jackson, B.P., 2003. Maternal transfer of contaminants to eggs of Common Grackles (*Quiscalus quiscala*) nesting on coal fly ash basins. Arch. Environ. Contam. Toxicol. 45, 273–277.

Bryan, A.L., Hopkins, W.A., Parikh, J.H., Jackson, B.P., Unrine, J.M., 2012. Coal fly ash basins as an attractive nuisance to birds: parental provisioning exposes nestlings to harmful trace elements. Environ. Pollut. 161, 170–177.

Caldas-Vieira, F., Feuerborn, H., 2013. Impact of political decuisions on production and use of coal combustion products in Europe. In: 2013 World of Coal Ash (WOCA) Conference April, 22–25, 2013, Lexington, KY <www.flyash.info>.

Carlson, C.L., Adriano, D.C., 1993. Environmental impacts of coal combustion residues. J. Environ. Qual. 22, 227–247.

Celik, M., Donbak, L., Unal, F., Yuzbasioglu, D., Aksoy, H., Yilmaz, S., 2007. Cytogenic damage in workers from a coal-fired power plant. Mutat. Res. 627, 158–163.

Central Electricity Authority, 2011. Report on fly ash generation at coal/lignite based thermal power stations and its utilization in the country for the year 2010–11, New Delhi, December 18, 2011.

Chakraborty, R., Mukherjee, A., 2009. Mutagenicity and genotoxicity of coal fly ash water leachate. Ecotoxicol. Environ. Saf. 72, 838–842.

Chaphekar, S.B., Madav, R.P., 1999. Thermal power plants and environmental management. J. Indian Assoc. Environ. Manag. 26, 48–53.

Chen, J., Liu, G., Kang, Y., Wu, B., Sun, R., Zhou, C., et al., 2013. Atmospheric emissions of F, As, Se, Hg, and Sb from coal-fired power and heat generation in China. Chemosphere 90 (6), 1925–1932.

Cheng, C.-M., Hack, P., Chu, P., Chang, Y.-N., Lin, T.-Y., Ko, C.-S., et al., 2009. Partitioning of mercury, arsenic, selenium, boron, and chloride in a full-scale coal combustion process equipped with selective catalytic reduction, electrostatic precipitation, and flue gas desulfurization systems. Energ. Fuel. 23, 4805–4816.

Chizhikov, V., Chikina, S., Gasparian, A., Zborovskaya, I., Steshina, E., Ungiadze, G., et al., 2002. Molecular follow-up of preneoplastic lesions in bronchial epithelium of former Chernobyl clean-up workers. Oncogene 21, 2398–2405.

CIRCA (Association of Canadian Industries Recycling Coal Ash), 2008. Production and use of coal combustion products (CCPS). <http://www.circainfo.ca/documents/ProductionandUseStatistics.pdf>.

Clarke, L.B., 1993. The fate of trace elements during coal combustion and gasification: an overview. Fuel 72, 731−736.

Costa, D.L., Dreher, K.L., 1997. Bioavailable transition metals in particulate matter mediate cardiopulmonary injury in healthy and compromised animal models. Environ. Health Perspect. 105, 1053−1060.

Cothern, C.R., Smith, J.E., 1987. Environmental Radon. Plenum Press, New York, p. 363.

Davison, R.L., Natusch, D.F.S., Wallace, J.R., Evans Jr., C.A., 1974. Trace elements in fly ash: dependence of concentration on particle size. Environ. Sci. Technol. 8, 1107−1113.

Dellantonio, A., Fitz, W.J., Repmann, F., Wenzel, W.W., 2010. Disposal of coal combustion residues in terrestrial systems: contamination and risk management. J. Environ. Qual. 39, 761−775.

Demir, I., Hughes, R.E., DeMaris, P.J., 2001. Formation and use of coal combustion residues from three types of power plants burning Illinois coals. Fuel 80, 1659−1673.

Dewey, G.R., Sutter, L.L., Sandell, J.F., 1996. Reactivity Based Approach for Classifying Fly Ash, 6. American Power Conference, Chicago, IL, pp. 1−4.

Dwivedi, S., Saquib, Q., Al-Khedhairy, A.A., Ali, A.S., Musarrat, J., 2012. Characterization of coal fly ash nanoparticles and induced oxidative DNA damage in human peripheral blood mononuclear cells. Sci. Total Environ. 437, 331−338.

El-Mogazi, D., Lisk, D.J., Weinstien, L.H., 1988. A review of physical, chemical and biological properties of fly-ash and effects on agricultural ecosystems. Sci. Total Environ. 74, 1−37.

Fairbrother, A., Brix, K.V., Toll, J.E., McKay, S., Adams, W.J., 1999. Egg selenium concentrations as predictors of avian toxicity. Hum. Ecol. Risk Assess. 5, 1229−1253.

Finkelman, R.B., 2007. Health impacts of coal: facts and fallacies. Ambio 36, 103−106.

Finkelman, R.B., Belkin, H.E., Zhang, B.S., Centeno, J.A., 2000. Arsenic poisoning caused by residential coal combustion. In: Proceedings of the 31st International Geological Congress, Guizhou Province, China.

Frandsen, F.J., 2009. Ash research from Palm Coast, Florida to Banff, Canada: entry of biomass in modern power boilers. Energ. Fuel. 23, 3347−3378.

GAINS, 2012. Greenhouse Gas and Air Pollution Interactions and Synergies e South Asia Program. International Institute of Applied Systems Analysis. Laxenburg, Austria.

Gohda, H., Hatano, H., Hanai, T., Miyaji, K., Takahashi, N., Sun, Z., et al., 1993. GC and GC-MS analysis of polychlorinated dioxins, dibenzofurans and aromatic hydrocarbons in fly ash from coal-burning works. Chemosphere 27, 9−15.

Goodarzi, F., 2006. Assessment of elemental content of milled coal, combustion residues, and stack emitted materials: possible environmental effects for a Canadian pulverized coal-fired power plant. Int. J. Coal Geol. 65, 17−25.

Gorman, J.M., Bhumbla, D.K., Sencindiver, J.C., 2000. Properties of fly ash used as a topsoil substitute in mineland reclamation. Proc. Am. Soc. Min. Reclam. 29 (3), 627−643.

Goyal, D., Kaur, K., Garg, R., Vijayan, V., Nanda, S.K., Nioding, A., et al., 2002. Industrial fly ash as a soil amendment agent for raising forestry plantations. In: Taylor, P.R. (Ed.), 2002 EPD Congress and Fundamental of Advanced Materials for Energy Conversion. TMS Publication, Warrendale, PA, pp. 251−260.

Guo, X.X., Zheng, C.C., Jia, X.X., Sun, T.T., 2003. The behavior of mercury, arsenic, selenium during coal combustion. J. Eng. Thermophys. 24, 793−795.

Guo, X.X., Zheng, C.C., Jia, X.X., 2004. Mobility of arsenic in coal fly ash of different particle size. J. Combust. Sci. Technol. 10, 299−302.

Guo, X., Zheng, C.-G., Xu, M., 2007. Characterization of mercury emissions from a coal-fired power plant. Energ. Fuel. 21, 898–902.

Guttikunda, S.K.S.K., Jawahar, P.P., 2014. Atmospheric emissions and pollution from the coal-fired thermal power plants in India. Atmos. Environ. 92, 449–460.

Gupta, D.K.D.K., Rai, U.N.U.N., Tripathi, R.D.R.D., Inouhe, M.M., 2002. Impacts of fly-ash on soil and plant responses. J. Plant Res. 115 (6), 401–409.

Han, J.J., Xu, M.M., Cheng, J.J., Qiao, Y.Y., Zeng, H.H., 2002. Study of trace element emission factor in coal-fired boilers. J. Eng. Thermophys. 23, 770–772.

Han, J.J., Wang, G.G., Xu, M.M., Yao, H.H., 2009. Experimental study of the vaporization of arsenic and selenium during coal combustion and pyrolysis. J. Huazhong Univ. Sci. Technol. (Nat. Sci. Ed.) 37, 113–116.

Haynes, R.J., 2009. Reclamation and revegetation of fly ash disposal sites – challenges and research needs. J. Environ. Manag. 90, 43–53.

Heinz, G.H., 1996. Selenium in birds. In: Beyer, W.N., Heinz, G.H., Redmon-Norwood, A.W. (Eds.), Environmental Contaminants in Wildlife: Interpreting Tissue Concentrations. CRC Press, Boca Raton, FL, pp. 447–458.

Helble, J.J., 2000. A model for the air emissions of trace metallic elements from coal combustors equipped with electrostatic precipitators. Fuel Process. Technol. 63, 125–147.

Hopkins, W.A., Mendonca, M.T., Rowe, C.L., Congdon, J.D., 1998. Elevated trace element concentrations in southern toads, *Bufo terrestris*, exposed to coal combustion wastes. Arch. Environ. Contam. Toxicol. 35, 325–329.

Hopkins, W.A., Rowe, C.L., Congdon, J.D., 1999. Elevated trace element concentrations and standard metabolic rate in banded water snakes (*Nerodia fasciata*) exposed to coal combustion wastes. Environ. Toxicol. Chem. 18, 1258–1263.

Hopkins, W.A., Congdon, J., Ray, J.K., 2000. Incidence and impact of axial malformations in larval bullfrogs (*Rana catesbeiana*) developing in sites polluted by a coal-burning power plant. Environ. Toxicol. Chem. 19, 862–868.

Hopkins, W.A., Snodgrass, J.W., Staub, B.P., Jackson, B.P., Congdon, J.D., 2003. Altered swimming performance of a benthic fish (*Erimyzon sucetta*) exposed to contaminated sediments. Arch. Environ. Contam. Toxicol. 44, 383–389. <http://www.kuenvbiotech.org/casestudy/.htm>.

IAEA, 2003. Extent of environmental contamination by naturally occurring radioactive material (NORM) and technological options for mitigation. Technical Report Series 419. International Atomic Energy Agency, Vienna, Austria.

IHME, 2013. The global burden of disease 2010: generating evidence and guiding policy. Institute for Health Metrics and Evaluation, Seattle, WA.

I.I.T. Kharagpur, 1999. Draft report of fly ash mission sponsored project "Utilization of Fly Ash and Organic Wastes in Restoration of Crop Land Ecosystem" submitted to fly ash mission.

International Energy Agency, 2013. Statistics: coal and peat. <http://www.iea.org/stats/prodresult.asp?PRODUCT = Coal%20and%20Peat>.

International Energy Outlook, 2009. DOE/EIA-0484. <www.eia.doe.gov/oiaf/ieo/index.html>.

Ito, S., Yokoyama, T., Asakura, K., 2006. Emissions of mercury and other trace elements from coal-fired power plants in Japan. Sci. Total Environ. 368, 397–402.

Iyer, R.S., Scott, J.A., 2001. Power station fly ash e a review of value-added utilization outside of the construction industry. Resour. Conserv. Recycl. 31, 217–228.

Izquierdo, M., Querol, X., 2012. Review article—leaching behavior of elements from coal combustion fly ash: an overview. Int. J. Coal Geol. 94, 54–66.

Jala, S., Goyal, D., 2006. Fly ash as a soil ameliorant for improving crop production—a review. Bioresour. Technol. 97, 1136–1147.

Jamil, S., Abhilash, P.C., Singh, A., Singh, N., Behl, H.B., 2009. Fly ash trapping and metal accumulating capacity of plants: Implication for green belt around thermal power plants. Landsc. Urban Plan. 92 (2), 136−147.

Janz, D.M., DeForest, D.K., Brooks, M.L., Chapman, P.M., Gilron, G., Hoff, D., et al., 2010. Ecological Assessment of Selenium in the Aquatic Environment. CRC Press, Boca Raton, FL, pp. 141−232.

Japan Coal Energy Centre, 2007. Status of coal ash production. <http://www.jcoal.or.jp/coaltech_en/coalash/ash01e.html>.

Jiao, J., 1998. The migration and concentration regularity of harmful trace elements in Permian coal seams of western Guizhou, China. J. Geol. Miner. Resour. North China 13, 236−242.

Jiménez, S., Pérez, M., Ballester, J., 2008. Vaporization of trace elements and their emission with submicrometer aerosols in biomass combustion. Energ. Fuel. 22, 2270−2277.

Kassim, T.A., Williams, K.J., 2005. Environmental impact assessment of recycled wastes on surface and ground waters. Concepts: Methodology and Chemical Analysis. Springer-Verlag, Berlin, Heidelberg, 94 pp.

King, K.A., 1988. Elevated selenium concentrations are detected in wildlife near a power plant. U.S. Department of Interior, Fish and Wildlife Service Research Information Bulletin 88-31.

King, K.A., Custer, T.W., Weaver, D.A., 1994. Reproductive success of barn swallows nesting near a selenium-contaminated lake in east Texas, USA. Environ. Pollut. 84, 53−58.

Kumari, A., Pandey, V.C., Rai, U.N., 2013. Feasibility of fern *Thelypteris dentata* for revegetation of coal fly ash landfills. J. Geochem. Explor. 128, 147−152.

Lemly, D.A., 1996. Selenium in aquatic organisms. In: Beyer, W.N., Heinz, G.H., Redmon-Norwood, A.W. (Eds.), Environmental Contaminants in Wildlife: Interpreting Tissue Concentrations. CRC Press, Boca Raton, FL, pp. 427−455.

Lemly, D.A., 2002. Symptoms and implications of selenium toxicity in fish: the Belews Lake case example. Aquat. Toxic. 57, 39−49.

Linak, W.P., Yoo, J.I., Wasson, S.J., Zhu, W., Wendt, J.O.L., Huggins, F.E., et al., 2007. Ultrafine ash aerosols from coal combustion: characterization and health effects. Proc. Combust. Inst. 31, 1929−1937.

Lior, N., 2010. Sustainable energy development: the present (2009) situation and possible paths future. Energy 35, 3976−3994.

Liu, K., Xie, W., Zhao, Z., Pan, W., Riley, J., 2000. Investigation of polycyclic aromatic hydrocarbons in fly ash from fluidized bed combustion systems. Environ. Sci. Technol. 34, 2273−2279.

Liu, K., Han, W., Pan, W.-P., Riley, J., 2001. Polycyclic aromatic hydrocarbons (PAH) emissions from a coal-fired pilot FBC system. J. Hazard. Mater. B84, 175−188.

Liu, K.L., Heltsley, R., Zou, D.H., Pan, W.P., Riley, J.T., 2002. Polyaromatic hydrocarbon emissions in fly ashes from an atmospheric fluidized bed combustor using thermal extraction coupled with GC/TOF-MS. Energ. Fuel. 16, 330−337.

Love, A., Tandon, R., Banerjee, B.D., Babu, C.R., 2009. Comparative study on elemental composition and DNA damage in leaves of a weedy plant species, *Cassia occidentalis*, growing wild on weathered fly ash and soil. Ecotoxicology 18, 791−801.

Luijk, R., Dorland, C., Kapteijn, F., Govers, H.A.J., 1993. The formation of PCDDs and PCDFs in the catalysed combustion of carbon: implications for coal combustion. Fuel 72, 343−347.

Luo, K., Xu, L., Li, R., Xiang, L., 2002. Fluorine emission from combustion of steam coal of north China plate and northwest China. Chin. Sci. Bull. 47, 1346−1350.

Luo, K., Zhang, X., Chen, C., Lu, Y., 2004. A preliminary estimation of atmospheric arsenic emission from coal-fired power plant in China. Chin. Sci. Bull. 49, 2014−2019.

Mandal, A., Sengupta, D., 2003. Radioelemental study of Kolaghat, thermal power plant, West Bengal, India: possible environmental hazards. Environ. Geol. 44, 180−186.

Markad, V.L., Kodam, K.M., Ghole, V.S., 2012. Effect of fly ash on biochemical responses and DNA damage in earthworm, *Dichogaster curgensis*. J. Hazard. Mater. 215−216, 191−198.

Mastral, A.M., Callén, M.S., 2000. A review on polycyclic aromatic hydrocarbon (PAH) emissions from energy generation. Environ. Sci. Technol. 34, 3051−3057.

Matsumoto, S., Hamanaka, A., Murakami, K., Shimada, H., Sasaoka, T., 2019. Securing topsoil for rehabilitation using fly ash in open-cast coal mines: effects of fly ash on plant growth. J. Pol. Miner. Eng. Soc. <https://doi.org/10.29227/IM-2019-01-02>.

Mattigod, S.V., Dhanpat, R., Eary, L.E., Ainsworth, C.C., 1990. Geochemical factors controlling the mobilization of inorganic constituents from fossil fuel combustion residues: I. Review of the major elements. J. Environ. Qual. 19, 188−201.

Meawad, A.S., Bojinova, D.Y., Pelovski, Y.G., 2010. An overview of metals recovery from thermal power plant solid wastes. Waste Manag. 30, 2548−2559.

Meij, R., te Winkel, H., 2007. The emissions of heavy metals and persistent organic pollutants from modern coal-fired power stations. Atmos. Environ. 41, 9262−9272.

Mishra, L.C., Shukla, K.N., 1986. Effects of fly-ash deposition on growth, metabolism and dry matter production of maize and soybean. Environ. Pollut. 42, 1−13.

Mittra, B.N., Karmakar, S., Swain, D.K., Ghosh, B.C., 2005. Fly ash a potential source of soil amendment and a component of integrated plant nutrient supply system. Fuel 84, 1447−1451.

Mohapatra, R., Rao, J.R., 2001. Some aspects of characterization, utilization and environmental effects of fly ash. J. Chem. Technol. Biotechnol. 76, 9−26.

Mukherjee, A.B., Zevenhoven, R., Bhattacharya, P., Sajwan, K.S., Kikuchi, R., 2008. Mercury flow via coal and coal utilization by-products: a global perspective. Resour. Conserv. Recycl. 52 (4), 571−591.

Nagle, R.D., Rowe, C.L., Congdon, J.D., 2001. Accumulation and selective maternal transfer of contaminants in the turtle *Trachemys scripta* associated with coal ash deposition. Arch. Environ. Contam. Toxicol. 40, 531−536.

NCAB, 2012. Israeli National Coal Ash Board. General information, coal ash production. <http://www.coal-ash.co.il/english/info.html>.

Nelson, P.F., Shah, P., Strezov, V., Halliburton, B., Carras, J.N., 2010. Environmental impacts of coal combustion: a risk approach to assessment of emissions. Fuel 89, 810−816.

Ondov, J.M., Ragaini, R.C., Biermann, A.H., 1979. Emissions and particle-size distributions of minor and trace elements at two western coal-fired power plants equipped with cold-side electrostatic precipitators. Environ. Sci. Technol. 13, 946−953.

Otero-Rey, J.R., Lopez-Vilarino, J.M., Moreda-Pineiro, J., Alonso-Rodriguez, E., Muniategui-Lorenzo, S., Lopez-Mahia, P., et al., 2003. As, Hg, and Se flue gas sampling in a coal-fired power plant and their fate during coal combustion. Environ. Sci. Technol. 37, 5262−5267.

Pacyna, J.M., Pacyna, E.G., Steenhuisen, F., Wilson, S., 2003. Mapping 1995 global anthropogenic emissions of mercury. Atmos. Environ. 37, 109−117.

Pacyna, E.G., Pacyna, J.M., Steenhuisen, F., Wilson, S., 2006. Global anthropogenic mercury emission inventory for 2000. Atmos. Environ. 40, 4048−4063.

Page, A.L., Elseevi, A.A., Straughan, I.R., 1979. Physical and chemical properties of fly ash from coal fired power plants with reference to environmental impact. Residue Rev. 71, 83−120.

Pandey, V.C., 2015. Assisted phytoremediation of fly ash dumps through naturally colonized plants. Ecol. Eng. 82, 1−5.

Pandey, V.C., 2017. Managing waste dumpsites through energy plantations. In: Bauddh, K., Singh, B., Korstad, J. (Eds.), Phytoremediation Potential of Bioenergy Plants. Springer, Singapore, pp. 371−386.

Pandey, V.C., 2012a. Invasive species based efficient green technology for phytoremediation of fly ash deposits. J. Geochem. Explor. 123, 13−18.

Pandey, V.C., 2012b. Phytoremediation of heavy metals from fly ash pond by *Azolla caroliniana*. Ecotoxicol. Environ. Saf. 82, 8−12.

Pandey, V.C., 2013. Suitability of *Ricinus communis* L. cultivation for phytoremediation of fly ash disposal sites. Ecol. Eng. 57, 336−341.

Pandey, V.C., Singh, N., 2010. Impact of fly ash incorporation in soil systems. Agr. Ecosyst. Environ. 136, 16−27.

Pandey, V.C., Singh, K., 2011. Is *Vigna radiata* suitable for the revegetation of fly ash basins? Ecol. Eng. 37, 2105−2106.

Pandey, V.C., Kumar, A., 2012. *Leucaena leucocephala*: an underutilized plant for pulp and paper production. Genet. Resour. Crop. Evol. 60, 1165−1171.

Pandey, V.C., Singh, B., 2012. Rehabilitation of coal fly ash basins: current need to use ecological engineering. Ecol. Eng. 49, 190−192.

Pandey, V.C., Singh, N., 2015. Aromatic plants versus arsenic hazards in soils. J. Geochem. Explor. 157, 77−80.

Pandey, V.C., Mishra, T., 2018. Assessment of *Ziziphus mauritiana* grown on fly ash dumps: prospects for phytoremediation but concerns with the use of edible fruit. Int. J. Phytoremediat. 20 (12), 1250−1256.

Pandey, V.C., Souza-Alonso, P., 2019. Market opportunities in sustainable phytoremediation. In: Pandey, V.C., Bauddh, K. (Eds.), Phytomanagement of Polluted Sites. Elsevier, Amsterdam, The Netherlands, pp. 51−82.

Pandey, V.C., Abhilash, P.C., Singh, N., 2009a. The Indian perspective of utilizing fly ash in phytoremediation, phytomanagement and biomass production. J. Environ. Manag. 90, 2943−2958.

Pandey, V.C., Abhilash, P.C., Upadhyay, R.N., Tewari, D.D., 2009b. Application of fly ash on the growth performance, translocation of toxic heavy metals within *Cajanus cajan* L.: implication for safe utilization of fly ash for agricultural production. J. Hazard. Mater. 166, 255−259.

Pandey, V.C., Singh, J.S., Kumar, A., Tewari, D.D., 2010. Accumulation of heavy metals by chickpea grown in fly ash treated soil: effect on antioxidants. Clean Soil Air Water 38, 1116−1123.

Pandey, V.C., Singh, J.S., Singh, R.P., Singh, N., Yunus, M., 2011. Arsenic hazards in coal fly ash and its fate in Indian scenario. Resour. Conserv. Recy. 55, 819−835.

Pandey, V.C., Singh, K., Singh, R.P., Singh, B., 2012a. Naturally growing *Saccharum munja* on the fly ash lagoons: a potential ecological engineer for the revegetation and stabilization. Ecol. Eng. 40, 95−99.

Pandey, V.C., Singh, N., Singh, R.P., Singh, D.P., 2014. Rhizoremediation potential of spontaneously grown *Typha latifolia* on fly ash basins: study from the field. Ecol. Eng. 71, 722−727.

Pandey, V.C., Prakash, P., Bajpai, O., Kumar, A., Singh, N., 2015a. Phytodiversity on fly ash deposits: evaluation of naturally colonized species for sustainable phytorestoration. Env. Sci. Pollut. Res. 22, 2776−2787.

Pandey, V.C., Bajpai, O., Pandey, D.N., Singh, N., 2015b. *Saccharum spontaneum*: an underutilized tall grass for revegetation and restoration programs. Genet. Resour. Crop. Evol. 62 (3), 443−450.

Pandey, V.C., Bajpal, O., Singh, N., 2016. Energy crops in sustainable phytoremediation. Renew. Sust. Energ. Rev. 54, 58–73.

Pandey, V.C., Rai, A., Korstad, J., 2019. Aromatic crops in phytoremediation: from contaminated to waste dumpsites. In: Pandey, V.C., Bauddh, K. (Eds.), Phytomanagement of Polluted Sites. Elsevier, Amsterdam, The Netherlands, pp. 255–275.

Papastefanou, C., 2008. Radioactivity of coals and fly ashes. J. Radioanal. Nucl. Chem. 275, 29–35.

Peles, J.D., Barrett, G.W., 1997. Assessment of metal uptake and genetic damage in small mammals inhabiting a fly ash basin. Bull. Environ. Contam. Toxicol. 59, 279–284.

Pirrone N., Cinnirella S., Feng X., Finkelman R.B., Friedli H.R., Leaner J., et al., 2009. Global mercury emissions to the atmosphere from natural and anthropogenic sources. Springer, New York, USA, chap. 1, 3–49.

Poliakova, V.A., Suchko, V.A., Tereshchenko, V.P., Bazyka, D.A., Golovinia, O.M., Rudavskaia, G.A., 2001. Invasion of microorganisms in the bronchial mucosa of liquidators of the Chernobyl accident consequences. Mikrobiology 63, 41–50.

Praharaj, T., Swain, S.P., Powell, M.A., Hart, B.R., Tripathi, S., 2002. Delineation of groundwater contamination around an ash pond geochemical and GIS approach. Environ. Int. 27, 631–638.

Prayas, 2011. Thermal Power Plants on the Anvil: Implications and Need for Rationalisation. Prayas Energy Group, Pune, India.

Prayas, 2013. Black and Dirty: The Real Challenges Facing India's Coal Sector. Prayas Energy Group, Pune, India.

Querol, X., Fernandez Turiel, J.L., Lopez Soler, A., Duran, M.E., 1992. Trace elements in high-S subbituminous coals from the Teruel mining district, northeast Spain. Appl. Geochem. 7, 547–561.

Ram, L.C., Masto, R.E., 2010. Review: an appraisal of the potential use of fly ash for reclaiming coal mine spoil. J. Environ. Manag. 91, 603–617.

Ram, L.C., Masto, R.E., 2014. Fly ash for soil amelioration: a review on the influence of ash blending with inorganic and organic amendments. Earth Sci. Rev. 128, 52–74.

Ram, L.C., Srivastava, N.K., Tripathi, R.C., Jha, S.K., Sinha, A.K., Singh, G., et al., 2006. Management of mine spoil for crop productivity with lignite fly ash and biological amendments. J. Environ. Manag. 79, 173–187.

Readman, J.W., Fillmann, G., Tolosa, I., Bartocci, J., Villeneuve, J.-P., Catinni, C., et al., 2002. Petroleum and PAH contamination of the Black Sea. Mar. Pollut. Bull. 44, 48–62.

Reash, R.J., Lohner, T.W., Wood, K.V., 2006. Selenium and other trace metals in fish inhabiting a fly ash stream: implications for regulatory tissue thresholds. Environ. Pollut. 142, 397–408.

Reddy, M.S., Basha, S., Joshi, H.V., Jha, B., 2005. Evaluation of the emission characteristics of trace metals from coal and fuel oil fired power plants and their fate during combustion. J. Hazard. Mater. B123, 242–249.

Rowe, C.L., Kinney, O.M., Fiori, A.P., Congdon, J.D., 1996. Oral deformities in tadpoles (*Rana catsbeiana*) associated with coal ash deposition: effects on grazing ability and growth. Freshw. Biol. 36, 723–730.

Rowe, C.L., 1998. Elevated standard metabolic rate in a freshwater shrimp (*Palaemonetes paludosus*) exposed to trace element-rich coal combustion waste. Comp. Biochem. Physiol. Part A 121 (4), 299–304.

Roy, W.R., Thiery, R.G., Schuller, R.M., Suloway, J.J., 1981. Coal fly ash: a review of the literature and proposed classification system with emphasis on environmental impacts. In: Environmental Geology Notes 96. Illinois State Geological Survey, Champaign, IL.

Ruhl, L., Vengosh, A., Dwyer, G.S., Hsu-Kim, H., Deonarine, A., Bergin, M., et al., 2009. Survey of the potential environmental and health impacts in the immediate aftermath of the coal ash spill in Kingston, Tennessee. Environ. Sci. Technol. 43, 6326–6333.

Ruhl, L., Vengosh, A., Dwyer, G.S., Hsu-Kim, H., Deonarine, A., 2010. Environmental impacts of the coal ash spill in Kingston, Tennessee: an 18-month survey. Environ. Sci. Technol. 44, 9272–9278.

Sahu, K.C., 1994. Power plant pollution: cost of coal combustion. Survey of the Environment. The Hindu, Madras, pp. 47–51.

Sahu, S.K., Bhangare, R.C., Ajmal, P.Y., Sharma, S., Pandit, G.G., Puranik, V.D., 2009. Characterization and quantification of persistent organic pollutants in fly ash from coal fueled thermal power station in India. Microchem. J. 92, 92–96.

Schilling, C.J., Tams, I.P., Schilling, R.S.F., Nevitt, A., Rossiter, C.E., Wilkinson, B., 1988. A survey into the respiratory effects of prolonged exposure to pulverized fuel ash. Br. J. Ind. Med. 45, 810–817.

Schroeder, W.H., Munthe, J., 1998. Atmospheric mercury – an overview. Atmos. Environ. 32, 809–822.

Schwarzbauer, J., 2006. Organic Contaminants in Riverine and Groundwater Systems – Aspects of the Anthropogenic Contribution. Springer-Verlag, Berlin, Heidelberg, p. 464.

Sekine, Y., Sakajiri, K., Kikuchi, E., Matsukata, M., 2008. Release behavior of trace elements from coal during high-temperature processing. Powder Technol. 180, 210–215.

Shaheen, S.M., Peter, S., Hooda, P.S., Tsadilas, C.D., 2014. Opportunities and challenges in the use of coal fly ash for soil improvements -a review. J. Environ. Manag. 145, 249–267.

Sharma, S., 1989. Fly ash dynamics in soil water systems. Crit. Rev. Environ. Control. 19 (3), 251–275.

Singh, D.N., Kolay, P.K., 2002. Simulation of ash–water interaction and its influence on ash characterization. J. Prog. Energ. Combust. Sci. 28, 267–299.

Singh, J.S., Pandey, V.C., 2013. Fly ash application in nutrient poor agriculture soils: impact on methanotrophs population dynamics and paddy yields. Ecotoxicol. Environ. Saf. 89, 43–51.

Singh, S., Ram, L.C., Masto, R.E., Verma, S.K., 2011a. A comparative evaluation of minerals and trace elements in the ashes from lignite, coal refuse, and biomass fired power plants. Int. J. Coal Geol. 87, 112–120.

Singh, J.S., Pandey, V.C., Singh, D.P., Singh, R.P., 2011b. Coal fly ash and farmyard manure amendments in dry-land paddy agriculture field: effect on N – dynamics and paddy productivity. Appl. Soil. Ecol. 47, 133–140.

Singh, K., Pandey, V.C., Singh, B., Patra, D.D., Singh, R.P., 2016. Effect of fly ash on crop yield and physico-chemical, microbial and enzyme activities of sodic soils. Environ. Eng. Manag. J. 15 (11), 2433–2440. <http://omicron.ch.tuiasi.ro/EEMJ/>.

Skousen, J., Ziemkiewicz, P., Yang, J.E., 2012. Use of coal combustion by-products mine reclamation: review case studies in the USA. Geosyst. Eng. 15, 71–83.

Srivastava, N.K., Ram, L.C., 2010. Reclamation of coal mine spoil dump through fly ash and biological amendments. Int. J. Ecol. Dev. 17 (F10), 17–33.

Staub, B.P., Hopkins, W.A., Novak, J., Congdon, J.D., 2004. Respiratory and reproductive characteristics of eastern mosquitofish (*Gambusia holbrooki*) inhabiting a coal ash settling basin. Arch. Environ. Contam. Toxicol. 46, 96–101.

Streets, D.G., Hao, J., Wu, Y., Jiang, J., Chan, M., Tian, H., et al., 2005. Anthropogenic mercury emissions in China. Atmos. Environ. 39, 7789–7806.

Sun, J., Yao, Q., Liu, H., Lu, J., Yin, G., Zhao, C., 2004. Distribution of arsenic in PM10 and PM2.5 caused by coal combustion and its enrichment mechanism. J. China Coal Soc. 29, 78−82.

Tadmore, J., 1986. Radioactivity from coal-fired power plants: a review. J. Environ. Radioact. 4, 177−244.

Tang, Z., Ma, S., Ding, J., Wang, Y., Zheng, S., Zhai, G., 2013. Current status and prospect of fly ash utilisation in China. In: World of Coal Ash (WOCA) Conference April, Lexington, KY, 22−25. <www.flyash.info>.

Taylor Jr., E.M., Schuman, G.E., 1988. Fly-ash and lime amendment of acidic coal spoil to aid revegetation. J. Environ. Qual. 17, 120−124.

Tian, H., Wang, Y., Xue, Z., Qu, Y., Chai, F., Hao, J., 2011. Atmospheric emissions estimation of Hg, As, and Se from coal-fired power plants in China, 2007. Sci. Total Environ. 409, 3078−3081.

Tolle, D.A., Arthur, M.F., Pomeroy, S.E., 1982. Fly Ash Use for Agriculture and Land −Reclamation: A Critical Literature Review and Identification of Additional Research Needs. RP. Battelle Columbus Laboratories, Columbus, OH, pp. 1224−1225.

UNEP, 2010. Study on mercury sources and emissions, and analysis of cost and effectiveness of control measures. <http://www.unep.org/hazardoussubstances/LinkClick. aspx?fileticket = BCcJWXkn0aQ%3D&tabid = 3593&mid = 6077>.

UNSCEAR, 1982. Ionizing radiation, sources and biological effects. Reportto Gen. Assembly, AnnexC 108−110.

US EPA, 2007. Human and Ecological Risk Assessment of Coal Combustion Wastes (Draft), Research Triangle Park, NC.

US EPA, 2008. Particulate matter. <http://www.epa.gov/air/particlepollution/health. html>

USGS. 2002. Characterization and modes of occurrence of elements and fly ash:An integrated approach. USGS Fact Sheet 0038-02. USGS, Denver, CO.

Uyanik, S., Topeli, M., 2012. Development fly ash utilization in Turkey and contribution of Isken to the market. In: EUROCOALASH 2012 Conference, Session I. CCP Production, Management and Marketing, Thessaloniki, Greece, September 25−27, 2012.

Vassilev, S.V., Vassileva, C.G., 2006. A new approach for the classification of coal fly ashes based on their origin, composition, properties, and behavior. Fuel 86, 1490−1512.

Vassilev, S.V., Vassileva, C.G., Karayigit, A.I., Bulut, Y., Alastuey, A., Querol, X., 2005. Phase mineral and chemical composition of fractions separated from composite fly ashes at the Soma Power Station, Turkey. Int. J. Coal Geol. 61, 65−85.

VGB Power Tech., 2010. Statistics on Production and Utilisation of By-products From Coal-Fired Power Plants, Germany.

Verma, S.K., Singh, K., Gupta, A.K., Pandey, V.C., Trivedi, P., Verma, R.K., et al., 2014. Aromatic grasses for phytomanagement of coal fly ash hazards. Ecol. Eng. 73, 425−428.

Wang, S., 2008. Application of solid ash based catalysts in heterogeneous catalysis. Environ. Sci. Technol. 42, 7055−7063.

Wang, Y., Ren, D., Xie, H., 1995. The distribution and migration of trace elements in coal combustion. Coal Mine Environ. Prot. 9, 25−28.

Wagner, N.J., Hlatshwayo, B., 2005. The occurrence of potentially hazardous trace elements in five Highveld coals, South Africa. Int. J. Coal Geol. 63, 228−246.

Ward, C.R., French, D., 2006. Determination of glass content and estimation of glass composition in fly ash using quantitative X-ray diffractometry. Fuel 85, 2268−2277.

Wild, S.R., Jones, K.C., 1995. Polynuclear aromatic hydrocarbons in the United Kingdom environment: a preliminary source inventory and budget. Environ. Pollut. 88, 91−108.

Yokoyama, T., Asakura, K., Matsuda, H., Ito, S., Noda, N., 2000. Mercury emissions from a coal-fired power plant in Japan. Sci. Total. Environ. 259, 97–103.

Yunusa, I.A.M., Eamus, D., DeSilva, D.L., Murray, B.R., Burchett, M.D., Skilbeck, G.C., et al., 2006. Fly-ash: an exploitable resource for management of Australian agricultural soils. Fuel 85, 2337–2344.

Zandi, M., Russell, N.V., 2007. Design of a leaching test framework for coal fly ash accounting for environmental conditions. Environ. Monit. Assess. 131, 509–526.

Zhang, L., Zhuo, Y., Chen, L., Xu, X., Chen, C., 2008. Mercury emissions from six coal-fired power plants in China. Fuel Process. Technol. 89, 1033–1040.

Zhou, Y., 1998. Distribution type and occurrence form of arsenic in anthracite of Laochang mining area. Coal Geol. Explor. 26, 8–13.

Zhou, J., Zhang, L., Luo, Z., Hu, C., He, S., Zheng, J., et al., 2008. Study on mercury emission and its control for boiler of 300MW unit. Therm. Power Gener. 37, 22–27.

Zielinski, R.A., Finkelman, R.B., 1997. Radioactive elements in coal and fly ash: abundance, forms and environmental significance. US Geological Survey Fact Sheet FS. USGS, Denver, CO, pp. 163–197, <http://www.acaa.usa.org/PDF/FS-163-97.pdf>.

## Further reading

ASTM standard specification for coal fly ash and raw or calcined natural pozzolan for use in concrete (C618-05). In: Annual Book of ASTM Standards, Concrete and Aggregates, vol. 04. 02. American Society for Testing Materials; 2005.

USGS, 1997. Radioactive elements in coal fly ash: abundance, forms, and environmental significance. USGS Fact Sheet FS-163-97. USGS, Denver, CO.

# CHAPTER 2

# Bioremediation of fly ash dumpsites—holistic approach

## Contents

## 2.1 The fly ash dumps

Fly ash (FA), a by-product of coal combustion, is generated from coal-based thermal power stations all over the world due to energy demand. The quantity of FA generation is increasing persistently in meeting the global energy demand, which could not yet find a safe and feasible option for its disposal (Pandey et al., 2009). As a result, FA dumps have encroached thousands of hectares of land adjacent to thermal power stations. Some countries with higher rates of FA generation include the United States, China, Europe, India, Australia, Japan, South Africa, Greece, and Italy. Thus most of the countries will continue this practice for a long time into the future in view of seeking energy demand due to high coal reserves. FA production in India is estimated to increase to 300 million tons yearly by 2016−17 (Skousen et al., 2012), while global FA production is projected to exceed 750 million tons yearly, but only <50% of world's production is utilized (Izquierdo and Querol, 2012). The World Bank has notified that the FA disposal in India would need about $1000 \text{ km}^2$ of land by 2015 (Maiti and Nandhini, 2006), while global FA disposal would require around $3235 \text{ km}^2$ of land by 2015. The

*Phytomanagement of Fly Ash*
ISBN: 978-0-12-818544-5
DOI: https://doi.org/10.1016/B978-0-12-818544-5.00002-X

FA contains different types of pollutants, such as heavy metals, metalloids, radioactive elements, and polycyclic aromatic hydrocarbons, which pollutes both terrestrial and aquatic ecosystems (Pandey and Singh, 2010; Pandey et al., 2010, 2011).

Huge quantities of FA are being disposed in ash ponds across the nations. FA generated in large amounts can be utilized in multiple applications such as cement production, concrete products, cover material, structural fill, roadway, lightweight aggregate, underground void filling, infiltration barrier, and soil, water, and environmental improvement. The worldwide average FA utilization is around 25% (Wang, 2008). Thus it is clear that a major amount of the world's FA is unutilized and requires urgent actions to recognize potential opportunities. Moreover, thousands of hectares of land have been occupied for the storage of FA across the nations, which could be reclaimed through different plant species (Pandey et al., 2012b). In general, two types of regions such as terrestrial and aquatic regions are present in FA dumping sites, and created due to wet disposal method of ashes. After many years of FA disposal, some plants show natural colonization on both regions. Thus terrestrial and aquatic plants have been identified for remediation of both sites.

## 2.2 Fly ash bioremediation

FA dumps rank highly in the most hazardous environments across the nations, creating pollution in the surrounding ecosystems through wind erosion, leaching, or seepage (Pandey et al., 2011; Pandey and Singh, 2012). Consequently, several heavy metals and metalloids of FA are added to environments, and thus pose a hazard to receiving habitats (Pandey et al., 2009). Even that a human health-related problem has also been noticed in inhabitants of coal-fired power stations (US EPA, 2007). Therefore the bioremediation of FA dumps is a pressing need all over the world. In this regard, several researches on phytoremediation and microbial remediation have been done worldwide to mitigate FA pollution. Both plant- and microbe-based approaches will be explored in detail in Sections 2.2.1 and 2.2.2, respectively.

### 2.2.1 Phytoremediation

Phytoremediation of FA dumps in order to stop environmental pollution is a pressing need across the nations. In general, the FA dumps do not support plant's growth and establishment due to its harsh conditions.

Some plant growth limiting factors of FA include extreme alkalinity, toxicity of metal(loid)s, unavailability of phosphorus, lack of nitrogen, and organic matter (Pandey and Singh, 2010). These situations play a major role to prevent the natural colonization of plants on abandoned FA dumps. Still some plants grow naturally on FA dumps after few years of ash deposition due to their inherent ability. In this regard, globally, several naturally growing vegetations have been noticed on derelict FA dumps (Maiti and Jaiswal, 2008; Pandey, 2012a,b, 2013; Pandey et al., 2012b, 2014a, 2015b; Gajić et al., 2016, 2019). A number of plant species have been identified from FA dumps to remediate metal(loid)s all over the world. These plant species can be characterized in two categories, that is, phytostabilization and phytoextraction on the basis of their phytoremediation potential. Phytostabilization approach reduces the mobility and bioavailability of metal(loid)s, phenols, and chlorinated solvents in the environment to inhibit their migration in the soil and water, and thus checks their entry in the food chain (Nwoko, 2010). This technique uses excluders that have the ability to arrest pollutants in the substrate or roots (Prasad, 2006). Therefore phytostabilization is an effective and important method, which can be applied on large areas with multielement contamination, for example, FA dumps. Appropriate plants for phytostabilization must have extended root systems and the ability to immobilize pollutants at the rhizosphere level with a restricted transport to aerial parts over wide-ranging substrate concentrations (Baker, 1981).

The phytostabilizers of FA dumps are *Ricinus communis* (Cd, Cu, Ni, Pb, Zn; Pandey, 2013), *Saccharum munja* (Cr, Cd, Cu, Pb, Fe, Mn, Zn, Ni; Pandey et al., 2012b), *Typha latifolia* (Cr, Cd, Pb, Cu, Ni, Zn; Pandey et al., 2014a), *Ipomea carnea* (Cu, Cr, Ni, Pb; Pandey, 2012b), *Thelypterys dentate* (As, Cd, Fe, Pb, Si; Kumari et al., 2013), *Festuca rubra* (As, B, Cu, Mn, Zn; Gajić et al., 2016), *Sida cardifolia* (Co, Ni, Pb; Gupta and Sinha, 2008), *Blumea lacera* (Co, Mn; Gupta and Sinha, 2008), *Cynodon dactylon* (Cd, Cu, Fe, Mn, Ni, Pb; Kumar et al., 2015), *Calotropis procera* (Co, Zn; Gupta and Sinha, 2008), *Cassia tora* (Cr, Cu, Fe; Gupta and Sinha, 2008), *Chenopodium album* (Cd, Cr, Mn; Gupta and Sinha, 2008), *Jatropha curcas* (As, Cd, Co, Cr, Cu, Fe, Li, Mn, Mo, Ni, Pb, Pt, Se, Sn, Zn; Vurayai et al., 2017), *Pennisatum clandestinum* (Cd, Co, Cu, Fe, Li, Mn, Mo, Ni, Pb, Sn; Vurayai et al., 2017). The phytostabilizer also improves the physicochemical and biological properties of the substrate of FA dumps by enhancing the organic matter content. Thus the important phytostabilizers of FA dumps belong to the genus of *Acer, Calotropis, Carex, Cassia,*

*Cynodon, Euphorbia, Festuca, Ipomea, Leucaena, Lolium, Lotus, Medicago, Miscanthus, Phragmites, Populus, Vicia, Saccharum, Salix, Sesbania, Sida, Thelypterys, Trifolm,* and *Typha* (Gupta and Sinha, 2008; Pandey and Singh, 2012 Pandey, 2013; Kumari et al., 2016; Gajić et al., 2016; 2019).

Phytoextraction approach uses plants that have the ability to absorb and accumulate metal(loid)s in below and above ground plant parts to recover valuable elements (Keller et al., 2005). The phytoextractor must have fast growth rate, high biomass, extensive root system, high root-shoot transfer, and high tolerance against metal(loid)s (Tong et al., 2004). The phytoextractors of FA dumps are *T. latifolia* (Mn; Pandey et al., 2014a), *I. carnea* (Cd; Pandey, 2013), *C. procera* (Cd, Cr, Cu, Fe, Mn, Ni, Pb; Gupta and Sinha, 2008), *C. tora* (Cd, Co, Ni, Pb, Zn; Gupta and Sinha, 2008), *B. lacera* (Cd, Fe, Zn; Gupta and Sinha, 2008), *C. dactylon* (Zn; Kumar et al., 2015), *S. cardifolia* (Cd, Cr, Cu, Fe, Zn; Gupta and Sinha, 2008), and *C. album* (Co, Cu, Fe, Ni, Pb, Zn; Gupta and Sinha, 2008).

Moreover, many aquatic species have been recognized as potential plant species for the remediation and management of aquatic sites of FA dumps. *Azolla caroliniana*, water fern, has been reported as a potential asset for phytoremediation of heavy metals (i.e., Fe, Zn, Ni, Cu, Mn, Cr, Pb, and Cd) of FA ponds due to their best adaptation and plant growth rate (Pandey, 2012a). A field-based research showed that naturally grown *Eichhornia crassipes* has the potential for the phytoremediation of heavy metals (i.e., Zn, Cu, Cr, Cd, and Pb) from aquatic regions of FA dumps (Pandey, 2016). Another field-based study exposed that *T. latifolia* is an appropriate plant for the rhizoremediation of heavy metals (i.e., Zn, Mn, Cu, Pb, Cd, Cr, and Ni) in FA ponds due to lower translocation factor values than one (Pandey et al., 2014a). Thus the natural and lush growth of *A. caroliniana*, *E. crassipes*, and *T. latifolia* on aquatic sites of FA dumps has shown some of their significant and desirable features such as easy and fast multiplication with high biomass, toxitolerant against alkalinity and heavy metals, and easily harvestable and colonizer nature (Table 2.1). Certain biofuel crop species with phytoremediation potential are value declaring, mainly with respect to FA deposits. These are *R. communis* (Huang et al., 2011; Pandey, 2013), *S. munja* and *Saccharum spontaneum* (Pandey et al., 2012b, 2015c), *Salix* species (Hilton, 2002; Volk et al., 2006), *Populus* species (Rockwood et al., 2004), *J. curcas* (Jamil et al., 2009; Pandey et al., 2012a, 2016b), *Panicum virgatum* (Awoyemi and Dzantor, 2017), and *Miscanthus giganteus* (Técher et al., 2012).

**Table 2.1** Dominant and naturally grown plant species on fly ash deposits at Feroze Gandhi Unchahar Thermal Power Station, Raebareli, Uttar Pradesh, India.

| S. no. | Species description | Images of species growing on fly ash dumps |
|---|---|---|
| 1. | *P. juliflora* (Sw.) DC.<br>*Origin*: native to Mexico, Central and northern South America<br>*Family*: Fabaceae<br>*Hindi name*: Vilayati babul<br>*Growth habit*: tree 3—12 m tall, sometimes shrubby with spreading branches, woody<br>*Leaves*: bipinnate, glabrous or pubescent, 1—3 pairs of pinnae, leaflets 6—23 mm long $\times$ 1.6—5.5 mm wide<br>*Phenology*: flowering occurs shortly after bud-bursting in the beginning of spring, when the weather becomes warmer<br>*Propagation*: seed<br>*Socioeconomic importance*: fuel wood, restoration |  |
| 2. | *S. spontaneum* L.<br>*Origin*: India. Widely distributed in the warm regions of old world tropics (Asia, Africa, and Australia)<br>*Family*: Poaceae<br>*Hindi name*: Kans grass<br>*Growth habit*: tufted and rhizomatous perennial grass, stems with inflorescence 2—6 m tall<br>*Leaves*: mostly glabrous with ensheating base, 1—2 cm wide<br>*Phenology*: flowering from July to September<br>*Propagation*: by rhizomes and stem fragments, also by seeds<br>*Socioeconomic importance*: medicinal value, ethanol production, and good fibers |  |
| 3. | *S. munja* Roxb.<br>*Origin*: Native to Indian subcontinent<br>*Family*: Poaceae<br>*Hindi name*: Munja grass<br>*Growth habit*: tufted and rhizomatous perennial grass, it grows up to 7 ft (2 m) in height<br>*Leaves*: leaf blades are long and sharp, and growing up to 3-ft long and 1-cm (10-mm) wide |  |

*(Continued)*

**Table 2.1** (Continued)

| S. no. | Species description | Images of species growing on fly ash dumps |
|---|---|---|
| | *Phenology*: flowering from October to December<br>*Propagation*: by rhizomes and stem fragments, also by seeds<br>*Socioeconomic importance*: making huts, thatching roofs, baskets, brooms, hand fans, mats, and ropes | |
| 4. | *C. dactylon* L. (Pers.)—$C_4$ grass<br>*Origin*: throughout the tropical and subtropical areas of the world<br>*Family*: Poaceae<br>*Hindi name*: Doob, *English name*: Bermuda grass<br>*Growth habit*: ground cover, it creeps lavishly by scaly rhizomes or strong stolons<br>*Leaves*: alternate-distal pattern, leaf blades (1—15 cm) are green to dull-green, and open up to the base<br>*Phenology*: flowering from August to October<br>*Propagation*: seeds, stolons, rooted runners, and rhizomes<br>*Socioeconomic importance*: grazing grass by livestock, medicinal use, sacred grass in Hindu rituals, and forage resource |  |
| 5. | *T. latifolia* L.—Aggressive colonizer<br>*Origin*: Native to Africa (northern and eastern), America (North), Europe, the Middle East, and Asia (western and northern)<br>*Family*: Typhaceae<br>*Hindi name*: Goam, *English name*: Cattail<br>*Growth habit*: Plant's height range normally from 1 to 3 m<br>*Habitat*: aquatic, wetlands<br>*Leaves*: broad and flat leaves (2—4 cm) with a sheath at the base<br>*Phenology*: flowering from June to August<br>*Propagation*: seed and rhizome (vigorous) |  |

*(Continued)*

**Table 2.1** (Continued)

| S. no. | Species description | Images of species growing on fly ash dumps |
|---|---|---|
| | *Socioeconomic importance*: building material (rafts, boats, mats, and huts), paper, fibers, biofuel, wild food source, etc. | |
| 6. | *E. crassipes* L.—Invasive Species<br>*Origin*: Native of South America, but is now naturalized in Africa, Australia, India, and many other countries<br>*Family*: Pontederiaceae<br>*Hindi name*: Jalkumbhi, *English name*: water hyacinth<br>*Growth habit*: It can grow up to 1 m in height<br>*Habitat*: aquatic<br>*Leaves*: it consists of petiole (swollen) and blade (shape-round, ovoid, or kidney)<br>*Phenology*: flowering from May to September<br>*Propagation*: seed and vegetative<br>*Socioeconomic importance*: bioenergy, phytoremediation of waste water, and medicine |  |
| 7. | *A. caroliniana* Willd.<br>*Origin*: North America; specific native range— Eastern North America and the Caribbean; widely distributed throughout the tropics and subtropics<br>*Family*: Azollaceae<br>*English name*: mosquito plant, water velvet<br>*Growth habit*: it can grow from 1 to 2 cm in height and forms mats to 4-cm thick<br>*Habitat*: still water—ponds, ditches, and marshes<br>*Leaves*: leaf like structure present<br>*Phenology*: no flowering, reproduces by spores<br>*Propagation*: runners<br>*Socioeconomic importance*: fertilizer, eradication of mosquito, and phytoremediation of contaminated water |  |

A number of plant species grow naturally on derelict and barren FA deposits across the nations. The selection of plant species for phytoremediation proposes of the FA deposits would be better, if we select potential plants after the assessment of phytodiversity on FA deposits, and then we can obtain adaptive plant species for FA phytoremediation. These adaptive plant species have higher capability to tolerate high radiation, temperatures, and drought along with FA toxicity and deficiency of essential plant nutrients such as N and P in FA (Pilon-Smits, 2005; Pandey et al., 2009; Pandey, 2012a; Kostić et al., 2012; Gajić et al., 2013, 2016). Therefore the study of ecological succession of FA deposits is very important in view of the remediation and management of the FA deposits all over the world. Several researchers have worked on ecological succession of the FA deposits worldwide (Pandey et al., 2015a; Chu, 2008; Chibrik et al., 2016; Jasionkowski et al., 2016). In general, native pioneer species (primary colonizer) are naturally adapted plant species especially grasses against multiple stress tolerances of FA deposits. In this regard, some plant species showed a great promise in view of their best adaptation and multiple uses. The ability of FA deposit revegetation has been explored in several countries worldwide: in Australia (Jusaitis and Pillman 1997), India (Raj and Mohan 2008; Babu and Reddy 2011a; Pandey et al., 2015b), China (Chu 2008), the United States (Bilski et al., 2011), Kosovo in Europe (Mustafa et al., 2012), and Poland (Jasionkowski et al., 2016). *Saccharum* is a genus of tall perennial grasses, which has the ability to colonize naturally on the barren FA deposits in India. Some important species of *Saccharum* are *S. munja*, *S. spontaneum*, and *S. reveneae*, which are reported as native colonizers of barren FA deposits across India (Pandey et al., 2012b, 2015b; Rau et al., 2009). *Saccharum* species have an extensive root system with tall thick clumps that bind FA particles, multiply with root suckers and profuse seeding, and produce high biomass (Pandey et al., 2015b, c). These grasses are widely used in making of ropes, brooms, baskets, hand fans, huts, and mats. The rational utilization of these grasses may support the rural socioeconomic development of the nearby communities of FA deposits. Therefore *S. munja* and *S. spontaneum* are recognized as greatest primary colonizers for the revegetation/rehabilitation of barren FA deposits (Table 2.1) due to having potential features such as perennial, self-propagation, extensive root system, good FA binder, unpalatable at mature stage to animals, and tolerance to a wide-ranging environmental and edaphic stresses (Sharma et al., 2011; Pandey and Singh, 2011, 2014).

*S. spontaneum* grass quickly colonizes on the harsh conditions of the FA substrate, establishes a thick green cover, increases FA-substrate fertility, and offers sustainable microclimates for starting secondary colonizers (Pandey and Singh, 2014). The massive fibrous root system of these grasses prevents FA erosion, controls leaching of metal(loid)s, increases soil carbon, and preserves soil moisture. After decay of their biomass, they supply organic carbon and nutrients to upper layer of the FA substrate (Maiti and Maiti, 2015). Certain spontaneously growing grass species on FA deposits have been assessed for their phytoremediation efficiency: *S. munja* (Pandey et al., 2012b), *S. spontaneum* (Pandey et al., 2015b), *S. reveneae* (Rau et al., 2009), *Vetiveria zizanioides* (Verma et al., 2014), and *Cymbopogon citratus* (Panda et al., 2018) toward environmental and socioeconomic benefits (Pandey, 2015). To enhance the economic return of FA dumps' rehabilitation programs, ecologically and socioeconomically valuable tree species should be carefully chosen as secondary colonizers (Pandey and Singh, 2011, 2012). In addition, the leguminous plant species have been noticed special significance in the revegetation program of FA deposits due to their soil improving properties and rapid proliferation. Fast-growing, drought-resistant, and carbon- and nitrogen-fixing species are mostly used, and they readily produce decomposable litter in soil that releases essential nutrients and thus contributes to soil fertility (Singh et al., 2002; Jambhulkar and Juwarkar, 2009). *Prosopis juliflora*, a leguminous plant, has been reported as a naturally growing tree on FA deposits with a luxurious growth and serves as a potential plant for carbon sequestration on FA deposits (Pandey et al., 2016a). *P. juliflora* is also suggested for phytoremediation of FA deposits (Rai et al., 2004).

Naturally colonized plants on abandoned FA deposits were noted by a number of researchers across the nations (Pavlović et al., 2004; Gupta and Sinha, 2008; Mitrović et al., 2008; Pandey et al., 2012b, 2015a, 2016a; Pandey, 2015; Kumari et al., 2016). Hodgson and Townsend (1973) revealed that the plant species tolerant to harsh conditions of FA lagoons are: *F. rubra*, *L. perenne*, *Medicago sativa*, *Melilotus* sp., *Onobrychis sativa*, *Oryza sativa*, *Trifolium pretense*, *Trifolium repens*, and *Triticum* sp. Mulhern et al. (1989) showed by their experiments that herbaceous plants, such as *Festuca* sp., *Melilotus* sp., and *Agropyron* sp., grow better on FA deposits than trees. Furthermore, Shaw (1996) documented several herbaceous plant species on the FA basin in England: *Achillea millefolium*, *Agrostis stolonifera*, *Arrenatherrum elatius*, *Atriplex prostrate*, *Chenopodium rubrum*, *Cirsium arvense*, *Dactylis glomerata*, *Epilobium hirsitum*, *F. rubra*, *Lotus corniculatus*,

*Lycopus europaeus, Melilotus officinalis, Dactylorhiza* sp., *Oenothera biennis, Plantago lanceolata, Rorippa sylvestris, Reseda lutea, Rumex obtusifolia, Sonchus oleraceus, Senecio vulgaris, Suaeda maritime, T. repens, Tussilago farfara, Vicia sativa*, etc. Moreover, a number of herbaceous plants have been reported, which spontaneously colonize FA deposits in India: *Amaranthus deflexus, C. tora, C. procera, C. dactylon, C. album, Fimbristylis dichotoma, I. carnea, Saccharum nunja, S. spontaneum, Sida cordifolia*, and *T. latifolia* (Maiti and Jaiswal, 2008; Gupta and Sinha, 2008; Pandey et al., 2012b, 2015b, c; Pandey and Singh, 2014; Pandey, 2015). Plants that are able to grow and survive on the barren FA dumps mostly belong to the families: *Amaranthaceae, Aizoaceae, Asteraceae, Apiaceae, Azollaceae, Asclepidaceae, Brassicaceae, Boraginaceae, Chenopodiaceae, Caesalpinaceae, Convoluvaceae, Caryophylaceae, Cyperaceae, Euphorbiaceae, Elegnaceae, Equisetaceae, Fabaceae, Geraniaceae, Gentiaceae, Hypericaceae, Lamiaceae, Lythraceae, Moraceae, Malvaceae, Onagraceae, Poaceae, Portulaceae, Papaveraceae, Plantaginaceae, Polygonaceae, Rosaceae, Rhamnaceae, Resedaceae, Rubiaceae, Sapindaceae, Salicaceae, Solanaceae, Scropulariaceae, Tamaricaceae*, and Verbenaceae (Jusaitis and Pillman, 1997; Morgenthal et al., 2001; Pandey et al., 2015b; Gajić et al., 2019). Fig. 2.1 represents some naturally colonized plant species and planted species on the FA deposits of the 11-year-old passive cassette at the largest thermal power plant Nikola Tesla (TENT−A), Obrenovac, Belgrade, Serbia. Fig. 2.2 also shows spontaneously colonizing plant species on the abandoned FA deposited sites at Feroze Gandhi Unchahar Thermal Power Station, Raebareli, Uttar Pradesh, India. A phytorestored site of FA deposits can be used for silviculture or in plantations of tree species with an economic value, and explored in Chapter 10, An appraisal on phytomanagement of fly ash with economic returns.

## 2.2.2 Microbial remediation

Microbial remediation is a cost-effective, eco-friendly, and sustainable approach that uses different types of microbes to detoxify, degrade, and mineralize or biotransform pollutants to a harmless state. It involves the ex situ or in situ application of microbes, including archaea, actinomycetes, bacteria, and fungus, which are capable of removing, neutralizing, or biotransforming pollutants from a contaminated site. Because metal(loid)s are a fundamental part of coal FA, several researches have been conducted to study the effectiveness of special metal-resistant microbes in remediating metal(loid)s for inhibition of further environmental damage. A lot of

**Figure 2.1** Naturally colonized and planted species on the FA deposits of the 11-year-old passive cassette on the largest thermal power plant Nikola Tesla (TENT−A), Obrenovac, Belgrade, Serbia. (A) *Calamagrostis epigejos* (naturally grown as a native grass colonizer). (B) *Populus alba* (spontaneous colonized native tree), *Tamarix tetandra* (planted nonnative shrub), *Amorpha fruticosa* (spontaneous colonized nonnative shrub), and *C. epigejos* (spontaneous native grass colonizer). (C) *P. alba* and *T. tetandra*. Courtesy: *G. Gajic*.

**Figure 2.2** Natural colonization of vegetation on the abandoned FA deposited sites at Feroze Gandhi Unchahar Thermal Power Station, Raebareli, Uttar Pradesh, India. (A) The aquatic zone of FA deposited site. (B) Naturally grown *P. juliflora* and *Saccharum* species on the terrestrial zone of FA deposited site. (C) *C. procera* (Aiton) Dryand. is a flowering plant of the family Apocynaceae, and has the best adaptable potential to harsh conditions of FA dumps. *Courtesy: V.C. Pandey.*

studies have been done on the phytoremediation of heavy metals and metalloids from FA dumps (Pandey et al., 2009). The microbe-assisted phytoremediation of metal(loid)s from FA dumps to check soil and ground water contamination is also a significant and effective approach toward environmental sustainability. The synergistic use of plants and microbes for remediation of polluted sites is proposed by Glick (2003, 2010). The FA microbes have been found appropriate to increase the bioavailability of metal(loid)s from FA to complement phytoremediation potential in many ways (Kumar et al., 2008, 2009; Tiwari et al., 2012, 2013). Microbes also help to carry out metal transformations through redox conversions of inorganic forms. Roychowdhury et al. (2016) isolated several Cr(VI)-resistant strains from FA of MTPS coal-fired power station with abilities to change the toxic forms of metals to nontoxic forms. The reductive transformation of more toxic and soluble Cr(VI) forms into less toxic and insoluble Cr(III) forms by *Methylococcus capsulatus* Bath (methanotrophic bacteria) as the Cr(III) forms has a tendency to become precipitated at high pH levels (Hasin et al., 2010). Pandey et al. (2014b) reviewed in detail about the bioremediation potential of Methanotrophs bacteria, having great potential for the remediation of inorganic and organic pollutants due to the presence of broad-spectrum methane monooxygenases enzyme. The bioleaching of Fe, Zn, and Al from incinerator FA was reported after acceleration by *Aspergillus niger* (Xu and Ting, 2004). Mycoremediation, fungi or root-colonizing mycorrhiza-based remediation, is a highly effective approach for FA remediation (Awoyemi and Dzantor, 2017; Pandey et al., 2009; Suhara et al., 2003). Achievements have been proved with fungi (for instance, white rot) in decontamination of dioxins present in FA (Suhara et al., 2003). The following subsections will explore the role of FA tolerant microbes in solublization and immobilization of metal(loid)s with two purposes: (1) enhanced phytoextraction through an increase in metal(loid) mobility by rhizospheric manipulation (bioaugmentation); and (2) check migration of metal(loid)s to water reservoirs from FA dumps through exploiting metal immobilizing bacterial strains. Therefore microbe-assisted FA phytoremediation will be more effective than phytoremediation only, and microbe-assisted phytoremediation technology for FA remediation has been suggested by Roy et al. (2006) and Jala and Goyal (2006).

*Fungal remediation*—Mycorrhizas are important in social forestry and land reclamation, because they are helpful in enhancing nutrient uptake and cycling. They are known for the survival and growth of majority of plant species in barren and disturbed soils, and in the conversion of unfertile soil to fertile soil (Raman and Mahadevan 1996). Its hyphae form soil aggregates by binding soil particles probably producing polysaccharides (Koske and Polson, 1984). Selvam and Mahadevan (2000) proved the usefulness of a native fungus, *Glomus mosseae* in enhancing the growth of 20 plant species in a derelict FA dumps. Selvam and Mahadevan (2002) also surveyed a derelict FA dump (58 ha) for the presence of arbuscular mycorrhizal (AM) species in planted species. The planted species were *Azadirachta indica*, *Acacia leucophloea*, *Anacardium occidentale*, *Tectona grandis*, *Tamarindus indica*, and *Casuarina equisetifolia*. For each species, five plants were sampled for mycorrhizal isolation and identification. They isolated 15 AM species through direct isolation and trap culture, such as *Acaulospora gerdemannii*, *Gigaspora decipiens*, *Gigaspora gigantea*, *Gigaspora margarita*, *Glomus citricola*, *Glomus fasciculatum*, *Glomus formosanum*, *Glomus fulvum*, *Glomus maculosum*, *Glomus magnicaule*, *G. mosseae*, *Glomus tenebrosum*, *Sclerocystis pachycaulis*, *Scutellospora erythropa*, and *Scutellospora Fulgida*, from the rhizosphere of plant species on FA dumps. *G. mosseae* was recorded as the dominant AM fungus on FA dumps. Furthermore, they also noticed *G. gigantea* and *G. mosseae* from the plant *A. leucophloea*; *G. mosseae* from the plant *A. occidentale*; *A. gerdemannii*, *G. decipiens*, and *G. mosseae* from the plant *A. indica*; *G. citricola*, *G. maculosum*, and *G. mosseae* from the plant *C. equisetifolia*; *G. fasciculatum* and *G. mosseae* from the plant *T. indica*; *G. formosanum* and *G. mosseae* from the plant *T. grandis* through the direct isolation method. Fig. 2.3 represents the AM colonization and number of AM spores from the rhizosphere of plant species on FA dumps.

Channabasava et al. (2015) studied the mycorrhizoremediation of different FA doses, such as 2%, 4%, and 6%, using *Paspalum scrobiculatum* inoculated with AM fungus *Rhizophagus fasciculatus*. FA dose at 2% significantly improved nutrient uptake, nutrient-use efficiencies, AM colonization, spore number, grass growth, and grain yield of *P. scrobiculatum*. However, inoculation of soils amended with 2% FA with the AM fungus further enhanced nutrient uptake, the AM fungal, grass growth, and yield. AM colonization reduced with an increasing level of FA dose; however, such a decrease was not linear. The results also showed higher grass growth, root/shoot ratios, and nutrient contents in shoots raised on 2% FA dose with the AM fungus inoculation. Both AM fungus inoculation and FA amendment also significantly influenced the grain weight and the grain number. Both treatments affected Na, K, Ca, and Mg use efficiencies. Grass growth and nutrient parameters were toughly associated with

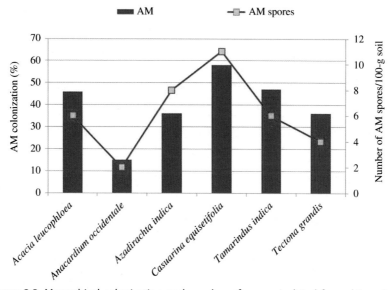

**Figure 2.3** Mycorrhizal colonization and number of spores isolated from rhizospheric zone of plants on FA dumps. *Modified from Selvam, A., Mahadevan, A., 2002. Distribution of mycorrhizas in an abandoned fly-ash pond and mined sites of Neyveli Lignite Corporation, Tamil Nadu, India.* Basic. Appl. Ecol. *3, 277—284.*

the level of AM fungal colonization in the root system. Overall, the results propose that the lower FA dose with AM fungi inoculation could be efficiently exploited for the reclamation of barren FA dumps. Likewise, mycorrhizal fungi are a direct association between soil and roots, and therefore of great importance in FA phytoremediation.

Babu and Reddy (2011b) studied on the isolation of AM fungi from the roots and the rhizospheres of six dominant plant species, *Acacia pennata, Calotropis gigantea, Cassia occidentalis, C. dactylon, Lantana camara,* and *Jatropha gossypifolia,* grown on FA deposit, Odisha, India. It was noticed that the AM fungal diversity differs among the rhizospheres of different plant species on FA deposit. All identified AM fungal species were found in the rhizosphere of *L. camara* except *G. magnicaule* fungus. *Glomus rosea* was the dominant fungus from the plant rhizospheres on FA deposit. The existence of AM fungi in small numbers in the roots and the rhizospheres of plants grown on FA is a sign of their FA adaptation. Inoculation of AM fungal consortia increased growth and nutrient uptake of *Eucalyptus tereticornis* seedlings and decreased translocation of Fe, Zn, Cu, and Al into the shoot part. Inoculated AM fungal consortia include *G. rosea,*

*Glomus multicaule, Glomus heterogama, Glomus etunicatum, G. maculosum, G. magnicaule, Scutellospora nigra,* and *Scutellospora heterogama.* Thus the results suggest that the application of FA adapted AM fungal species may aid in the remediation and management of FA deposits toward greening across the nations.

A case study was done by TERI's researchers in reclamation of FA dumps through the mycorrhizal-based approach. Different mycorrhizal fungal strains were collected from diverse zones of India and overseas. These strains were then isolated, selected, multiplied, and tested under greenhouse/nursery conditions to find out their growth pattern on FA dumps.

Strains help in survival and nutrient uptake to plants because of its high tolerance potential. TERI has successfully established three major examples of FA reclamation on FA dumps. Therefore, mycorrhizoremediation technology when put to application, life sprouted on the barren and derelict FA dumps in the form of green cover. In long term, such a green vegetation discloses a series of physiological changes through a biogeochemical cycle, leading to enrichment of the surface and subsurface ash to promote natural ground cover (Das et al., 2013; Das et al., 2017). In Romanian research, it is suggested that mycorrhizal fungi are potential bioremediation agents, and could be used in phytoremediation of FA deposits toward stabilization and revegetation (Popa et al., 2010; Popa et al., 2013). In addition, it can be used as a pollution bioindicator on FA deposits. Therefore it is well proved by numerous researchers across the nations that mycorrhizal colonization is a promising, effective, and low-prized biotechnology, and must be used in rehabilitation programs of FA dumps throughout the world.

*Bacterial remediation*—Several researches have been done in the bioremediation of FA dumps with the help of bacterial strains. In an investigation, Tiwari et al. (2008a) isolated 11 aerobic bacterial strains (4 gram–positive and others gram–negative) from the rhizospheric zone of *T. latifolia,* which was naturally grown on FA dumps. They separately inoculated in FA to observe the bioavailability or immobilization of heavy metals in FA. These strains enhanced the mobility of Zn, Fe, and Mn, or immobilized Cu and Cd in FA. In contrast, few exceptions are also observed. For example, NBRFT6 increased immobility of Zn and Fe and NBRFT2 of Mn. NBRFT8 and NBRFT9 boosted the Cu bioavailability. Likewise, Tiwari et al. (2008b) studied on the FA tolerant microbe-induced changes in metal(loid) extractability from FA. They observed that most of the FA tolerant bacterial strains either induced the solubilization of Ni, Fe, and Zn, or immobilization of Cr, Cd, Pd, and Cu in FA. The results also suggest that metal(loid) solublization (mobility) and immobilization (immobility) are specific functions to bacterial strains depending on the several environmental and edaphic factors. Thus these FA tolerant bacterial strains responsible for immobilization of FA-metal(loid)s may be employed for arresting their leaching to water systems, while induced bioavailability of metals by potential and FA tolerant bacterial strains may be exploited to boost the metal extraction from FA by metal phytoaccumulators. Thus FA tolerant bacteria strains offer an opportunity in bioremediation of FA with a dual purpose, and may be used in the microbe-assisted phytoremediation of FA dumps or FA contaminated sites. It is a low-cost, ecofriendly, sustainable, and holistic approach to bioremediate FA–metal(loid)s that pollute surface and ground water reservoirs in and around FA dumping ponds.

Kumar et al. (2008) isolated metal tolerant bacteria (*Enterobacter* sp. NBRI K28), having plant growth-promoting (PGP) potential from FA polluted soils. This strain and its siderophore overproducing mutant (NBRI K28 SD1) were noticed the ability to enhance plant biomass and to increase phytoextraction of Cr, Ni, and Zn from FA by *Brassica juncea* plants. In maximum cases, the mutant NBRI K28 SD1 showed more marked impact on growth performance and metal accumulation of *B. juncea* than wild type. Likewise, Kumar et al. (2009) reported bacterial strains NBRI K3 and NBRI K24, having PGP ability from FA polluted soil, which stimulate plant growth and minimize the Cr and Ni toxicity in *B. juncea*. These isolated strains were considered on the basis of 16 S rDNA sequencing and identified as *Rahnella aquatilis* and *Enterobacter aerogenes*, respectively. Siderophore producing ability was also observed in these strains. In both the studies of Kumar et al. (2008, 2009), simultaneous production of indole acetic acid, phosphate solubilization, 1-aminocyclopropane-1-carboxylic acid deaminase, and siderophores showed its PGP potential.

Tiwari et al. (2012) studied on the different combinations of four FA tolerant bacterial strains to test their potential for increased metal uptake by *B. juncea* grown in FA amended with press mud. After metal analysis in plants, it showed that out of eleven bacterial consortia set from the various groups of four bacterial strains, such as *Bacillus endophyticus* NBRFT4 (MTCC 9021), *Bacillus pumilus* NBRFT9 (MTCC 8913), *Micrococcus roseus* NBRFT2 (MTCC 9018), and *Paenibacillus macerans* NBRFT5 (MTCC 8912), a combination of NBRFT4, NBRFT9 (ST3), and NBRFT5 was noticed to have induced the maximum metal accumulation than other consortia. The bioaugmentation of the ST3 consortium increased the accumulation of Fe (247%), Ni (231%), and Zn (223%) in *B. juncea* than control plants.

In addition, siderophores and bacteria are also capable to produce enzymes, protons, and organic acids, which promote the phytoextraction process and increase the metal mobilization. The metal translocation from root to stem was always greater than from stem to leaf. Therefore ST3 was declared the best consortium to be used in the field application to boost the metal phytoextraction from FA through *B. juncea*. Likewise, Tiwari et al. (2013) studied on the metal phytoextraction through *S. munja* with the help of growth-promoting bacteria obtained from FA. They found that, when a consortium of *B. pumilus* NBRFT9, *P. macerans* NBRFT5, and *B. endophyticus* NBRFT4 inoculated in the rhizosphere of *S. munja*, it not only promoted the plant growth but also enhanced metal uptake through mobilization. The pooled effect of both factors enhanced the phytoextraction of Zn, Fe, and Ni by twofold or threefold. The synthesis of indole acetic acid, siderophore, cytokinins, and gibberellic acid by bacteria promoted both plant growth and metal bioavailability. Hence, a microbe-based phytoremediation in order to metal phytoextract from FA dumps may be suggested for use as a holistic approach.

Rau et al. (2009) studied the rhizobacteria of *Saccharum ravennae*, which was naturally grown on an Indraprastha and Badarpur FA dumps in Delhi urban ecosystems. They identified some potential rhizobacteria with the ability of high metal(loid) tolerance and good PGP traits. These are *Pseudomonas aeruginosa* BPSr43, *S. marcescens* IPSr82 and IPSr90, *Arthrobacter ureafaciens* BPSr55, *Enterococcus casseliflavus* BPSr32, *Paenibacillus azotofixans* BPSr107, and *Paenibacillus larvae* BPSr106. The significant improvement in the seedling establishment, shoot length, and plant weight in rhizobacterial inoculated grasses of *S. ravennae* in the FA situation showed the importance of rhizobacteria in its colonization and spread to the dumps. Therefore *S. ravennae* and its rhizobacteria may be potentially beneficial to develop potential inoculation technologies for the transformation of derelict FA deposits into ecologically and socioeconomically fruitful habitats. Raja and Omine (2013) isolated and characterized 17 boron-tolerant bacteria from an FA dumping site, Nagasaki prefecture, Japan. These boron-tolerant strains also showed resistance to As (III), Cd, Cr (VI), Cu, Ni, Pb, Se (III), Zn, and NaCl (15%). Moreover, these strains were observed to be arsenic oxidizing bacteria confirmed by silver nitrate test. The boron-tolerant strains were capable of removing $0.1-2.0$ and $2.7-3.7 \, \mathrm{mg \, L^{-1}}$ boron from the medium and FA at 168 h. The results conclude that boron-tolerant strains As14, KU22, and Cr-D can be important in the remediation and management of FA dumps either through microbial-induced metal mobilization in phytoextraction or microbial-induced metal immobilization in the phytostabilization process. Mukherjee et al. (2017) studied on the phytoremediation potential of rhizobacterial isolates from *S. spontaneum* grown on FA ponds. They isolated phosphate-solubilizing bacterial strains from the rhizosphere of *S. spontaneum* grown on bare FA pond of Mejia Thermal Power Station, and identified three strains, such as *Bacillus* sp. strain MHR4, *Staphylococcus* sp. strain MHR3, and *Bacillus anthracis* strain MHR2, with their plant growth-promoting abilities. These strains also showed plant growth-promoting features such as siderophore, production of ammonia, IAA, and hydrocyanide. These strains enhanced metal phytoextraction ability and increased the growth of *B. juncea* plants. Therefore the isolated native and stress-adapted rhizobacteria may serve as a potential and low-prized tool for the fruitful restoration of FA deposited sites.

## 2.3 The limitations of fly ash to plant growth

FA is just like a soil substrate, but it has harsh conditions that do not support to plant growth. The FA physiochemical and biological factors limit plant establishment and growth on FA dumpsites. These are high pH, formation of a compact layer, unfavorable mechanical composition, high level of soluble salts, absence of nitrogen, available phosphorus, presence of toxic metal(loid)s such as Cr, Cd, Pb, As, Mn, Cu, Mo, B, Ni, and Se, and reduced microbial activities (Adriano et al., 1980; Carlson and Adriano, 1993; Jusaitis and Pillman, 1997; Pavlović et al., 2004; Mitrović et al., 2008; Haynes, 2009; Pandey et al., 2009; Pandey and Singh, 2010, 2014; Gajić et al., 2013, 2016). Physical barriers include restriction of root growth on FA substrate because of natural compaction of fine ash particles/formation of hard cemented layers due to its pozzalanic nature. In general, microbial factors include a lack of microbial activity, resultant low nutrient turnover, and a lack of inoculum of symbiotic microbes, that is, *Rhizobium* and *mycorrhizae*. Therefore two approaches are available to overcome these constraints: (1) organic waste amendments are often mixed with the surface layers of FA substrate during plantation; and (2) selection of potential plants from the naturally restored FA deposits for the phytoremediation/revegetation of newly deposited FA sites, because naturally grown plants have inherent characteristics to grow on the harsh conditions of fresh deposited FA sites with minimum inputs and low maintenance (Pandey and Singh, 2011).

## 2.4 The limitations of microbial remediation in fly ash deposits

The limitations of the use of microbial remediation for metal(loid)s are that, although the metal(loid)s are bound to microorganisms, they can be released back into the FA soon after their death and decay of the microorganisms and still be present in the FA substrate. Thus the in situ microbial remediation approaches must be combined with phytoremediation approaches by appropriate hyperaccumulator plants that can effectively uptake the metal(loid)s from the FA substrate and accumulate them in their below and above ground parts, thus inhibiting their recycling in the FA substrate.

*Factors affecting microbial remediation*—Microbial remediation depends upon the presence of suitable microbes in the accurate extents, combinations, and proper environmental and edaphic conditions. Biostimulation and bioaugmentation are two essential factors influencing microbial remediation. Biostimulation is the addition of nutrients, oxygen, or other electron donors or acceptors that assist to increase the number and activity of microbes available for remediation, while bioaugmentation is the addition of microbes that can biotransform or biodegrade a certain pollutant.

## 2.5 Multiple benefits of fly ash phytoremediation

Phytoremediation is a holistic approach that uses plants to remediate metal(loid)s from FA contaminated sites or FA dumpsites by accumulating in plant parts or immobilizing in the soil system/plant roots, thereby inhibiting erosion and decreasing the adverse effect on human health (Pandey and Bajpai, 2019). The establishment of vegetation cover on the FA dumps results in the control of wind erosion, stabilization of ash, reduced mobility and toxicity of metal(loid)s, the dispersion of toxic metal(loid)s into the environment, and providing organic substance to bind pollutants (Pavlović et al., 2004, 2007; Djurdjević et al., 2006; Mitrović et al., 2008; Haynes, 2009; Pandey, 2012b; Kostić et al., 2012; Gajić et al., 2013, 2016; Pandey and Singh, 2014; Pandey et al., 2015b, 2016a). In addition, the other expected outcome of the phytoremediation of FA involves the following: (1) the utilization of abandoned FA dumpsite; (2) saving the ecosystem and its fauna, including human being from FA toxicity; (3) the reduction of greenhouse gas emissions and enhancement of carbon sequestration; (4) providing ecosystem services; (5) the potential opportunities in FA phytoremediation that also yield biofuel, biodiesel, aromatic oil, biofortified products, and pulp–paper biomass, which would provide an alternate source of income and would foster socioeconomic balance and development. In addition, the edible plants should be discouraged in the FA phytoremediation programs to avoid metal toxicity in the food chain (Pandey and Mishra, 2018). The environmental and socioeconomic developmental considerations using economically valuable plants in the phytoremediation of FA deposits are expected to be universally lauded for contributing to green cover development with economic returns. Proper research and commercialization of phytoremediated biomass and phytoproducts would likely enhance socioeconomic viability by decreasing input costs in phytoremediation and increasing productivity through a suitable crop selection (Pandey and Souza-Alonso, 2019b).

The most significant drawback is the lack of awareness on the benefits of phytoremediation/revegetation of FA deposits between producers and endusers, and specific and tough directives from governments are normally missing across the nations, including India, while giving importance to enhanced power production, authorities neglect how to beneficially use the resulting FA dumpsites. Future orders should include policies that every abandoned FA dumpsites must be rehabilitated through revegetation programs. Governmental efforts are needed for encouraging FA beneficiation through reclamation, and should involve coal-based industries, practitioners, policy makers, researchers, and stakeholders to promote livelihood generation by framing directives and policies (Roy et al., 2018).

## 2.6 Conclusion and future prospects

The future directions regarding phytomanagement/phytoremediation of FA deposits for the remediation and management need a holistic approach of a fusion of naturally grown and introduced native species that boost sustainable growth of plant communities. The development of vegetation through assisted phytoremediation of FA dumpsites is a potential approach for maintaining ecosystem services and quality of human life. Although reports by various researchers are evident of potential role of plants and microbes in bioremediation of FA deposits, still there is a need to completely understand the plant—microbe—FA interactions and their better exploitation in the remediation and management of FA dumps/contaminated sites. The knowledge and insights on the usefulness of ecologically and socioeconomically valuable native grasses and tress that stabilizes ash dumps, reduces erosion and leaching, enhances the fertility of the ash substrate, initiates carbon sequestration, and incurs gradual restoration of the ash site can be linked to action on the ground by scientists and practitioners for the remediation and management of FA dumps.

# References

Adriano, D.C., Page, A.L., Elseewi, A.A., Chang, A., Straughan, I.A., 1980. Utilization and disposal of fly ash and other coal residues in terrestrial ecosystem: a review. J. Environ. Qual. 9, 333—344.

Awoyemi, O.M., Dzantor, E.K., 2017. Toxicity of coal fly ash (CFA) and toxicological response of switchgrass in mycorrhizamediated CFA-soil admixtures. Ecotoxicol. Environ. Saf. 144, 438—444.

Babu, A.G., Reddy, M.S., 2011a. Dual inoculation of arbuscular mycorrhizal and phosphate solubilizing fungi contributes in sustainable maintenance of plant health in fly ash ponds. Water Air Soil Pollut. 219, 3—10.

Babu, A.G., Reddy, M.S., 2011b. Diversity of arbuscular mycorrhizal fungi associated with plants growing in fly ash pond and their potential role in ecological restoration. Curr. Microbiol. 63, 273—280.

Baker, A.J.M., 1981. Accumulators and excluders - strategies in the response of plants to heavy metals. J. Plant Nutr. 3, 643—654. Available from: https://doi.org/10.1080/01904168109362867.

Bilski, J., McLean, K., McLean, E., Soumaila, F., Lander, M., 2011. Environmental health aspects of coal ash phytoremediation by selected crops. Int. J. Environ. Sci. 1, 2028—2036.

Carlson, C.L., Adriano, D.C., 1993. Environmental impacts of coal combustion residues. J. Environ. Qual. 22, 227—247.

Channabasava, A., Lakshman, H.C., Muthukumar, T., 2015. Fly ash mycorrhizoremediation through Paspalum scrobiculatum L., inoculated with Rhizophagus fasciculatus. C. R. Biol. 338, 29—39. Available from: https://doi.org/10.1016/j.crvi.2014.11.002.

Chibrik, T.S., Lukina, N.V., Filimonova, E.I., Glazyrina, M.A., Rakov, E.A., Maleva, M. G., et al., 2016. Biological recultivation of mine industry deserts: facilitating the formation of phytocoenosis in the middle Ural region, Russia. In: Prasad, M.N.V. (Ed.), Bioremediation and Bioeconomy. Elsevier, pp. 389—418.

Chu, M., 2008. Natural revegetation of coal fly ash in a highly saline disposal lagoon in Hong Kong. Appl. Veg. Sci. 11, 297—306.

Das, M., Agarwal, P., Singh, R., Adholeya, A., 2013. A study of abandoned ash ponds reclaimed through green cover development. Int. J. Phytorem. 15, 320—329.

Das, M., Jakkula, V.S., Alok Adholeya, A., 2017. Role of mycorrhiza in phytoremediation processes: a review. In: Varma, A., et al., (Eds.), Mycorrhiza - Nutrient Uptake, Biocontrol, Ecorestoration. Springer, pp. 271—286. Available from: https://doi.org/10.1007/978-3-319-68867-1-14.

Djurdjević, L., Mitrović, M., Pavlović, P., Gajić, G., Kostić, O., 2006. Phenolic acids as bioindicators of fly ash deposit revegetation. Arch. Environ. Contam. Toxicol. 50, 488—495.

Gajić, G., Pavlović, P., Kostić, O., Jarić, S., Djurdjević, L., Pavlović, D., et al., 2013. Ecophysiological and biochemical traits of three herbaceous plants growing of the disposed coal combustion fly ash of different weathering stage. Arch. Biol. Sci. 65 (1), 1651—1667.

Gajić, G., Djurdjević, L., Kostić, O., Jarić, S., Mitrović, M., Stevanović, B., et al., 2016. Assessment of the phytoremediation potential and an adaptive response of *Festuca rubra* L. sown on fly ash deposits: native grass has a pivotal role in ecorestoration management. Ecol. Eng. 93, 250—261.

Gajić, G., Mitrović, M., Pavlović, P., 2019. Ecorestoration of fly ash deposits by native plant species at thermal power stations in Serbia. In: Pandey, V.C., Bauddh, K. (Eds.), Phytomanagement of Polluted Sites: Market Opportunities in Sustainable Phytoremediation. Elsevier, Amsterdam, The Netherlands, pp. 113—177.

Glick, B.R., 2003. Phytoremediation: synergistic use of plants and bacteria to clean up the environment. Biotechnol. Adv. 21, 383—393.

Glick, B.R., 2010. Using soil bacteria to facilitate phytoremediation. Biotechnol. Adv. 28, 367—374.

Gupta, A.K., Sinha, S., 2008. Decontamination and/or revegetation of fly ash dykes through naturally growing plants. J. Hazard. Mater. 153 (3), 1078—1087.

Hasin, A.A.L., Gurman, S.J., Murphy, L.M., Perry, A., Simth, T.J., Gardiner, P.H.E., 2010. Remediation of chromium (VI) by a methaneoxidizing bacterium. Environ. Sci. Technol. 44, 400—405.

Haynes, R.J., 2009. Reclamation and revegetation of fly ash disposal sites: challenges and research needs. J. Environ. Manag. 90, 43—53.

Hilton, B., 2002. Growing Short Rotation Coppice. Best Practice Guidelines. Department of Environment. Food and Rural Affairs, London.

Hodgson, D.R., Townsend, W.N., 1973. The amelioration and revegetation of pulverized fuel ash. In: Chadwick, M.J., Goodman, G.T. (Eds.), Ecology and Reclamation of Devastated Land. Gordon and Breach, London, pp. 247—270.

Huang, H., Yu, N., Wang, L., Gupta, D.K., He, Z., Wang, K., et al., 2011. The phytoremediation potential of bioenergy crop *Ricinus communis* for DDTs and cadmium co-contaminated soil. Bioresour. Technol. 102, 11034—11038.

Izquierdo, M., Querol, X., 2012. Review article—leaching behavior of elements from coal combustion fly ash: an overview. Int. J. Coal Geol. 94, 54—66.

Jala, S., Goyal, D., 2006. Fly ash as a soil ameliorant for improving crop production—a review. Bioresour. Technol. 97, 1136—1147.

Jambhulkar, H.P., Juwarkar, A.A., 2009. Assessment of bioaccumulation of heavy metals by different plant species grown on fly ash dump. Ecotoxicol. Environ. Saf. 72, 1122—1128.

Jamil, S., Abhilash, P.C., Singh, N., Sharma, P.N., 2009. *Jatropha curcas*: a potential crop for phytoremediation of coal fly ash. J. Hazard. Mater. 172, 269—275.

Jasionkowski, R., Wojciechowska, A., Kamiński, D., Piernik, A., 2016. Meadow species in the early stages of succession on the ash settler of power plant EDF Toruń SA in Toruń, Poland. Ecol. Quest. 23, 79—86.

Jusaitis, M., Pillman, A., 1997. Revegetation of waste fly ash lagoons. I. Plant selection and surface amelioration. Waste Manage. Res. 15, 307—321.

Keller, C., Marchetti, M., Rossi, L., Lugon-Moulin, N., 2005. Reduction of cadmium availability to tobacco (*Nicotiana tabacum*) plants using soil amendments in low cadmium-contaminated agricultural soils: a pot experiment. Plant Soil 276, 69—84.

Koske, R.E., Polson, W.R., 1984. Are VA mycorrhizae required for sand dune stabilization? Bioscience 34, 420—425.

Kostić, O., Mitrović, M., Knězević, M., Jarić, S., Gajić, G., Djurdjević, L., et al., 2012. The potential of four woody species for the revegetation of fly ash deposits from the 'Nikola Tesla-A' thermoelectric plant (Obenovac, Serbia). Arch. Biol. Sci. 64 (1), 145—158.

Kumar, K.V., Singh, N., Behl, H.M., Srivastava, S., 2008. Influence of plant growth promoting bacteria and its mutant on heavy metal toxicity in *Brassica juncea* grown in fly ash amended soil. Chemosphere 72, 678—683.

Kumar, K.V., Srivastava, S., Singh, N., Behl, H.M., 2009. Role of metal resistant plant growth promoting bacteria in ameliorating fly ash to the growth of *Brassica juncea*. J. Hazard. Mater. 170, 51—57.

Kumar, A., Ahirwal, J., Maiti, S.K., Das, R., 2015. An assessment of metal in fly ash and their translocation and bioaccumulation in perennial grasses growing at the reclaimed opencast mines. Int. J. Environ. Res. 9 (3), 1089—1096.

Kumari, A., Pandey, V.C., Rai, U.N., 2013. Feasibility of fern *Thelypteris dentata* for revegetation of coal fly ash landfills. J. Geochem. Explor. 128, 147—152.

Kumari, A., Lal, B., Rai, U.N., 2016. Assessment of native plant species for phytoremediation of heavy metals growing in the vicinity of NTPC sites, Kahalgaon, India. Int. J. Phytoremediat. 18, 592—597.

Maiti, S.K., Nandhini, S., 2006. Bioavailability of metals in fly ash and their bioaccumulation in naturally occurring vegetation: a Pilot Scale Study. Environ. Monit. Assess. 116, 263—273.

Maiti, S.K., Jaiswal, S., 2008. Bioaccumulation and translocation of metals in the natural vegetation growing on fly ash deposits: a field study from Santaldih Thermal Power Plant, West Bengal, India. Environ. Monit. Assess. 136, 355—370.

Maiti, S.K., Maiti, D., 2015. Ecological restoration of waste dumps by topsoil blanketing, coir-matting and seeding with grass—legume mixture. Ecol. Eng. 77, 74—84.

Mitrović, M., Pavlović, P., Lakušić, D., Djurdjevic, L., Stevanović, B., Kostić, O., et al., 2008. The potential of *Festuca rubra* and *Calamagrostis epigejos* for the revegetation on fly ash deposits. Sci. Total Environ. 72, 1090—1101.

Morgenthal, T.L., Cilliers, S.S., Kellner, K., van Hamburg, H., Michael, M.D., 2001. The vegetation of ash disposal sites at Hendrina power station I: Phytosociology. SA. J. Bot 67, 506—519.

Mukherjee, P., Roychowdhury, R., Roy, M., 2017. Phytoremediation potential of rhizobacterial isolates from Kans grass (*Saccharum spontaneum*) of fly ash ponds. Clean. Technol. Environ. Policy 19, 1373—1385.

Mulhern, D.W., Robel, R.J., Furness, J.C., Hensley, D.L., 1989. Vegetation of waste disposal areas at a coal-fired power plant in Kansas. J. Environ. Qual. 18, 285—292.

Mustafa, B., Hajdari, A., Krasniqi, F., Morina, I., Riesbeck, F., Sokoli, A., 2012. Vegetation of the ash dump of the "Kosova A" power plant and the slag dump of the "Ferronikeli" Smelter in Kosovo. Res. J. Environ. Earth Sci. 4 (9), 823—834.

Nwoko, C.O., 2010. Trends in phytoremediation of toxic elemental and organic pollutants. Afr. J. Biotechnol. 9, 6010—6016. Available from: https://doi.org/10.5897/AJB09.061.

Panda, D., Panda, D., Padhan, B., Biswas, M., 2018. Growth and physiological response of lemongrass (*Cymbopogon citratus* (D.C.) Stapf.) under different levels of fly ash-amended soil. Int. J. Phytoremediat. 12 20 (6), 538—544.

Pandey, V.C., 2012a. Phytoremediation of heavy metals from fly ash pond by *Azolla caroliniana*. Ecotoxicol. Environ. Saf. 82, 8—12.

Pandey, V.C., 2012b. Invasive species based efficient green technology for phytoremediation of fly ash deposits. J. Geochem. Explor. 123, 13—18.

Pandey, V.C., 2013. Suitability of *Ricinus communis* L. cultivation for phytoremediation of fly ash disposal sites. Ecol. Eng. 57, 336−341.

Pandey, V.C., 2015. Assisted phytoremediation of fly ash dumps through naturally colonized plants. Ecol. Eng. 82, 1−5.

Pandey, V.C., 2016. Phytoremediation efficiency of *Eichhornia crassipes* in fly ash pond. Int. J. Phytoremediat 18 (5), 450−452.

Pandey, V.C., Singh, N., 2010. Impact of fly ash incorporation in soil systems. Agric. Ecosyst. Environ. 136, 16−27.

Pandey, V.C., Singh, K., 2011. Is *Vigna radiata* suitable for the revegetation of fly ash landfills? Ecol. Eng. 37, 2105−2106.

Pandey, V.C., Singh, B., 2012. Rehabilitation of coal fly ash basins: current need to use ecological engineering. Ecol. Eng. 49, 190−192.

Pandey, V.C., Singh, N., 2014. Fast green capping on coal fly ash basins through ecological engineering. Ecol. Eng. 73, 671−675.

Pandey, V.C., Mishra, T., 2018. Assessment of *Ziziphus mauritiana* grown on fly ash dumps: prospects for phytoremediation but concerns with the use of edible fruit. Int. J. Phytoremediat. 20 (12), 1250−1256.

Pandey, V.C., Bajpai, O., 2019. Phytoremediation: from theory toward practice. In: Pandey, V.C., Bauddh, K. (Eds.), Phytomanagement of Polluted Sites. Elsevier, Amsterdam, The Netherlands, pp. 1−49.

Pandey, V.C., Abhilash, V.C., Singh, N., 2009. The Indian perspective of utilizing fly ash in phytoremediation, photomanagement and biomass production. J. Environ. Manage. 90, 2943−2958.

Pandey, V.C., Singh, J.S., Kumar, A., Tewari, D.D., 2010. Accumulation of heavy metals by chickpea grown in fly ash treated soil: effect on antioxidants. Clean Soil Air Water 38, 1116−1123.

Pandey, V.C., Singh, J.S., Singh, R.P., Singh, N., Yunus, M., 2011. Arsenic hazards in coal fly ash and its fate in Indian scenario. Resour. Conserv. Recycling 55, 819−835.

Pandey, V.C., Singh, K., Singh, J.S., Kumar, A., Singh, B., Singh, R.P., 2012a. Jatropha curcas: a potential biofuel plant for sustainable environmental development. Renew. Sust. Energ. Rev. 16, 2870−2883.

Pandey, V.C., Singh, K., Singh, R.P., Singh, B., 2012b. Naturally growing *Saccharum munja* on the fly ash lagoons: a potential ecological engineer for the revegetation and stabilization. Ecol. Eng. 40, 95−99.

Pandey, V.C., Singh, N., Singh, R.P., Singh, D.P., 2014a. Rhizoremediation potential of spontaneously grown *Typha latifolia* on fly ash basins: study from the field. Ecol. Eng. 71, 722−727.

Pandey, V.C., Singh, J.S., Singh, D.P., Singh, R.P., 2014b. Methanotrophs: promising bacteria for environmental remediation. Int. J. Environ. Sci. Technol. (2014) 11, 241−250.

Pandey, V.C., Pandey, D.N., Singh, N., 2015a. Sustainable phytoremediation based on naturally colonizing and economically valuable plants. J. Clean. Prod. 86, 37−39.

Pandey, V.C., Prakash, P., Bajpai, O., Kumar, A., Singh, N., 2015b. Phytodiversity on fly ash deposits: evaluation of naturally colonized species for sustainable phytorestoration. Environ. Sci. Pollut. Res. 22, 2776−2787.

Pandey, V.C., Bajpai, O., Pandey, D.N., Singh, N., 2015c. *Saccharum spontaneum*: an underutilized tall grass for revegetation and restoration programs. Genet. Resour. Crop. Evol. 62 (3), 443−450.

Pandey, V.C., Sahu, N., Behera, S.K., Singh, N., 2016a. Carbon sequestration in fly ash dumps: comparative assessment of three plant association. Ecol. Eng. 95, 198−205.

Pandey, V.C., Bajpal, O., Singh, N., 2016b. Energy crops in sustainable phytoremediation. Renew. Sust. Energ. Rev. 54, 58−73.

Pandey, V.C., Souza-Alonso, P., 2019b. Market opportunities in sustainable phytoremediation. In: Pandey, V.C., Bauddh, K. (Eds.), Phytomanagement of Polluted Sites. Elsevier, Amsterdam, The Netherlands, pp. 51–82.

Pavlović, P., Mitrović, M., Djurdjević, L., 2004. An ecophysiological study of plants growing on the fly ash deposits from the "Nikola Tesla-A" thermal power station in Serbia. Environ. Manage. 33 (5), 654–663.

Pavlović, P., Mitrović, M., Djurdjević, L., Gaijić, G., Kostić, O., Bojović, O., 2007. The ecological potential of Spiracea van-houttei (Briot.) zabel for urban (the city of Belgrade) and fly ash deposit (Obrenovac) landscaping in Servia. Pol. J. Environ. Stud. 3, 427–431.

Pilon-Smits, E., 2005. Phytoremediation. Annu. Rev. Plant Biol. 56, 15–39.

Popa, D., Hanescu, V., Constantin, C., Codreanu, M., 2010. The concept of plant–fungi–soil partnership—a realistic approach of the symbiotic system developed on the dumps of ashes from Isalnita area—Romania. J. Environ. Prot. Ecol. 11 (3), 958.

Popa, D., Codreanub, M.D., Babeanua, C., 2013. Mycorrhizal colonisation—a possible alternative biotechnology recommended to the fly ashes dumps. J. Environ. Prot. Ecol. 14 (1), 256–262.

Prasad, M.N.V., 2006. Stabilization, remediation, and integrated management of metal-contaminated ecosystems by grasses (Poaceae). In: Prasad, M.N.V., Sajwan, K.S., Naidu, R. (Eds.), Trace Elements in the Environment (Biogeochemistry, Biotechnology, and Bioremediation). CRC Taylor and Francis, Boca Raton, FL, pp. 405–424.

Rai, U.N., Pandey, K., Sinha, S., Singh, A., Saxena, R., Gupta, D.K., 2004. Revegetating fly ash landfills with *Prosopis juliflora* L.: impact of different amendments and *Rhizobium* inoculation. Environ. Int. 30, 293–300.

Raj, S., Mohan, S., 2008. Approach for improved plant growth using fly-ash amended soil. Int. J. Emerging Technol. Adv. Eng. 4, 709–715.

Raja, C.E., Omine, K., 2013. Characterization of boron tolerant bacteria isolated from a fly ash dumping site for bacterial boron remediation. Environ. Geochem. Health 35, 431–438.

Raman, N., Mahadevan, A., 1996. Mycorrhizal research—a priority in agriculture. In: Mukerji (Ed.), Concepts of Mycorrhizal Research. Kluwar, The Netherlands, pp. 41–75.

Rau, N., Mishra, V., Sharma, M., Das, M.K., Ahaluwalia, K., Sharma, R.S., 2009. Evaluation of functional diversity in rhizobacterial taxa of a wild grass (*Saccharum ravennae*) colonizing abandoned fly ash dumps in Delhi urban ecosystem. Soil Biol. Biochem. 41, 813–821.

Rockwood, D.L., Naidu, C.V., Carter, D.R., Rahmani, M., Spriggs, T.A., Lin, C., et al., 2004. Shortrotation woody crops and phytoremediation: opportunities for agroforestry? In: Nair, P.K.R., Rao, M.R., Buck, L.E. (Eds.), New Vistas in Agroforestry. Kluwer Academic Publishers, Dordrecht, The Netherlands, pp. 51–63.

Roy, S.V., Vanbroekhoven, K., Dejonghe, W., Diels, L., 2006. Immobilization of heavy metals in the saturated zone by sorption and in situ bioprecipitation processes. Hydrometallurgy 83, 195–203.

Roy, M., Roychowdhury, R., Mukherjee, P., 2018. Remediation of fly ash dumpsites through bioenergy crop plantation and generation: a review. Pedosphere 28 (4), 561–580.

Roychowdhury, R., Mukherjee, P., Roy, M., 2016. Identification of chromium resistant bacteria from dry fly ash sample of Mejia MTPS Thermal Power Plant, West Bengal, India. Bull. Environ. Contam. Toxicol. 96, 210–216.

Selvam, A., Mahadevan, A., 2000. Reclamation of ash pond of Neyveli Lignite Corporation, Neyveli, India. Minetech 21, 81–89.

Selvam, A., Mahadevan, A., 2002. Distribution of mycorrhizas in an abandoned fly ash pond and mined sites of Neyveli Lignite Corporation, Tamil Nadu, India. Basic. Appl. Ecol. 3, 277–284.

Sharma, M., Mishra, V., Rau, N., Sharma, R.S., 2011. Functionally diverse rhizobacteria of *Saccharum munja* (a native wild grass) colonizing abandoned morrum mine in Aravalli hills (Delhi). Plant Soil 341, 447–459.

Shaw, P.J.A., 1996. Role of seedbank substrates in the revegetation of fly ash and gypsum in the United Kingdom. Restor. Ecol. 4, 61–70.

Singh, A.N., Raghubanshi, A.S., Singh, J.S., 2002. Plantations as a tool for mine spoil restoration. Curr. Sci. 82 (12), 1436–1441.

Skousen, J., Ziemkiewicz, P., Yang, J.E., 2012. Use of coal combustion by-products in mine reclamation: review of case studies in the USA. Geosyst. Eng. 15, 71–83.

Suhara, H., Daikoku, C., Takata, H., Suzuki, S., Matsufuji, Y., Sakai, K., et al., 2003. Monitoring of white-rot fungus during bioremediation of polychlorinated dioxin-contaminated fly ash. Appl. Microbiol. Biotechnol. 62, 601–607.

Técher, D., Laval-Gilly, P., Bennasroune, A., Henry, S., Martinez-Chois, C., D'Innocenzo, M., et al., 2012. An appraisal of Miscanthus x giganteus cultivation for fly ash revegetation and soil Restoration. Ind. Crop. Prod. 36, 427–433.

Tiwari, S., Kumari, B., Singh, S.N., 2008b. Microbe-induced changes in metal extractability from fly ash. Chemosphere 71, 1284–1294.

Tiwari, S., Kumari, B., Singh, S.N., 2008a. Evaluation of metal mobility/immobility in fly ash induced by bacterial strains isolated from rhizospheric zone of *Typha latifolia* growing on fly ash dumps. Bioresour. Technol. 99, 1305–1310.

Tiwari, S., Kumari, B., Garg, S.K., 2012. Stimulated phytoextraction of metals from fly ash by microbial interventions. Environ. Technol. 33 (21), 2405–2413.

Tiwari, S., Singh, S.N., Garg, S.K., 2013. Microbially enhanced phytoextraction of heavy-metal fly-ash amended soil. Commun. Soil Sci. Plant Anal. 44 (21), 3161–3176.

Tong, Y.P., Kneer, R., Zhu, Y.G., 2004. Vacuolar compartmentalization: a second-generation approach to engineering plants foe phytoremediation. Trends Plant Sci. 9 (1), 8–9.

US EPA, 2007. Human and Ecological Risk Assessment of Coal Combustion Wastes (Draft), Research Triangle Park, NC.

Verma, S.K., Singh, K., Gupta, A.K., Pandey, V.C., Trevedi, P., Verma, S.K., et al., 2014. Aromatic grasses for phytomanagement of coal fly ash hazards. Ecol. Eng. 73, 425–428.

Volk, T.A., Abrahamson, L.P., Nowak, C.A., Smart, L.B., Tharakan, P.J., White, E.H., 2006. The development of short-rotation willow in the northeastern United States for bioenergy and bioproducts, agroforestry and phytoremediation. Biomass Bioenerg. 30, 715–727.

Vurayai, R., Nkoane, B., Moseki, B., Chartuvedi, P., 2017. Phytoremediation potential of *Jatropha curcas* and *Pennisetum clandestinum* grown in polluted soil with and without coal fly ash: a case of BCL Cu/Ni mine, Selibe-Phikwe, Botswana. J. Biodivers. Environ. Sci. 10 (5), 193–206.

Wang, S., 2008. Application of solid ash based catalysts in heterogeneous catalysis. Environ. Sci. Technol. 42 (19), 7055–7063.

Xu, T.J., Ting, Y.P., 2004. Optimisation on bioleaching of incinerator fly ash by *Aspergillus niger*. Use of central composite design. Enzyme Microb. Technol. 35 (5), 444–454.

# Further reading

Pandey, V.C., 2017. Managing waste dumpsites through energy plantations. In: Bauddh, K., Singh, B., Korstad, J. (Eds.), Phytoremediation Potential of Bioenergy Plants. Springer, Singapore, pp. 371–386.

Pandey, V.C., Kumar, A., 2013. *Leucaena leucocephala*: an underutilized plant for pulp andpaper production. Genet. Resour. Crop. Evol. 60, 1165–1171.

Pandey, V.C., Singh, N., 2015. Aromatic plants versus arsenic hazards in soils. J. Geochem. Explor. 157, 77–80.

Pandey, V.C., Rai, A., Korstad, J., 2019a. Aromatic crops in phytoremediation: from contaminated to waste dumpsites. In: Pandey, V.C., Bauddh, K. (Eds.), Phytomanagement of Polluted Sites. Elsevier, Amsterdam, The Netherlands, pp. 255–275.

# CHAPTER 3

# Scope of fly ash use in agriculture: prospects and challenges

## Contents

## 3.1 Introduction

Fly ash (FA), a by-product of coal combustion, is generated in the coal-based thermal power stations. Nutrient-enriched FA has been accepted as a potential soil amendment for plant growth (Pandey and Singh, 2010). In most cases, FA consists of plant macronutrients Ca, Mg, K, P, and S as well as micronutrients Fe, Co, B, Zn, Cu, and Mn. FA also contains heavy metals such as Pb, Ni, As, Cr, and Cd (Pandey and Singh, 2010). Toxic effect of FA is insignificant, and concentration of toxic elements within permissible limits at its low doses in some agricultural work (Pandey et al., 2009b). Hence, major initiatives have been taken in India and elsewhere to use such a cost-effective resource in large

*Phytomanagement of Fly Ash*
ISBN: 978-0-12-818544-5
DOI: https://doi.org/10.1016/B978-0-12-818544-5.00003-1

volumes in agriculture (Pandey et al., 2009a, b, 2010; Pandey and Singh, 2010; Singh et al., 2011, 2016; Singh and Pandey, 2013; Shaheen et al., 2014; Verma et al., 2014). Presently, over 600 million tons of FA is produced worldwide. Globally, FA utilization rate is 25%, and unutilized FA requires an important action. Therefore FA needs eco-friendly use in a sustainable mode, which is a major challenge. In this regard, FA has been utilized in a wide range of construction works such as cement production, concrete manufacturing, structural fills, brick making, and road construction. Besides, FA works in soil stabilization, soil improvement, mine restoration, and agriculture. However, there are no clear regulations, specifications, or guidelines for its use in agriculture. The absence of guidelines or enactment is the main barrier to the FA utilization for improving soil properties in the agriculture (Ahmaruzzaman, 2010). Different challenges to the FA utilization in the crop cultivation are the distance of thermal power station from the agricultural land and its acceptability by the farmers. The above-mentioned facts created significant curiosity among scientists for the finding of alternative environmentally safe possibilities for the FA use.

The agronomic advantages of FA amendment are mainly related to the enhanced physicochemical and biological features of the soil. Ca−Si minerals are present in FA with the pozzolanic nature, which improves soil porosity, bulk density (BD), available water capacity, and water-holding capacity (WHC; Pandey and Singh, 2010). Nutrient-enriched FA supports crop growth and increases crop yield, as it has a good amount of phosphorus but nitrogen is absent. This is the main reason for the selection of the leguminous crops for the utilization of FA in agriculture due to its capability of nitrogen fixation (Pandey et al., 2009b), as nitrogen is absent in FA, because it is oxidized into gaseous constituents during the combustion. Therefore leguminous crops should be considered in agriculture, while utilizing FA as a soil ameliorant. The core focus of the present chapter is to explore the probability of FA as a fertilizer or soil conditioner in the agriculture sector.

## 3.2 Impact of fly ash incorporation in the soil−plant system

The impact of FA incorporation in the soil−plant system has been explored by several researchers across the nations. Numerous studies suggested that FA can be utilized as a soil additive for the improvement of physicochemical, biological, and nutritional qualities of degraded soils. A number of amendments have been applied for the reducing FA toxicity during the FA application in agriculture (Pandey and Singh, 2010; Ram and Masto, 2014), although the higher FA dose results in metal (loid)s contamination and hinders the microbial activities. Besides, it is also noticed that, at lower doses, the nutrient-enriched FA-aided soil improves the growth and yield of the plants (Pandey et al., 2009a, b, 2010; Singh and Pandey, 2013; Singh et al., 2011, 2016). Overall, FA studies reveal that FA could be successfully used in the degraded soil to improve the soil fertility and the crop yields. The impact of FA incorporation on soil amelioration and plant response will be explored in detail in the next sections.

## 3.3 Fly ash and soil amelioration

### 3.3.1 Soil pH

The FA liming property has been examined by numerous researchers. FA may be acidic or alkaline, depending on some aspects such as coal source, particularly the S content of the used coal, and the operating condition of the thermal power plant (Ram and Masto, 2010; Pandey and Singh, 2010). Most of the globally produced FAs are alkaline in nature. The degree of soil pH change on the application of FA mostly depends on the difference among the FA pH and soil pH, the soil buffering capacity, and the FA neutralizing capacity as determined by the quantity of $Al_2SiO_5$, MgO, and CaO present (Truter et al., 2005). FAs may have significant amounts of silicate minerals, for instance, mullite (Ram et al., 1995), which can take up $H^+$, lead to neutralization via silicic acid formation. Therefore dissolving even a lesser quantity of silicate minerals with the FA application would considerably enhance the soil pH (Jankowski et al., 2006), while dolomite and lime can be used for the acidic soil reclamation, but they are costly and long-term processes in improving soil characteristics (particularly structural and physical characteristics), and release a huge amount of $CO_2$. Most of the acidic soils are not appropriate for commercial cultivation due to lack of proper management. The range of optimum pH for most of the crops is 6.5−7.5, on which maximum nutrients are available to crops. Soil fertility decreases at very low pH (<5.0), while the solubility of toxic metals such as Mn and Al increases, which harms plant roots and hampers the plant growth. The bioavailability of essential nutrients to plant is adversely affected at lower pH (Mengel and Kirkby, 1987). Alkaline FAs with high calcium contents can be replaced for lime to lessen soil acidity to a certain level that is appropriate for cultivation (Matsi and Keramidas, 1999; Pathan et al., 2003). It is noted that alkaline FA is chemically similar to around 20% reagent-grade $CaCO_3$ to supply calcium to plants and increase soil pH (Phung et al., 1979a, b). The application of FAs to acidic soils adjusted their pH (Lee et al., 2006). FA use in place of lime in crop cultivation can decrease net $CO_2$ emission, and consequently lessen global warming too. Furthermore, the application of FA with the amendments [i.e., paper mill sludge, sewage sludge (SS), lime mud, and farm yard manure] at different doses has raised the pH of wide-ranging acidic soils (Wong and Lai, 1996; Jiang et al., 1999; Ulrichs et al., 2005; Ram et al., 2007). The high-pH FA can be used as a feasible alternative to lime for decreasing SS acidity, and ameliorating acidic mine spoils and acidic soils during the eco-remediation (Taylor and Schuman, 1988; Stout et al., 1999; Zhang et al., 2008a). In a field study of mine spoil, an enhancement in pH of mine spoil has been noticed after lignite FA addition (Ram et al., 2006). In a long-term field study, it was observed that sulfo-calcic FA and silico-aluminous FA buffered natural soil acidification and regulated trace element bioavailability and mobility (Lopareva-Pohu et al., 2011). There was a minor reduction in sodic-soil pH, while electrical conductivity (EC) remarkably decreased with the FA application during the sodic-soil reclamation. It was suggested that gypsum can be replaced (up to 40% of its requirement) with acidic FA (Kumar and Singh, 2003). FA was effective for ameliorating the sodic soil. The initial pH value of the sodic soil was 10.0, and it reduced to 8.3 after the FA application and the pyrite amendment (Tiwari et al., 1992). Moreover, the FA application with press mud can be used for an effective reclamation of sodic soils. This combination had a favorable effect on the soil physicochemical, microbial, and enzymatic activities and the crop growth (Singh et al., 2016). Consequently, it is now well proved that FA is a potential soil pH ameliorator and has a fantastic role in bringing back a wide range of soils (i.e., mine spoil, acidic soil, sodic soil, and degraded soil) for agriculture, floriculture, and forestry purposes (Pandey et al., 2009b). The corrected pH has multiple benefits for the establishment and growth of plants via nutrient availability, reducing metal toxicity, etc. Fig. 3.1 shows that FA is a potential ameliorator for improving the soil system in terms of the physicochemical and biological properties.

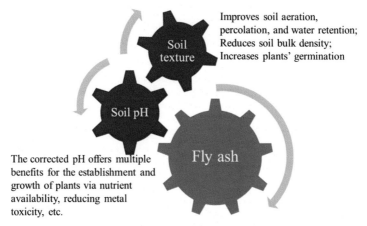

**Figure 3.1** Fly ash (gray circle) works as an ameliorator for the soil system (dark brown circles) toward improving the soil physicochemical and biological properties.

## 3.3.2 Soil physical properties

The impact of FA on the soil physical characteristics is mainly attributable to the changes in the soil texture. The change in the soil texture is associated with porosity, BD, WHC, hydraulic conductivity, and void ratio, which have a direct effect on the soil biological activity, the nutrient retention, and the plant growth (Fang et al., 2001; Guest et al., 2001; Pandey and Singh, 2010; Jayasinghe and Tokashiki, 2012; Ram and Masto, 2014). Direct FA amendment modified the soil texture from sandy clay to loamy and sandy loam to silt loam, besides ameliorating their physicochemical characteristics (Stout et al., 1999; Truter et al., 2005). FA hollow spheres accumulate small silt particles in voids, which modifies the soil texture (Ram et al., 2007) and the soil pore structure (Yunusa et al., 2006). FAs with high soluble calcium content and suitable particle-size distribution, particularly a higher percentage of particles ranging $2-200 \, \mu m$, may support to enhance the soil structural properties (Yunusa et al., 2006). FA incorporation also decreases surface encrustation, and increases soil aeration and plant seed germination. FA addition even with higher dose ($600 \, t \, ha^{-1}$) in acid clay soils upgraded their physicochemical characteristics (Fulekar, 1993), with a favorable alteration in the soil workability and significant decreasing of clayey properties. The BD of the FA has been reported to be lower than the normal cultivable soil

(Sikka and Kansal, 1994). A wide-ranging BD $(0.81-1.16\,\mathrm{Mg\,m}^{-3})$ of FAs is observed by several researchers (Chatterjee et al., 1988; Sarkar et al., 2005). Mostly, a noticeable decrease in the BD of a range of agricultural soils $(1.25-1.65\,\mathrm{Mg\,m}^{-3})$ on the FA incorporation is observed (Page et al., 1979). The improvement in the soil porosity, WHC, permeability, and workability of different types of soil after the reduction in their BD on FA addition is also well recognized (Phung et al., 1978; Adriano and Weber, 2001). The FA application (10%−30%) to coarse-textured soils improved the WHC manifold depending on the FA dose and soil texture (Campbell et al., 1983; Sims et al., 1995). Soil moisture is a main variable of the climate system that has effects on the biogeochemical cycles, energy, and water. FA is helpful to conserve the soil moisture (Seneviratne et al., 2010). On the FA incorporation $(200-400\,\mathrm{t\,ha}^{-1})$ to sandy loam soil, a significant improvement in the soil permeability and soil moisture-holding capacity along with a reduction in BD and acidity was determined to the benefit of crop production (Gracia et al., 1995). Moreover, Pandey and Kumar (2013) noticed that 10% of FA amended sand is an appropriate rooting medium for the vegetative propagation of Subabul plant (*Leucaena leucocephala*). Fig. 3.1 depicts the potential of FA to improve the physical properties of the soil system.

### 3.3.3 Soil fertility

The effect of FA on soil fertility mainly depends on the coal and soil properties. FA may be acidic or alkaline nature depending on the source, and can be employed to buffer the soil pH (Elseewi et al., 1978). FA increases the nutrient status of soil (Rautaray et al., 2003). Nutrient-enriched FA has been suggested for use as a source of plant nutrients (Pichtel et al., 1994; CFRI (Central Fuel Research Institute), 2003; Palumbo et al., 2004; Pandey et al., 2009a, b; Tripathi et al., 2009, 2010; Ram and Masto, 2010; Srivastava and Ram, 2010; Singh et al., 2011, Singh and Pandey, 2013; Shaheen et al., 2014). The FA application significantly increased the soil contents such as K, Ca, P, Mg, B, Mn, Zn, carbonates, bicarbonates, and sulfates (Khan and Singh, 2001). FA addition showed positive results in acidic and neutral soils toward the increase in the P adsorption and enhanced the P retention capacity of the soil, while alkaline soils exhibited the least response (Seshadri et al., 2013a). Several studies have shown that nutrient-enriched FA may improve the agronomic soil properties and support plant growth (Adriano et al., 1980).

The N, P, and K contents in the soil and their uptake by paddy crop at all the growth stages on 20% of the FA application (w/w) were noticed (Wong and Lai, 1996). An increase in the soil contents (i.e., P, K, Mg, and B) was observed on various FA doses ranging 5%—10% (w/w) (Lai et al., 1999). A significant increase in the nutrient content of oil seed crops (Selvakumari and Jayanthi, 1999) and *Brassica* (Menon et al., 1993) upon FA addition was recorded. FA incorporation to soil enhanced the Si content of paddy crops (Lee et al., 2006). FA applications have been witnessed to correct plant nutritional deficiencies of Zn (Mulford and Martens, 1971; Schnappinger et al., 1975), Mo (Elseewi et al., 1980), S, Mg (Hill and Camp, 1984), B (Martens, 1971; Ransome and Dowdy, 1987), P, and Mn (Carlson and Adriano, 1993). The FA application is beneficial to the correction of S and B deficiencies in acid soils (Chang et al., 1977). Nutrient deficiencies were ameliorated by either direct FA incorporation or increasing the bioavailability in soil through pH corrections. The maximum crops prefer optimum pH values between 6.5 and 7.0, on which the availability of the majority of nutrients to plants is maximized. Fertility is impaired at very low pH levels as dissolution, and the bioavailability of Mn and Al that are toxic to plants increases. The soil pH can be increased after the FA incorporation, which is explained by the rapid release of Ca, Na, Al, and OH ions from FA (Wong and Wong, 1990).

FA addition on acidic mine spoils at different places increased the yield of many crops, which was attributed to increased availability of $Ca^{2+}$ and $Mg^{2+}$ in soil and preventing toxic effects of $Al^{3+}$, $Mn^{2+}$, and other metallic ions by neutralizing the soil acidity (Fail and Wochok, 1977). The activity of certain metals may increase with an increase in the pH of soil. For instance, Al is relatively insoluble as $Al(OH)_3$ at neutral pH, but it mainly exists as highly soluble and toxic aluminate anions above a soil pH of 8.0. $Al^{3+}$ is the toxic species for monocots, namely wheat roots, when $Al^{3+}$ activities were increased; the activities of the hydroxyl-Al species were decreased. For dicots, either $Al(OH)^{2+}$ or $Al(OH)_2^+$ is the phytotoxic species, and $Al^{3+}$ is much less toxic (Kochian, 1995). Al is the most abundant metal in FA. Although higher B availability limits the use of FA in crop production (Page et al., 1979), the problem can be overcome by proper weathering of the FA, which reduces B availability to below toxic level.

## 3.3.4 Soil biota

Precious soil and its biota are vital components of the uppermost fertile layer of the earth, where plants grow and are helpful to make earth's ecosystem that support life (Wilkinson et al., 2009). The study on the impact of FA on soil biota is not more extensive compared with the physicochemical characteristics of soil. The appropriate value of the FA application for the amelioration of soil can be determined only after repetitively assessing the soil quality bioindicators (viz., microbial diversity, microbial biomass, microbial quotients, soil enzymes, earthworms, respiration, etc.). It is well known that FA is not a source of soil biota, but its positive effect on the soil physicochemical properties enhances microbial activity (Truter et al., 2005; Pandey and Singh, 2010; Ram and Masto, 2010; Singh et al., 2011, 2016; Singh and Pandey, 2013). The FA application with acidic nature at $100 \, t \, ha^{-1}$ in soil had no significant effect on the activity of microbes, but higher doses from 400 to $7000 \, t \, ha^{-1}$ harmfully affected the soil microbes (Arthur et al., 1984). An enhancement effect was noticed in the microbial activity after the FA addition in soil up to 5%, while an inhibitory effect was found at higher FA doses (Kalra et al., 1997). Likewise, the FA application of up to $100 \, t \, ha^{-1}$ was safe for microbial characteristics of tropical red laterite soil (Roy and Joy, 2011). The number of P-solubilizing bacteria and Rhizobium spp. was increased when soil was amended with either FA or FYM alone or in mixture (Sen, 1997). In contrast, it was reported that the lignite FA application reduced the growth of soil microbes (Karpagavalli and Ramabadran, 1997). The numbers of gram-negative bacteria and arbuscular mycorrhizal fungi were increased by the FA application in soil at $505 \, t \, ha^{-1}$ (Schutter and Fuhrmann, 2001). An enhancement in the numbers of microbial community in degraded subsoil modified with FA was noticed (Schutter and Fuhrmann, 2001). Muir et al. (2007) observed that adult populations of a species of *Aporrectodea trapezoides* (an exotic species) and megascolecid worm (a native species) were not harmfully affected when acidic loamy soil was treated with FA of up to $25 \, t \, ha^{-1}$. Maity et al. (2009) reported that up to 50% of FA dose does not apparently harm the *Lampito mauritii* earthworm for their growth and survival. A significant enhancement in the concentration of tissue metallothionein was observed in *L. mauritii* without tissue metal accumulation, showing that metallothionein is involved in scavenging of free radicals and reactive oxygen species metabolites. It is determined that the biochemical response observed in *L. mauritii* exposed to FA amended soil could be used as a vital tool for eco-toxicological field monitoring. Yunusa et al. (2009) revealed that native earthworms were mainly sensitive to FA addition ($>5 \, t \, ha^{-1}$); however, even the more strong exotic species exhibited marks of stress at high FA doses. Pandey and Singh (2010) suggested that extensive studies are needed to know the FA effect on the fundamental features of the earthworms' burrows, such as their diameter, length, tortuosity, sinuosity, orientation, connectivity, branching, and volume that differ among species and the functional groups of earthworms along with the behaviors of earthworms in FA amended soil.

The soil enzymatic activity is also a significant factor for assessing soil biological properties in FA amended soil. Pati and Sahu (2004) studied the enzyme activities (i.e., dehydrogenase, amylase, and protease) and $CO_2$ evolution, and reported little or no inhibition of enzyme activities and soil respiration of up to 2.5% of FA addition. Lal et al. (1996) also found that FA incorporation to soil at 16% (w/w) enhanced cellulose and urease activities, while acid phosphatase activity was decreased with FA addition. In a thorough study, alkaline FA did not hinder microbial activity and respiration at 10% and 20% of FA with swine manure (Vincini et al., 1994). A field-based study showed higher dehydrogenase, protease amylase, and invertase activities over the control when using FA at $20 \, t \, ha^{-1}$ (Sarangi et al., 2001). Altering soil with suitable doses of FA-filtered mud combination (1:1, w/w) augmented soil microbes and soil enzyme activities, namely phosphatase, cellulase, and urease (Xing et al., 2001). Kumpiene et al. (2009) observed an enhancement in microbial biomass, respiration, and soil enzyme activities, and a reduction in microbial stress when FA was incorporated in a Cu−Pb-contaminated soil. FA addition in soils can also inhibit microbial respiration, soil nitrification, N mineralization, and enzyme activities such as dehydrogenase, amylase, invertase, and protease. The inhibition of microbial activity was noticed due to FA [attributed to its high pH, high alkalinity, higher amount of metal(loid)s, and soluble salts], and a limited amount of organic matters in amended soils (Pichtel and Hayes, 1990; Klubek et al., 1992; McCarty et al., 1994; Sims et al., 1995; Pandey and Singh, 2010; Ram and Masto, 2014; Pandey et al., 2017). Overall, the FA impact on microbial activities is beneficial in the majority case of FA applications at lower doses or in the presence of organic amendments, while at higher doses, the microbial activity is decreased, mainly with high-pH ashes.

## 3.4 Fly ash and plant response

### 3.4.1 Plant morphology, physiology, and growth

The fine FA dust particles discharged by coal-based thermal power stations are mostly alkaline in nature. Being fine particles, FA dust has tendency to exist in atmosphere for a long period, and its impact on plant physiology may be positive and negative, because alkaline dust particles usually form a surface crust on plant parts with the formation of water films. Therefore, in the presence of humidity, it sticks to the plant parts (such as leaf and fruit) and creates injuries (i.e., physical and chemical), and small necrotic spots with dark brown appear on the leaves of several vegetables, for instance, turnip, cabbage, tomato, and green beans (Singh and Yunus, 2000). These surface films can also have detrimental effects, probably by blocking stomata, thereby reducing gaseous exchange and photosynthesis due to shading effect (Ots et al., 2011; Grantz et al.,2003). At lower FA deposition rates, FA particles gather on the leaf pores especially the guard cells, induce the stomatal mechanism that regulates opening and closing, and stop the stomatal closing (Fluckiger et al., 1979, Krajickova and Majstrick, 1984), thus impeding increased transpiration rates. In addition, higher FA deposition rates ($8 \text{ g m}^{-1} \text{ day}^{-1}$) as a foliar application resulted in reduced transpiration rates owing to the hurdle formed by a thicker layer, and consequently decreased vapor loss from the plant leaves (Mishra and Shukra, 1986). FA thick layers restrict the light that is essential for photosynthesis, and therefore decrease the rate of photosynthesis. A plant leaf loaded with FA absorbs more heat in comparison with an unloaded leaf, and thus the enhanced leaf temperature results in increased transpiration rates. Numerous studies have been performed by researchers to assess the effect of FA deposition on foliar morphology, physiology, plant growth, and survival near coal-based thermal power stations due to fine dust particle pollution. The main infirmities triggered by FA pollutants include necrosis, chlorosis, and epinasty (Sharma and Tripathi, 2009). The toxicity of B was observed at the leaf margin of *Agropyron elongatum* (Wong and Su, 1997). In addition, airborne FA dust particles may affect many other functions of plant parts such as root, shoot, and leaf.

In a study (Qadir et al., 2016), the impact of FA dust particle deposition on the foliar and biochemical characters of *Azadirachta indica* near an ash dumping site in coal-based Thermal Power Station, Badarpur, New Delhi, India, was assessed. Airborne FA dust particle and plant responses, such as stomatal index, stomatal conductance, leaf pigments, net photosynthetic rate, intercellular carbon dioxide concentration, sulfur content, nitrogen, and proline, were evaluated. A significant decrease in leaf pigments was detected at the FA dumpsite. FA dust stress showed the stimulatory impact on the stomatal index, antioxidants, sulfur, nitrogen, and proline content in the leaves. In FA dust stress, stomatal conductance was low, leading to a decrease in the photosynthetic rate and an increase in the internal carbon dioxide concentration of the leaf. The leaf length, width, and area also revealed a decreasing trend from control to the FA polluted site. Antioxidant enzymes enhanced in the leaves of *A. indica* reflect FA stress and mitigate reactive oxygen species (Qadir et al., 2016). Several greenhouse and field-based studies proved that nutrient-enriched FA increases the growth of plant species via the improvement of the soil physicochemical and biological properties (Singh et al., 1997, 2011, 2016; Matsi and Keramidas, 1999; Singh and Siddiqui, 2003; Tripathi et al., 2009; Pandey et al., 2009a, b; Pandey et al., 2010; Singh and Pandey, 2013).

## 3.4.2 Plant yield

There are numerous problematic soils, which have inherent structural and nutritional limitations, the use of weathered FA can increase crop yields and enhance food security, and it is realized that weathered FA is more beneficial than unweathered FA because of the FA attributed decreased pH, salinity, and phytotoxicity (Pandey et al., 2009a; Pandey and Singh, 2010; Ukwattage et al., 2013; Ram and Masto, 2014). Exhaustive studies over the world have demonstrated that FA applications in soil increased the crop yield of *Medicago sativa, Hordeum vulgare, Cynodon dactylon, Trifolium repens, Helianthus* sp., *Arachis hypogaea, Cymbopogon martini, Cymbopogon nardus, Cajanus cajan, Cicer arietinum* (Elseewi et al., 1980; Hill and Camp, 1984; Weinstein et al., 1989; Asokan et al., 1995; Sajwan et al., 1995; Jala and Goyal, 2006; Pandey et al., 2009b, 2010). Sikka and Kansal (1995) reported that lower FA dose (2%−4%, w/w) increased the rice yield, while higher FA dose (8%, w/w) reduced the rice yield. Kim et al. (1997) also found similar increases in the Chinese cabbage yield due to FA applications. Matsi and Keramidas (1999) studied the influence of FA applications on the total biomass yield of ryegrass in two acidic soils, and the highest yield was found when FA was applied at 5% (w/w, $50 \, g \, kg^{-1}$ soil). A greenhouse study (Singh and Siddiqui, 2003) showed the effects of various FA doses on the yield of three cultivars of *Oryza sativa*. They reported that the 20% and 40% of FA application rates increase in the rice yield of all the three cultivars. Similar results of FA on the rice yield were found in the studies of Lee et al. (2006) and Dwivedi et al. (2007). In contrast, Singh et al. (2008) studied the effects of various doses of the FA (i.e., 0%, 5%, 10%, 15%, and 20%) application on the yield of

*Beta vulgaris.* They noticed that the higher doses of the FA application (15% and 20%) caused significant declines in the biomass yield. Tsadilas et al. (2009) conducted a field study on *Triticum vulgare*, which was grown in FA-treated soil (Alfisol), with two doses (5.5 and 11 t ha$^{-1}$). They observed that FA doses enhanced the biomass and grain yield, and the yield was increased with the FA rate. In a pot study, *C. cajan* was grown in FA-treated garden soil at the doses of 0%, 25%, 50%, and 100% (w/w), and the 25% of FA amended soil showed positive results in most of the yield parameters (Pandey et al., 2009b). They also confirmed that the lower FA dose (25%) was safe for the *C. cajan* leguminous crop. However, the yield parameters were adversely affected at higher FA doses (50% and 100%, w/w) compared with the control (Pandey et al., 2009b).

A field study was done by Tripathi et al. (2009) to evaluate the impact of the FA use on the nutrient composition and the yield of wheat, maize, and eggplant crops grown in order on a formerly uncultivated land. FA was amended with soil at 100 t ha$^{-1}$. In the wheat case, the increases in the yields of grain and straw were recorded as 29.4% and 26.6%, respectively, compared with the untreated control. The FA application also increased the yield of maize (33.1%) and eggplant (18.4%). This noticeably proves that, when FA is applied in suitable amounts as a soil amendment, the FA application can support to increase and sustain the crop yield. Similarly, a field investigation was performed by Singh and Agrawal (2010) to assess the impact of FA on the yield of three cultivars of *Vigna radiata* on a soil treated with various FA doses (0%, 5%, 10%, 15%, and 20%, w/w). A significant positive result in all the growth factors was noticed at 10% of FA amendment for all the taken cultivars. An experiment was done to assess the impact of the FA application rate (0%, 25%, 50%, 75%, and 100%) on the yield of the *Mentha piperita* crop. They reported that the FA application at >50% decreased the crop oil yield, while the FA incorporation at the dose of ≤50% showed positive responses (Kumar and Patra, 2012).

## 3.5 Coapplication of fly ash with amendments

*Inorganic amendments*—Acidic FA in mixture with gypsum or alone enhanced the paddy and wheat yield via the improvement of the sodic-soil physicochemical properties (Kumar and Singh, 2003). The application of FA and lime showed a significant increase in the plant production (Taylor and Schuman, 1988). Lee et al. (2005) used a combination of FA and gypsum (50:50, wt./wt.) to assess the Si and Ca demands of rice, and noticed positive impacts on the rice growth by increasing the soil fertility. The combination of FA and gypsum at $25 \, t \, ha^{-1}$ increased the paddy grain yield up to 8%, besides enhancing the Si and N uptake of rice. Lee et al. (2003) assessed the FA—gypsum mixture to reestablish the nutrient equilibrium in paddy soils and to minimize the nitrogen application dose of rice. Moreover, Lee et al. (2008) appraised the potential of FA and phosphogypsum for decreasing dissolved P in arable soil, and observed that this mixture reduced P loss from the paddy field and improved the paddy soil fertility. Therefore it is proved that the FA—gypsum mixture is a good paddy soil ameliorant. Gutenmann and Lisk (1996) detected higher Se uptake in onions that were grown on 50% of FA media comprising vermiculite—sphagnum peat moss. The application of FA and CaO (quicklime) was used to immobilize Pb, $Cr^{3+}$, and $Cr^{6+}$ in contaminated clayey sand soils. They observed that this combination reduced toxic metal leachability well under the safe regulatory limits. The mechanism of Pb and $Cr^{6+}$ immobilization is surface adsorption, while for $Cr^{3+}$ it is hydroxide precipitation (Dermatas and Meng, 2003). Daniels and Das (2006) also reported that the application of FA and lime mixture reduced the leachability of Cd, Se, and to certain extent of As. FA-based granular aggregate contains FA, gypsum, paper waste, and lime developed by Jayasinghe et al. (2009), who observed their potential as a soil ameliorant for the improvement of soil physicochemical properties and toxic metal contents in the soil under the acceptable limits. A pot study showed that the application of FA and steel-slag mixture increased soil pH, reduced the phytoavailability of toxic metals, and further inhibited metal uptake by the rice crop (Gu et al., 2011).

*Farmyard manure*—Historically, FYM is known as a vital manure used by the ancestors for growing crops due to easy accessibility of all the nutrients needed by the plants. Several studies revealed that the application of FA and organic manure combination enhanced the soil physicochemical and biological properties (Rautaray et al., 2003; Jala and Goyal, 2006; Lee et al., 2006; Pandey and Singh, 2010; Ram and Masto, 2014). Singh and Agrawal (2010) showed that the application of FA and FYM significantly decreased the BD in the field condition and increased the *V. radiata* yield. Singh et al. (2011) also noticed a reduction in the BD with the use of FA and FYM. The positive response of the application of FA along with FYM has been observed in the yield of several crops such as sunflower, groundnut, maize, paddy, wheat, garlic, carrot, radish, onion, turnip, potato, eggplant, and beetroot (Yeledhalli et al., 2005; Tripathi et al., 2009, 2010; Singh et al., 2011), and soil amelioration in terms of the nutrient availability, solubility of P, N mineralization, nitrification rates, WHC, and suppressed heavy metal (Ghodrati et al., 1995; Bhattacharya and Chattopadhyay, 2002; Pierzynski et al., 2004; Sinha et al., 2005, 2011; Urvashi et al., 2007; Tripathi et al., 2009; Roy and Joy, 2011). Alkaline stabilization of swine manure along with lime (4%) and FA (50%) was helpful in devitalizing pathogens and decreasing their poststabilization regrowth (Wong and Selvam, 2009). In addition, P in FA could be mobilized by incorporating swine manure (Vincini et al., 1994). The application of combinations of FA and poultry manure was beneficial to the improvement in the yield of crops such as *Brassica parachinensis*, *Brassica chinensis*, sunflower, mustard, and grass (Wong and Wong, 1987; Phunshon et al., 2002; Seshadri et al., 2013b), and the soil amelioration (Salingar et al., 1994; Dao et al., 2001; Phunshon et al., 2002; Seshadri et al., 2013b). In a study, it was reported that the application of biofertilizer and FYM assisted in lessening the metal toxicity (i.e., Cr, Zn, Cu, Pb, Ni, and Cd) of FA deposits in the range 25%–48% (Juwarkar and Jambhulkar, 2008). Therefore the application of FA with FYM is helpful to improve the soil physicochemical and microbial properties such as WHC, porosity, pH, nutrient availability, stressed or decreased metal toxicity, size, and activity of microbes. These all improvements are attributed to an increase in the organic matter content in the FA. This combination is helpful in agriculture to use massive and biologically inert FA (Pandey and Singh, 2010).

*Sewage sludge*—SS is a main solid organic waste of sewage treatment plants in cities across the nations. SS application on land is the most largely used reasonable way for its disposal worldwide (Metcalf and Eddy, 2003), because it is a source of easily available essential nutrients (i.e., N, P, S, and Mg), micronutrients (such as Cu, Zn, Mo, B, and Fe), organic matters, and microorganisms, which are helpful for plantations and landscaping (Abbott et al., 2001). It improves the soil physicochemical and biological properties (Aggelides and Londra, 2000; Sajwan et al., 2003). The risk of heavy metal toxicity by SS during its application to soil can be reduced with the combined use of FA and SS. The alkaline FA has a large quantity of CaO that plays as a stabilization agent for SS to reduce the pathogen load and the availability of heavy metals (Wong, 1995; Jiang et al., 1999). Thus FA can be used as a sludge stabilizing agent (Samaras et al., 2008). In addition, the mixture of FA and SS can be used as artificial soil media (Wong, 1995; Kelley et al., 2002), and contains high nutrient contents (mg kg$^{-1}$) with the availability of K (762), P (375), and N (109) (Zhang et al., 2008b). The application of the FA—SS mixture has led to well-balanced nutrient contents (Guest et al., 2001; Shen et al., 2004). The physical quality of wide-ranging soils (i.e., sandy soil, saline—sodic soil, barren red soil, black soil, and sodium-rich soil) can be improved by incorporating the FA—SS mixture (Lai et al., 1999; Veeresh et al., 2003; Pierzynski et al., 2004; Sahin et al., 2008; Mataix-Solera et al., 2011).

In general, the FA—SS combinations had higher pH levels than the normal levels for the suitable growth and establishment of plant species (Fang et al., 1999; Papadimitriou et al., 2008). Increased pH of the mixture due to the FA addition is the most important factor in devitalizing the pathogens of SS and in ameliorating SS. The coapplication of the FA—SS mixture significantly decreased the total coliform population of the sludge (in most of the cases the decrease was approximately 100%), apart from the reduced metal leaching from the SS (Papadimitriou et al., 2008). Alkaline stabilization of SS with 10% of FA and 8.5% of lime (dry weight basis) resulted in satisfactory amounts of salmonella and coliforms, and was attributable to the increase in pH (Wong et al., 2001). The application of the FA—SS mixture in a wide range of soils (acid soil, neutral soil, Alfisol, acidic loamy soil, loamy sand, red acidic soil, loamy soil, silty loam, and calcific soil) significantly increased the yields of several crops (*Sorghum vulgare*, *Triticale*, *Helianthus annuus*, *Zoysia matrella*, *A. elongatum*, *B. chinensis*, *Zea mays*, *Sorghum bicolor*, and *Glycine max*) by up to many folds (Adriano et al., 1982; Reynolds et al., 2002; Kallesha et al., 2001; Wong et al., 1998; Sajwan et al., 2003; Xu et al., 2012; Wong and Su, 1997; Wong and Selvam, 2009; Su and Wong, 2002; Schumann and Sumner, 2004; Huapeng et al., 2012). FA stabilized SS exhibited a positive role on crop production, where no harmful impacts on crop growth was noticed (Ghuman et al., 1994). In an extensive study, the FA—SS mixture (1:1) showed a positive impact on biomass production and soil fertility, without any significant risk of soil—plant contamination (Sajwan et al., 2003). The FA—SS mixture in suitable ratios increased the root growth of *Z. mays* (Su andWong, 2002). In addition, the phytotoxicity studies of *Z. mays* exhibited the best performance at 30% of FA incorporation to SS (Masto et al., 2012). FAs work as a low-cost adsorbent for Cd and Pb and a soil ameliorant, and could be used for SS amended acidic soils (Shaheen and Tsadilas, 2010). There was no too much uptake of heavy metals by the paddy crop on the amendment of silty clay loam soil (rich in B) with FA—SS (Lee et al., 2003). An enhancing FA amendment dose significantly decreased diethylenetriaminepentaacetic acid-extractable Cd, Ni, Zn, and Cu concentrations in loamy soil of Hong Kong (Su and Wong, 2004). The coapplication of the FA—SS mixture is suggested for decreasing the bioavailability of heavy metals and for increasing the ambient amounts of K, Ca, S, and B (Lau et al., 2001; Su and Wong, 2004; Xu et al., 2012).

*Biochar*—Biochar is a carbon rich material and made from biomass via pyrolysis. It is gradually being recognized by researchers, policy makers, and practitioners for its potential role in carbon sequestration, reducing greenhouse gas emissions, waste mitigation, and renewable energy, and as a soil amendment because of its relatively stable and inert nature (Lehmann, 2007; Kookana et al., 2011). Recently, biochar is famous worldwide as an important and potential asset to increase carbon sequestration that results in the enhancement of terrestrial soil carbon pool. It is an eco-friendly material for creating a stable carbon sink in the upper layer of soils, particularly in agricultural soils because of its agronomic benefits such as nutrient retention, water retention, detoxification of contaminants, and high cation exchange capacity (CEC). Biochar and FA have the potential to enhance the soil quality, soil carbon pool, and crop yield (Palumbo et al., 2004; Lehmann, 2007). Masto et al. (2013) notified the highest increase in the grain yield of maize with biochar + FA (4.6 t ha$^{-1}$) followed by biochar (4.0 t ha$^{-1}$) and the lowest increase with the control (3.59 t ha$^{-1}$) and FA (3.42 t ha$^{-1}$) amendments. These treatments did not affect the crop height, but lower shoot yield was observed in the FA amendment (5.9 t ha$^{-1}$) than that in the others (6.6—6.7 t ha$^{-1}$). The grain yield was increased by 28.1% for biochar + FA and 11.4% for biochar, but decreased by 4.7% under the FA amendment. In addition, the soil quality factors such as soil organic carbon (SOC), pH, BD, WHC, EC, biological properties, and nutrient contents have noticeably improved owing to the joint effect of biochar and FA. Biochar incorporation to FA enhanced extractable P, K, Ca, Zn, Mn, and Cu, and improved WHC, mesoporosity, and CEC, but it showed a little or no stimulatory impact on N fertility, plant growth, and the soil microbial community's size (Belyaeva and Haynes, 2012).

*Press mud*—Press mud is a waste product of the sugar industry, and contains MgO (0.5%—1.5%), PO$_4$ (1%—3%), CaO (1%—4%), SiO (4%—10%), total ash (9%—10%), crude wax (5%—14%), sugar (5%—15%), crude protein (5%—15%), and fiber (15%—30%; Partha and Sivasubramanian, 2006). In addition, it is a rich and important source of enzymes, higher microbes, organic carbon, organic matters, micronutrients (viz., Zn, Fe, Mn, Cu, etc.), and macronutrients (viz., N, P, and K; Prakash and Karmegam, 2010). The application of press mud and FA combination enhanced dry matter production, panicle numbers, tiller numbers, and straw and paddy grain yields (Jeyabal et al., 2000). Press mud incorporation to FA improved the onion crop yield (Selvam and Mahadevan, 2002). Press mud and FA amendment increased the amount of amino acids and soluble protein in *Cicer arithenium* (Gupta et al., 2006). Gupta et al. (2000) observed increased chlorophyll, protein contents, and growth rates in *L. leucocephala* on a combined treatment of press mud + FA than FA alone treatment. It was noticed that press mud is the most appropriate amendment for the revegetation of FA deposits through *Cassia siameia* (Kumar et al., 2002). Mixing press mud with FA at 25% considerably reduced the cysteine, malondialdehyde, and stress protein contents in *Prosopis juliflora* than the control, thus indicating the ameliorating effect of press mud (Sinha et al., 2005). Selvam and Mahadevan (2002) observed that the negative impact of FA alone could be improved by using the FA and press mud mixture. Field-based studies with mine spoil and red soils showed a positive impact of FA and press mud on the rice yield and soil quality (Ram et al., 2006). Moreover, increasing the soil organic matter content, the organic treatments immobilized toxic metals (Ram et al., 2007). On the basis of pot and field study, the impact of press mud on FA-affected soil showed a significant improvement in the quality of the produce apart from increasing the growth rate of *Lycopersicum esculentum*. Besides, an increase in the available Ca, S, Mg, Mn, Fe, K, N, Cu, and Zn contents and an enhancement in the physical characteristics of the amended soil were noticed (Rani et al., 2012). The cocoon and biomass of *Eisenia fetida* significantly enhanced with the increasing quantities of press mud incorporation to FA (Sarojini et al., 2009). Likewise, Anbalagan and Manivannan (2012) stated that the mixture of FA, press mud, and cow dung (1:1:1 ratio) had no adverse impact on the reproduction and growth of *L. mauritii* and *Eudrilus eugeniae*.

*Other organic industrial waste*—Guttila et al. (2009) noticed a significant increase in the growth and yield due to the incorporation of FA and paper waste to less productive red soil in Okinawa, Japan. In a field experiment, the addition of paper mill sludge—FA combinations at the rates of $24-48 \, t \, ha^{-1}$ improved crop yields without any significant impact on heavy metal uptake by crops (Simpson et al., 1983). The incorporation of the FA and pine wood peeling mixture increased the flower diameter of marigolds than the other treatments. Furthermore, this mixture revealed higher Ca, Mg, and K concentrations than the other treatments (Woodard et al., 1993). In a field investigation, the addition of FA and crop residues (left over stubbles) on rice—wheat cropping systems improved the productivity and soil condition over no residue fusion (Kachroo et al., 2006). The FA and tannery sludge blend increased the oil content and the yield of *Brassica campestris* at 1:4 ratio (Gupta et al., 2010). The blend of FA and diluted spent wash could be used as a biofertilizer (Kumari et al., 2012). The use of FA and jatropha oil cake is reported as a promising ameliorant to support the essential oil yield of *M. piperita* than the other treatments (Kumar and Patra, 2012).

## 3.6 Vermicomposted fly ash for potential agricultural practices

Vermicomposting is a potential approach with the support of earthworms to obtain fertilizer substances from waste materials that can be used for the improvement of soil (Sinha et al., 2010). It includes the stabilization and biooxidation of organic matters depending on the pooled impact of microbes and earthworms. The FA impact on the growth and survival of worms was due to the positive effect of FA on soil pH (Muir et al., 2007). The FA application during the vermicomposting was positive for enhancing the bioavailability of nutrients, namely Mn, B, and N, and releasing more P from FA (Bhattacharya and Chattopadhyay, 2002). Saxena et al. (1998) studied on vermicomposted FA and reported that 25% of FA mixed with *Parthenium* cutting, green grass cutting, and sisal green pulp formed superb feed for the earthworm *E. fetida*, and observed higher NKP contents in vermicomposted FA than other manures. A high existence of nitrogen-fixing bacteria has been detected in vermicomposted FA (Bhattacharya and Chattopadhyay, 2004). The possibility of large-scale production of vermicomposted FA using cow dung, crop residue, and press mud with the earthworms *L. mauritii* and *E. eugeniae* has been studied in laboratory conditions (Bhattacharya et al., 2012). The FA application of 100% showed the incapability of earthworms to survive. However, the highest growth and reproduction in terms of biomass, and the cocoon and hatchling production were reported in the mixture of FA, press mud, and cow dung (at 1:1:1 ratios) without no adverse effect on the earthworms. Bhattacharya et al. (2012) concluded that vermicomposted FA contains increased nutrient availability with the decreased solubility of Cd, Cr, and Pb. Vermicomposted FA with rock phosphate and SS showed a significant reduction in metal bioavailability (Wang et al., 2013). Vermicomposted FA enhanced the soil fertility and replaced the chemical fertilizer to a significant extent. Likewise, the vermicomposting of other wide-ranging organic wastes, for instance, animal–vegetable wastes (Loher et al., 1985), sludge from pulp–paper industries (Elvira et al., 1998), sludge–horse manure (Neuhauser et al., 1979), paper wastes (Gajalakshmi et al., 2001), kitchen wastes (Adi and Noor, 2009), and press mud (Parthasarathi and Ranganathan, 2000; Khawairakpam and Bhargava, 2009), with FA can be advantageous. Increased microbial biomass was noticed owing to vermicomposting, and more metabolic quotient advised the rejuvenation of microbial spores in vermicompost (Pramanik and Chung, 2011). Vermicompost has been used for the revegetation of FA deposits using *Eucalyptus citriodora*, *Casuarina equisetifolia*, and *Anacardium occidentale*, which obtained higher biomass (Manivannan, 2005). Eijsackers (2010) concluded earthworms as one of the first colonizers in FA deposits.

## 3.7 Prospects and challenges

Now, FA has been well recognized as an important nutrient-enriched source for agricultural practices released from coal-based thermal power stations over the world. Numerous studies of FA applications on a wide range of crops showed that FA has the potential to increase plant growth/ yield by improving soil fertility. However, several studies also revealed that leafy vegetables are the most sensitive crops for the FA utilization as a nutrient-enriched source due to the presence of heavy metals. More studies are needed on evaluating the large-scale FA applications in agricultural areas especially cropwise and soilwise areas, but it requires policy interventions for enhancing the FA utilization. Table 3.1 represents the strategies for market opportunities of the FA use in agriculture. This chapter is a focused challenge to contribute to a better understanding of the FA utilization in agriculture. It also highlights the reclamation of degraded lands using nutrient-enriched FA along with organic waste as an emerging field, focusing on the enhancement of agricultural and rural development through improved crop yield and soil fertility. This chapter concludes that "researchers, practitioners, stakeholders, and farmers" are the key factors

**Table 3.1** Approaches for safe and bulk utilization of fly ash in agriculture.

| Approach | Action | References |
|---|---|---|
| **Market opportunities** | | |
| Fly ash use in modification of soil health | The fly ash has vast potential in modification of soil health due to its special physicochemical properties but its application and use must be done through research, assessment, and validation process. These improvements/modifications of problematic soils are: (1) neutralise soil acidity, (2) ameliorate soil sodicity, (3) nutrient delivery and minimization of nutrient loss, (4) carbon sequestration, and (5) improvement of soil structural and hydrological properties. Therefore, the fly ash has great potential to modify the health of problematic soil systems toward ecological and socioeconomic sustainability | Kumar and Singh (2003); Pandey et al. (2009a); Pandey and Singh (2010); Ram and Masto (2014); Shaheen et al. (2014); Singh et al. (2016) |
| Nutrient-enriched fly ash to enhance agricultural productivity | Technological forecasting and assessment of fly ash use in agriculture, investment into wider fly ash testing related to agriculture, expanding the safe use of nutrient-enriched fly ash as a potential asset for enhancing agricultural productivity, and the environmental aspects of fly ash associated heavy metals, organic pollutants, and radioactive elements. Extend the knowledge of agricultural use of fly ash among local farmers | Inthasan et al. (2002); Haynes (2009); Pandey et al. (2009a,b); Pandey et al. (2011); Singh and Pandey (2013); Sheoran et al. (2014) |

| | | |
|---|---|---|
| Fly ash use and application in agricultural disease control | Fly ash has been effectively used in agricultural diseases control. Several studies confirmed the insecticidal properties of lignite fly ash against pests such as Epilachna, Spodoptera, Lepidoptera, and Coleoptera. Fly ash could also be used as an active carrier in certain insecticide formulations like dust, wettable powder, and granules | Narayanasamy and Gnanakumar (1989); Mendki et al. (2001); Sankari and Narayanasamy (2007); Eswaran and Manivannan (2007); Pandey et al. (2009a) |
| **Policy implementation** | | |
| Environmental law/regulation framework | Advocacy within the regulatory sphere to recognize responsibilities for stakeholders, practitioners, and consumers, with the outcome that legal certainty can be explained. In India, the Ministry of Environment, Forests and Climate Change (MoEFCC) has revised norms for fly ash usage and disposal by granting permission to use it for agriculture | Aiken and Heidrich (2014) https://www.downtoearth.org.in/news/agriculture/moefcc-revises-fly-ash-notification-53260 |
| Permissible limit of the fly ash doses | The long-term impacts of fly ash application on soil quality and environment are not well studied in detail. Hence, the lack of information on this topic is highly important owing to risk associated. Therefore, the permissible limit of fly ash doses should be mentioned clearly in the agricultural department by the government of each state for each type of soil and crops. In addition, the soil quality monitoring programs should be undertaken to avoid metal toxicity to consumer who used fly ash grown agricultural food products. Several opportunities exist to improve fly ash quality by using different types of organic amendments | Pandey et al. (2009a,b); Pandey and Mishra (2018); Ram and Masto (2014) |

(Continued)

**Table 3.1** (Continued)

| Approach | Action | References |
|---|---|---|
| Memorandum of understanding (MoU) | MOUs should be established between two or more countries, between stakeholders and practitioners for better utilization of fly ash in agriculture. It will express a convergence between them for fly ash utilization, and will indicate an intended common line of action in agriculture | http://www.psuconnect.in/news/NTPC-Signs-Pact-with-Indian-Railways-for-Fly-Ash-Transportation/17593 |
| Cost sharing in fly ash utilization | Cost sharing by the three major stakeholders such as farmers, thermal power plants, and government for fly ash utilization in agricultural sector should be done | Rama Rao et al. (2012) |
| Incentives and subsidies for fly ash based agriculture | Incentives (monetary and support based) should be given to state and private nodal agencies involved in fly ash promotion in agriculture. Farmers should be given subsidies on the utilization of fly ash in agriculture | Parab et al. (2012) |
| **Fly ash knowledge and education** | | |
| Extension of the fly ash knowledge | Conduct an extension process, supporting publications on fly ash use and application in agriculture sector across the scientific journals, technical literature, reports, and books. In addition, the fly ash use and application in agriculture must be included in the course of postgraduate of relevant subjects | Aiken and Heidrich (2014) https://www.manage.gov.in/pgdaem/studymaterial/aem205.pdf |

| | | |
|---|---|---|
| Exchange of fly ash knowledge and skills | Supporting knowledge on fly ash use and application in agriculture should be exchange across the nations. At local level, the transfer of knowledge and skills regarding fly ash use and application in agriculture to farmers and their organization is an important criterion to achieve the goal of "fly ash-assisted agriculture" | http://dst.gov.in/sites/default/files/fly_ash_signing_brief_with_photo.pdf |
| Awareness programs on fly ash–assisted agriculture | Awareness programs on the safe and effective use of fly ash in agriculture as a soil additive must be done among the farmers, stakeholders, practitioners, and policy makers | Pandey et al. (2009b); Parab et al. (2012) |

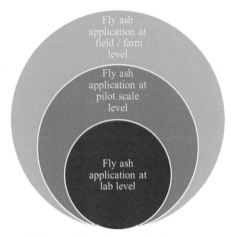

**Figure 3.2** Steps at which fly ash—based agricultural practices must be implemented before recommending to farmers.

for the application of FA as a nutrient-enriched source on large scale in field level. Fig. 3.2 reveals that three step trials (lab scale, pilot scale, and field scale) are necessary, at which FA-based agricultural practices must be implemented before advising to farmers.

## 3.8 Discussion and conclusions

FA, a by-product of coal-based thermal power stations, is generated in gigantic amounts across the nations around 750 million tons year$^{-1}$ (Blissett and Rowson, 2012). The use of coal for electricity production is increasing year on year at a significant rate, and is expected to increase more in future due to power demand in society. Therefore FA generation in massive amounts across the nations has been regarded as a waste that needs proper disposal. Hazardous pollutants are present in FA. However, FA is not listed as a hazardous waste in the US or EU member states, because FA has not been found to show any of the characteristics of hazardous wastes: corrosivity, reactivity, ignitability, and toxicity (Feuerborn, 2011). This has allowed an extensive FA use in several types of industries (i.e., brick, cement, concrete, road construction, embankments, and structural fills). In addition, FA is also used for land restoration and geotechnical (soil stabilization and road construction) purposes, although there is no clear cut legislation for the use of FA in agriculture, for example, some states in United States govern FA but their guidelines differ. This lack of

**Need of fly ash–based agricultural practices**

Increase in nutrient poor soil
Availability of degraded lands
Overpopulation
Rising food demand
Due to higher cost of chemical fertilizer
Need of carbon sequestration
Mitigate costs related to the agricultural practices
Stabilization of heavy metal contaminated agricultural soils

**Advantages of fly ash–based agricultural practices**

Improves soil texture, soil aeration, percolation, and water retention
Reduces bulk density of soil and crust formation
Optimizes pH value
Increases soil buffering capacity
Provides micro- and macronutrients to plants
Reduces the consumption of soil ameliorants (fertilizers and lime)
Uses as insecticidal purposes
Decreases the heavy metal mobility and availability in soil

**Figure 3.3** Pictorial representation of the need and the benefits of fly ash–based agricultural practices.

legislation is one of the main obstacles to the use of FA for the soil amelioration (Ahmaruzzaman, 2010). Other barricades to the FA use in agriculture include the distance between FA lagoons and farmers' fields. As a result, a huge volume of FA still remains unutilized and stored near coal-based thermal power stations. Such a situation causes air, soil, and water pollution. The FA storage in lagoons form is not environmentally safe and sustainable, because FA generation is an ever-increasing process. Nutrient-enriched FA also contains potentially hazardous pollutants (e.g., toxic metals, radioactivity, polychlorinated biphenyls (PCBs), and polycyclic aromatic hydrocarbons (PAHs)). FA is a potential resource for soil amelioration, but it can pollute soil and water through potentially hazardous pollutants. Consequently, the use of FA for soil improvement has been suggested much attention (Adriano et al., 1980; Basu et al., 2009; Yunusa et al., 2006; Pandey and Singh, 2010; Ukwattage et al., 2013; Skousen et al., 2012; Ram and Masto, 2014). Fig. 3.3 represents the need and the benefits of FA-based agricultural practices.

It is well proved that the application of FA as a soil additive is useful in several ways, as FA improves soil physical conditions, for example, BD, water infiltration, soil aggregation, soil WHC, and hydraulic conductivity (Adriano et al.,1980; Pathan et al., 2003; Singh and Agrawal, 2010;

Pandey and Singh, 2010; Skousen et al., 2012; Ukwattage et al., 2013). In sodic soil, FA is also helpful to lessen soil compaction and improve soil aggregate stability (Kumar and Singh, 2003; Singh et al., 2016), can offer micro and macronutrients to plants (Sajwan et al., 2003; Pandey et al., 2009a), and can correct acidic pH (Ciccu et al., 2003; Shaheen and Tsadilas, 2010). It is important to note that FA and problematic soil properties must be carefully considered as improper use of FA may not show advantages and probably may cause toxicity/adverse effects (viz., the use of acidic FA is not beneficial in acidic soil, or too large quantities of FA are applied; Jala and Goyal, 2006; Skousen et al., 2012; Ukwattage et al., 2013; Ram and Masto, 2014). The potential of FA applications for soil amelioration can be enhanced by blending organic amendments, because nitrogen and phosphorus are absent in FA (Pandey and Singh, 2010; Ram and Masto, 2014). Nitrogen is oxidized into a gaseous constitute during the coal combustion, and the presence of excessive iron and aluminum converts soluble phosphorus to insoluble phosphorus compounds (Adriano et al., 1980). Undoubtedly, the coapplication of FA with organic and inorganic amendments offers multiple benefits, that is, organic matter addition, nutrient availability, pH buffering, and overall soil health improvement.

The coapplication of FA and SS is a beneficial blending as soil amelioration offers a disposal problem for the huge amounts of both wastes. SS also contains nutrients and contaminants like FA. The application of SS is controlled by metal loadings and the mobility, and therefore the ecotoxicity risk of SS-borne metals increases after its break (Sajwan et al., 2003; Hooda, 2010). Alkaline FA is very helpful in such a combination to mitigate toxicity risk through alkaline reaction. At the same time, SS delivers desired nitrogen, phosphorus, and carbon, as it is well known that FA is unable to supply such an offer. Such blending is mainly suitable for the restoration of acid mine drainage-affected soils through establishing an effective plant cover on this site (Iyer and Scott, 2001; Xenidis et al., 2002). In addition, alkaline FA is used to remediate contaminated sites by metals through an immobilization mechanism by increasing soil pH and forming stable metal phases (Pandey and Singh, 2010; Lopareva-Pohu et al., 2011; Skousen et al., 2012; Ram and Masto, 2014). However, such an FA use should not lead to further rise of total metal concentrations in the soils, because FA is a strong sorbent and has been cost-effective in the treatment of wastewater for the removal of toxic metals and pesticides (Ahmaruzzaman, 2010).

Problems to the FA use as a soil additive could be either inadequate or too much quantities of CaO, $CaCO_3$, and/or $Ca(OH)_2$ in increasing soil pH too little or extremely; its excessive B content can create B toxicity in plants (Aitken et al., 1984) and hinder microbial respiration and enzyme activity, due to its negative effects on soil conditions (Pandey and Singh, 2010). Excessive FA-borne sulfur contents can be toxic to plants, and excessive amounts of unwanted toxic elements, namely Cd, Cr, Pb, As, Ni, B, and Se, (Sajwan et al., 2003; Ram and Masto, 2014), radioactive elements, and organic pollutants (PAHs and PCBs) can contaminate eco-systems and cause toxicity to flora and fauna, including human (Sahu et al., 2009; Pandey et al., 2011). However, all these problems related to FA applications can be mitigated by using a suitable FA dose, considering crop to crop along with site-specific conditions. Hence, most of the problems can eliminate by applying the above strategies. Appropriate FA application rates in several types of crops and soils have been worked out worldwide, but it still needs more research on the agricultural use of FA. The main constraint of the FA use on large scale in agriculture is a lack of guidelines and regulations. Thus it is a pressing need to work more for the sustainable and profitable use of FA in agriculture.

## 3.9 Recommendations

- Overall, this study shows that FA at lower concentration can ameliorate plant growth and yield, and it can be used as a supplement to fertilizers.
- FA can be directly used as an economic fertilizer and soil amendment in the small-scale cultivation of plants that have ornamental, medicinal, and agricultural or forestry potentials.
- However, the recommendation for the large-scale application of FA to the agricultural soils in a region cannot be made, unless extensive trials are made to find out a proper combination of FA with each type of soil and each crop to be grown in the area.
- Before using the FA as an economic fertilizer by farmers for agricultural purposes, such as field crops, vegetables and other edible plants must await more extensive field experiments to establish its quality and safety.
- The potential and FA-adaptive microbes should be isolated and screened for nitrogen fixation, carbon sequestration, and phosphorus

solubilization from old FA deposits, as we know that usually FA is deficient in nitrogen, carbon, and the available phosphorus.
- Use of FA as soil amendments in mitigation of climate change via soil carbon sequestration.
- Long-term and continuous monitoring of FA use in the agricultural field to check soil quality and risk assessment to ecosystems.

# References

http://dst.gov.in/sites/default/files/fly_ash_signing_brief_with_photo.pdf
https://www.manage.gov.in/pgdaem/studymaterial/aem205.pdf
http://www.psuconnect.in/news/NTPC-Signs-Pact-with-Indian-Railways-for-Fly-Ash-Transportation/17593
https://www.downtoearth.org.in/news/agriculture/moefcc-revises-fly-ash-notification-53260
Abbott, D.E., Essington, M.E., Mullen, M.D., Ammons, J.T., 2001. Fly ash and lime-stabilized biosolid mixtures in mine spoil reclamation: simulated weathering. J. Environ. Qual. 30, 608–616.
Adi, A.J., Noor, Z.M., 2009. Waste recycling: utilization of coffee grounds and kitchen waste in vermicomposting. Bioresour. Technol. 1027–1030.
Adriano, D.C., Weber, J.T., 2001. Influence of fly ash on soil physical properties and turf grass establishment. J. Environ. Qual. 30, 596–601.
Adriano, D.C., Page, A.L., Elseewi, A.A., Chang, A., Straughan, I.A., 1980. Utilization and disposal of fly ash and other coal residues in terrestrial ecosystem: a review. J. Environ. Qual. 9, 333–344.
Adriano, D.C., Page, A.L., Chang, A.C., Elsweei, A.A., 1982. Co-recycling of sewage sludge and **fly** ash: cadmium accumulation by crop. Environ. Technol. Lett. 3, 145–150.
Aggelides, S.M., Londra, P.A., 2000. Effects of compost produced from town wastes and sewage sludge on the physical properties of a loamy and a clay soil. Bioresour. Technol. 71, 253–259.
Ahmaruzzaman, M., 2010. A review of the utilization of fly ash. Prog. Energy Combust. Sci. 36, 327–363.
Aiken, J.T., Heidrich, C.R., 2014. Market opportunities for coal ash within Australian agriculture. In: Patti, A., Tang, C., Wong, V. (Eds.), Proceedings of the Soil Science Australia National Soil Science Conference: Securing Australia's Soils - For Profitable Industries and Healthy Landscapes, Melbourne. Australian Society of Soil Science Inc. ISBN: 978-0-09586-595-2-9.
Aitken, R.L., Campbell, D.J., Bell, L.C., 1984. Properties of Australian fly ash relevant to their agronomic utilization. Aust. J. Soil. Res. 22, 443–453.
Anbalagan, M., Manivannan, S., 2012. Capacity of fly ash and organic additives to support adequate earthworm biomass for large scale vermicompost production. J. Res. Ecol. 1, 001–005.
Arthur, M.F., Zwick, T.C., Tolle, D.A., VanVoris, P., 1984. Effects of fly ash on microbial $CO_2$ evolution from an agricultural soil. Water Air Soil Pollut. 22, 209–216.
Asokan, P., Saxena, M., Bose, S.K.Z., Khazenchi, A.C., 1995. In: Proceedings of Workshop on Fly Ash Management in the State of Orissa. R.R.L., Bhubaneswar, India, pp. 64–75.

Basu, M., Pande, M., Bhadoria, P.B.S., Mahapatra, S.C., 2009. Potential fly-ash utilization in agriculture: a global review. Prog. Nat. Sci. 19, 1173—1186.

Belyaeva, O.N., Haynes, R.J., 2012. Comparison of the effects of conventional organic amendments and biochar on the chemical, physical and microbial properties of coal fly ash as a plant growth medium. Environ. Earth Sci. 66, 1987—1997.

Bhattacharya, S.S., Chattopadhyay, G.N., 2002. Increasing bioavailability of phosphorus from fly ash through vermicomposting. J. Environ. Qual. 31, 2116—2119.

Bhattacharya, S.S., Chattopadhyay, G.N., 2004. Transformation of nitrogen during vermicomposting in fly ash. Waste Manag. Res. 22, 488—491.

Bhattacharya, S.S., Iftikar, W., Sahariah, B., Chattopadhyay, G.N., 2012. Vermicomposting converts fly ash to enrich soil fertility and sustain crop growth in red and lateritic soils. Resour. Conserv. Recycl. 65, 100—106.

Blissett, R.S., Rowson, N.A., 2012. A review of the multi-component utilisation of coal fly ash. Fuel 97, 1—23.

Campbell, D.J., Fox, W.E., Aitken, R.L., Bell, L.C., 1983. Physical characterization of sands amended with fly ash. Aust. J. Soil Res. 21, 147—154.

Carlson, C.L., Adriano, D.C., 1993. Environmental impacts of coal combustion residues. J. Environ. Qual. 22, 227—247.

CFRI (Central Fuel Research Institute), 2003. Demonstration trials in farmer's field for the popularization of bulk use of fly-ash from Anpara, Obra and Hardugang TPPs of UPRVUNL in agriculture and for reclamation of degraded/wasteland. CFRI, Dhanbad, Jharkhand (CFRI TR/CFRI/1.08/2002—2003).

Chang, A.C., Lund, L.J., Page, A.L., Warneke, J.E., 1977. Physical properties of fly ash amended soils. J. Environ. Qual. 6, 267—270.

Chatterjee, T., Mukhopadhya, M., Dutta, M., Gupta, S.K., 1988. Studies on some agrochemical properties of fly ash. Clay Res. 7, 19—23.

Ciccu, R., Ghiani, M., Serci, A., Fadda, S., Peretti, R., Zucca, A., 2003. Heavy metal immobilization in the mining-contaminated soils using various industrial wastes. Miner. Eng. 16, 187—192.

Daniels, J.L., Das, G.P., 2006. Leaching behavior of lime—fly ash mixtures. Environ. Eng. Sci. 23, 42—52.

Dao, T.H., Sikora, L.J., Hamasaki, A., Chaney, R.L., 2001. Manure phosphorus extractability as affected by aluminium and iron by-products and aerobic composting. J. Environ. Qual. 30, 1693—1698.

Dermatas, D., Meng, X., 2003. Utilization of fly ash for stabilization/solidification of heavy metal contaminated soils. Eng. Geol. 70, 377—394.

Dwivedi, D., Tripathi, R.D., Srivastava, S., Mishra, S., Shukla, M.K., Tiwari, K.K., et al., 2007. Growth performance and biochemical responses of three rice (*Oryza sativa* L.) cultivars grown in fly-ash amended soil. Chemosphere 67, 140—151.

Eswaran, A., Manivannan, K., 2007. Effect of foliar application of lignite fly ash on the management of papaya leaf curl disease. Acta Hortic. (ISHS) 740, 271—275.

Eijsackers, H., 2010. Earthworms as colonisers: primary colonisation of contaminated land, and sediment and soil waste deposits. Sci. Total Environ. 408, 1759—1769.

Elseewi, A.A., Bingham, F.T., Page, A.L., 1978. Availability of sulphur in fly-ash to plants. J. Environ. Qual. 7, 69—73.

Elseewi, A.A., Straughan, I.R., Page, A.L., 1980. Sequential cropping of fly ash amended soils: effects on soil chemical properties and yield and elemental composition of plants. Sci. Total Environ. 15, 247—259.

Elvira, C., Sampedro, L., Beritez, E., Nogales, R., 1998. Vermicomposting of sludge from paper mill and dairy industries with *Eisenia andrei*: a pilot scale study. Bioresour. Technol. 63, 211—218.

Fail Jr., J.L., Wochok, Z.S., 1977. Soyabean growth on fly ash amended strip mine spoils. Plant Soil 48, 473–484.

Fang, M., Wong, M.H., Wong, J.W.C., 2001. Digestion activity of thermophilic bacteria isolated from ash amended sewage sludge compost. Water Air Soil Pollut. 126, 1–12.

Fang, M., Wong, J.W.C., Ma, K.K., Wong, M.H., 1999. Co-composting of sewage sludge and coal fly ash: nutrient transformations. Bioresour. Technol. 67, 19–24.

Feuerborn, H.J., 2011. Coal combustion products in Europe: an update on production and utilization, standardisation and regulation. In: World Coal Ash Conference, May 9–12, 2011, Dever, CO. <http://www.flyash.info>.

Fluckiger, W., Oertli, J.J., Fluckiger-Keller, H., 1979. Relationship between stomatal diffusive resistance and various applied particle sizes on leaf surfaces. Z. Pflanzenphysiol. 91, 173–175.

Fulekar, M.H., 1993. The pH effects on leachability of fly-ash heavy metals: laboratory experiment. Indian J. Environ. Prot. 13, 185–192.

Gajalakshmi, S., Ramasamy, E.V., Abbasi, S.A., 2001. Screening of four species of detritivorous (humus-former) earthworms for sustainable vermicomposting of paper waste. Environ. Technol. 22, 679–685.

Ghodrati, M., Sims, J.T., Vasilas, B.L., 1995. Evaluation of fly ash as a soil amendment for the Atlantic Coastal Plain: I. Soil hydraulic properties and elemental leaching. Water Air Soil Pollut. 81, 349–361.

Ghuman, G.S., Menon, M.P., Chandra, K., James, J., Adriano, D.C., Sajwan, K.S., 1994. Uptake of multielements by corn from fly-ash compost amended soil. Water Air Soil Pollut. 72, 285–329.

Gracia, G., Zabaleta, I., Canibano, J.G., Gyeyo, M.A., 1995. Use of coal ash/mine waste in agriculture; their properties from agriculture point of view. Coal Abstr. 95, 32–46.

Grantz, D.A., Garner, J.H.B., Johnson, D.W., 2003. Ecological effects of particulate matter. Environ. Int. 29, 213–239. Available from: https://doi.org/10.1016/S0160-4120(02)00181-2.

Gu, H., Qiu, H., Tian, T., Zhan, S., Deng, T., Chaney, R.L., et al., 2011. Mitigation effects of silicon rich amendments on heavy metal accumulation in rice (*Oryza sativa* L.) planted on multi-metal contaminated acidic soil. Chemosphere 83, 1234–1240.

Guest, C.A., Johnston, C.T., King, J.J., Alleman, J.E., Tishmack, J.K., Norton, L.D., 2001. Chemical characterization of synthetic soil from composting coal combustion and pharmaceutical by-product. J. Environ. Qual. 30, 25–246.

Gupta, M., Kumar, A., Yunus, M., 2000. Effect of fly-ash on metal composition and physiological responses in *Leucaena leucocephala* (Lamk.) De Wit. Environ. Monit. Assess. 61, 399–406.

Gupta, D.K., Tripathi, R.D., Rai, U.N., Dwivedi, S., Mishra, S., Srivastava, S., et al., 2006. Changes in amino acid profile and amino acid content in seeds of Cicer arietinum L. (chick –pea) grown under various fly ash amendments. Chemosphere 65, 939–945.

Gupta, A.K., Mishra, R.K., Sinha, S., Lee, B., 2010. Growth, metal accumulation and yield performance of *Brassica campestris* L. (cv. Pusa Jaikisan) grown on soil amended with tannery sludge/fly ash mixture. Ecol. Eng. 36, 981–991.

Gutenmann, W.H., Lisk, D.J., 1996. Increasing selenium in field-grown onions by planting in peat moss pots containing coal fly ash. Chemosphere 32, 1851–1853.

Guttila, Y., Jayasinghe, Y.T., Kazuthoshi, K., 2009. Recycling of coal fly ash and paper waste to improve low productive red soil in Okinawa, Japan. Clean. 37, 687–695.

Haynes, R.J., 2009. Reclamation and revegetation of fly ash disposal sites – challenges and research needs. J. Environ. Manag 90, 43–53.

Hill, M.J., Camp, C.A., 1984. Use of pulverized fuel ash from Victoria brown coal as a source of nutrient for pasture species. Aust. J. Exp. Agric. Anim. Husbandary 30, 377–384.

Hooda, P.S., 2010. Trace Elements in Soils, first ed. John Wiley & Sons Ltd, Chichester, UK.

Huapeng, N., Shengrong, L., Junfeng, S., Meijuan, Y., 2012. Characteristics of heavy metal accumulation on fly ash- and sewage sludge-amended calcific soil. Chin. J. Geochem. 31, 181–186.

Inthasan, J., Hirunburana, N., Herrmann, L., Stahr, K., 2002. Effects of fly ash applications on soil properties, nutrient status and environment in Northern Thailand. In: 17th World Congress of Soil Science.

Iyer, R.S., Scott, J.A., 2001. Power station fly ash: a review of value-added utilization outside of the construction industry. Resour. Conserv. Recyc. 31, 217–228.

Jala, S., Goyal, D., 2006. Fly ash as a soil ameliorant for improving crop production a review. Bioresour. Technol. 97, 1136–1147.

Jankowski, J., Ward, C.R., French, D., Groves, S., 2006. Mobility of trace elements from selected Australian fly ashes and its potential impact on aquatic ecosystems. Fuel 85, 243–256.

Jayasinghe, G.Y., Tokashiki, Y., 2012. Influence of coal fly ash pellet aggregates on the growth and nutrient composition of Brassica campestris and physicochemical properties of grey soils in Okinawa. J. Plant. Nutr. 35, 47–453.

Jayasinghe, G.Y., Tokashiki, Y., Kinjo, K., 2009. Recycling of coal fly ash and paper waste to improve low productive red soil in Okinawa, Japan. Clean Soil Air Water 37, 687–695.

Jeyabal, A., Arivazhagan, K., Thanunathan, K., 2000. Utilisation of lignite fly ash as source of plant nutrient for rice. Fertiliser N. 45, 55–58.

Jiang, R.F., Yang, C.G., Su, D.C., Wong, J.W.C., 1999. Coal fly ash and lime stabilized biosolids as an ameliorant for boron deficient acidic soils. Environ. Technol. 2, 645–649.

Juwarkar, A.A., Jambhulkar, H.P., 2008. Restoration of fly ash dump through biological interventions. Environ. Monit. Assess. 139, 355–365.

Kachroo, D., Dixit, A., Balp, A., 2006. Influence of crop residue, fly ash and varying starter dosages on growth, yield and soil characteristics in rice (Oryza sativa)–wheat (Triticum aestivum) cropping system under irrigated conditions of Jammu region. Indian J. Agric. Sci. 76, 3–6.

Kallesha, R.S., Yeledhalli, N.A., Patil, C.V., Prakash, S.S., 2001. Effect of fly ash and sewage sludge levels on crop yield and soil properties. In: Ram, L.C. (Ed.), Proceedings of the National Seminar on Bulk Utilization of Fly Ash in Agriculture and for Value-Added Products. Technical Session, V. CFRI, Dhanbad, India, pp. 187–194.

Kalra, N., Joshi, H.C., Chaudhary, A., Chaudhary, R., Sharma, S.K., 1997. Impact of fly ash incorporation in soil on germination of crops. Bioresour. Technol. 61, 39–41.

Karpagavalli, S., Ramabadran, R., 1997. Effect of lignite fly ash on the growth and dry matter production (DMP) of soil borne pathogens. In: Souvenir and Abstract: National Seminar on Bio-Utilization of Fly Ash. Khallikote Autonomous College, Berhampur, Orissa, India, April 4–5, p. 11.

Kelley, R., Richard, K., Norman, R., Source, T.W., 2002. The production of an artificial soil from sewage sludge and fly ash and from the subsequent evaluation of growth enhancement, heavy metal translocation and leaching potential. Water Res. 28, 73–77.

Khan, M.R., Singh, W.N., 2001. Effects of soil application of fly ash on the fusarial wilt on tomato cultivars. Int. J. Pest. Manag. 47, 293–297.

Khawairakpam, M., Bhargava, R., 2009. Bioconversion of filter mud using vermicompost-ing employing two exotic and one local earthworm species. Biores. Technol. 100, 5846–5852.

Kim, B.J., Back, J.H., Kim, Y.S., 1997. Effect of fly ash on the yield of Chinese cabbage and chemical properties of soil. J. Korean Soc. Soil Sci. Fert. 30, 161–167.

Klubek, B., Carlson, C.L., Oliver, J., Adriano, D.C., 1992. Characterization of microbial abundance and activity from three coal ash basins. Soil Biol. Biochem. 24, 1119–1125.

Kochian, L.V., 1995. Cellular mechanisms of aluminum toxicity and resistance in plants. Ann. Rev. Plant. Physiol. Plant Mol. Biol. 47, 237–260.

Kookana, R.S., Sarmah, A.K., Zwieten, L.V., Krull, E., Singh, B., 2011. Chapter three - Biochar application to soil: agronomic and environmental benefits and unintended consequences. Adv. Agron. 112, 103–143.

Krajickova, A., Majstrick, V., 1984. The effect of fly-ash particles on the plugging of sto-mata. Env. Pollut. Ser. A 36, 83–93.

Kumar, D., Singh, B., 2003. The use of coal fly ash in sodic soil reclamation. Land. Deg. Dev. 14, 285–299.

Kumar, K.V., Patra, D.D., 2012. Alteration in yield and chemical composition of essential oil of Mentha piperita L. plant: effect of fly ash amendments and organic wastes. Ecol. Eng. 47, 237–241.

Kumar, A., Vajpayee, P., Ali, M.B., Tripathi, R.D., Singh, N., Rai, U.N., et al., 2002. Biochemical responses of Cassia siamea Lamk. grown on coal combustion residue (fly-ash). Bull. Environ. Contam. Toxicol. 68, 675–683.

Kumari, K., Ranjan, N., Sharma, J.P., Agarwal, P.K., Sinha, R.C., 2012. Integrated man-agement of diluted distillery effluent and fly ash as a potential biofertiliser: a case study on the vegetative growth and chlorophyll content of the marigold plant, Tagetes patu-la. Int. J. Environ. Technol. Manag. 15, 275–290.

Kumpiene, J., Guerri, G., Landi, L., Pietramellara, G., Nannipieri, P., Renella, G., 2009. Microbial biomass, respiration and enzyme activities after in situ aided phytostabiliza-tion of a Pb and Cu-contaminated soil. Ecotoxicol. Environ. Saf. 72, 115–119.

Lai, K.M., Ye, D.Y., Wong, J.W.C., 1999. Enzyme activities in sandy soil amended with sewage sludge and coal fly ash. Water Air Soil Pollut. 113, 261–272.

Lal, J.K., Mishra, B., Sarkar, A.K., 1996. Effect of fly ash on soil microbial and enzymatic activity. J. Indian Soc. Soil Sci. 44, 77–80.

Lau, S.S.S., Fang, M., Wong, J.W.C., 2001. Effects of composting process and fly ash amendment on phytotoxicity of sewage sludge. Arch. Environ. Contam. Toxicol. 40, 184–191.

Lee, Y.B., Ha, H.S., Lee, K.D., Park, K.D., Cho, Ys, Kim, P.J., 2003. Evaluation of use of fly ash–gypsum mixture for rice production at different nitrogen rates. Soil Sci. Plant Nutr. 49, 69–76.

Lee, Y.B., Ha, H.S., Lee, C.H., Lee, H., Ha, B.H., Kim, P.J., 2005. Improving rice pro-ductivity and soil quality by coal ash–phospho gypsum mixture application. J. Korea Soc. Soil Sci. Fert. 38, 45–50.

Lee, H., Ho, S.H., Lee, C.H., Lee, Y.B., Kim, P.J., 2006. Fly ash effect on improving soil properties and rice productivity in Korean paddy soils. Bioresour. Technol. 97, 1490–1497.

Lehmann, J., 2007. A handful of carbon. Nature 447, 143–144.

Loher, R.C., Neuhauser, E.F., Melecki, M.R., 1985. Factors affecting the vermistabiliza-tion process temperature, moisture content and polyculture. Water Res. 19, 1311–1317.

Lopareva-Pohu, A., Pourrut, B., Waterlot, C., Garçon, G., Bidar, G., Pruvot, C., et al., 2011. Assessment of fly ash-aided phytostabilisation of highly contaminated soils after an 8-year field trial part 1. Influence on soil parameters and metal extractability. Sci. Total Environ. 409, 647−654.

Maity, S., Bhattacharya, S., Chaudhury, S., 2009. Metallothionein response in earthworms Lampito mauritii (Kinberg) exposed to fly ash. Chemosphere . Available from: https://doi.org/10.1016/j.chemosphere.2009.07.011.

Manivannan, K., 2005. Reclamation and re-vegetation through horticultural crops in the lignite ash pond at Neyveli in Tamilnadu. In: Fly Ash India New Delhi, Fly Ash Utilization Programme. Technical Session, XII. TIFAC, DST, New Delhi, pp. 33.1−33.6.

Martens, D.C., 1971. Availability of plant nutrients in fly ash. Compos. Sci. 12, 15−19.

Masto, R.E., Sunar, K.K., Sengupta, T., Ram, L.C., Rout, T.K., Selvi, V.A., et al., 2012. Evaluation of the co-application of fly ash and sewage sludge on soil biological and biochemical quality. Environ. Technol. 33, 897−905.

Masto, R.E., Ansari, Md.A., George, J., Selvi, V.A., Ram, L.C., 2013. Co-application of Biochar and lignite fly ash on soil nutrients and biological parameters at different crop growth stages of Zea mays. Ecol. Eng. 58, 314−322.

Mataix-Solera, J., Cerdà, A., Arcenegui, V., Jordán, A., Zavala, L.M., 2011. Fire effects on soil aggregation: a review. Earth Sci. Rev. 109, 44−60.

Matsi, T., Keramidas, V., 1999. Fly ash application on two acid soils and its effect on soil salinity, pH, B, P and on ryegrass growth and composition. Environ. Pollut. 104, 107−112.

McCarty, G.W.R., Siddaramappa, R.J., Wright, E.E., Gao, G., 1994. Evaluation of coal combustion byproducts as soil liming materials: their influence on soil pH and enzyme activities. Biol. Fertil. Soil. 17, 147−172.

Mendki, P.S., Maheshwari, V.L., Kothari, R.M., 2001. Fly-ash as a post-harvest preservative for five commonly utilized pulses. Crop Prot. 20, 241−245.

Mengel, K., Kirkby, E.A., 1987. Principles of Plant Nutrition. International Potash Institute, Bern, Switzerland.

Menon, M.P., Sajwan, K.S., Ghuman, G.S., James, J., Chandra, K., 1993. Fly ash-amended compost as manure for agricultural crops. J. Environ. Sci. Health A Environ. Sci. Eng. 28, 2167−2182.

Metcalf, Eddy, 2003. Wastewater Engineering: Treatment, Disposal and Reuse, fourth ed. McGraw-Hill Publishing Company Ltd, New York.

Mishra, L.C., Shukra, K.N., 1986. Effects of fly ash deposition on growth, metabolism, and dry matter production of maize and soyabean. Environ. Pollut. 42, 1−13.

Muir, M.A., Yunusa, I.A.M., Burchett, M.D., Lawrie, R., Chan, K.Y., Manoharan, V., 2007. Short-term responses of contrasting species of earthworms in an agricultural soil amended with coal fly-ash. Soil Biol. Biochem. 39, 987−992.

Mulford, F.R., Martens, D.C., 1971. Response of alfalfa to boron in fly ash. Soil Sci. Soc. Am. Proc. 35, 296−300.

Narayanasamy, P., Gnanakumar, D., 1989. A Lignite fly-ash: A nonpolluting substance for tackling pest problems. In: Devaraj, K.V. (Ed.), 1989: Progress in Pollution Research. University of Agricultural Sciences, Bangalore, pp. 201−206.

Neuhauser, E.F., Kaplan, D.L., Hartenstein, R., 1979. Life history of the earthwom Eudrilus eugeniae (Kinberg). Rev. Ecol. Biol. Soil. 16, 524−534.

Ots, K., Indriksonsb, A., Kabasinskiene, I.V., Mandre, M., Kuznetsova, T., Klõšeiko, J., et al., 2011. Changes in canopies of Pinus sylvestris and Picea abies under alkaline dust impact in the industrial region of Northeast Estonia. For. Ecol. Manag. 262, 82−87. Available from: https://doi.org/10.1016/j.foreco.2010.07.031.

Page, A.L., Elseevi, A.A., Straughan, I.R., 1979. Physical and chemical properties of fly ash from coal fired power plants with reference to environmental impact. Residue Rev. 71, 83−120.

Palumbo, A.V., McCarthy, J.E., Amonette, J.F., Fisher, L.S., Wullschlege, S.D., Daniels, W.L., 2004. Properties for enhancing carbon sequestration and reclamation of degraded lands with fossil fuel combustion by-products. Adv. Environ. Res. 8, 425−438.

Pandey, V.C., Mishra, T., 2018. Assessment of *Ziziphus mauritiana* grown on fly ash dumps: Prospects for phytoremediation but concerns with the use of edible fruit. Int. J.Phytoremediat. 20 (12), 1250−1256.

Pandey, V.C., Singh, N., 2010. Impact of fly ash incorporation in soil systems. Agric. Ecosyst. Environ. 136, 16−27.

Pandey, V.C., Kumar, A., 2013. *Leucaena leucocephala*: an underutilized plant for pulp and paper production. Genet. Resour. Crop. Evol. 60, 1165−1171.

Pandey, V.C., Abhilash, V.C., Singh, N., 2009a. The Indian perspective of utilizing fly ash in phytoremediation, photomanagement and biomass production. J. Environ. Manage. 90, 2943−2958.

Pandey, V.C., Abhilash, P.C., Upadhyay, R.N., Tewari, D.D., 2009b. Application of fly ash on the growth performance and translocation of toxic, heavy metals within *Cajanus cajan* L.: implication for safe utilization of fly ash for agricultural production. J. Hazard. Mater. 166, 255−259.

Pandey, V.C., Singh, J.S., Kumar, A., Tewari, D.D., 2010. Accumulation of heavy metals by chickpea grown in fly ash treated soil: effect on antioxidants. Clean Soil Air Water 38, 1116−1123.

Pandey, V.C., Singh, J.S., Singh, R.P., Singh, N., Yunus, M., 2011. Arsenic hazards in coal fly ash and its fate in Indian scenario. Resour. Conserv. Recy. 55, 819−835.

Pandey, V.C., Patel, D., Thakare, P.V., Ram, L.C., Singh, N., 2017. Microbial and enzymatic activities of fly ash amended soil. In: Singh, J.S., Singh, D.P. (Eds.), Microbes and Environmental Management. Studium Press LLC, USA, , ISBN: 9789380012834pp. 142−167.

Papadimitriou, C.A., Haritou, I., Samaras, P., Zouboulis, A.I., 2008. Evaluation of leaching and ecotoxicological properties of sewage sludge−fly ash mixtures. Environ. Res. 106, 340−348.

Partha, N., Sivasubramanian, V., 2006. Recovery of chemicals from press mud—a sugar industry waste. Indian Chem. Eng. 48, 160−163.

Parthasarathi, K., Ranganathan, L.S., 2000. Influence of press mud on the development of the ovary, oogenesis and the neurosecretory cells of the earthworm *Eudrilus eugeniae* (Kinberg). Afr. Zool. 35, 281−286.

Pathan, S.M., Aylmore, L.A.G., Colmer, T.D., 2003. Properties of several fly ash materials in relation to use as soil amendments. J. Environ. Qual. 32, 687−693.

Pati, S.S., Sahu, S.K., 2004. $CO_2$ evolution and enzyme activities (dehydrogenase, protease and amylase) of fly ash amended soil in the presence and absence of earthworms (Drawida willsi, Michaelsen) under laboratory conditions. Geoderma 118, 289−301.

Phung, H.T., Lam, H.V., Page, A.L., Lund, L.J., 1979a. The practice of leaching boron and soluble salts from fly ash-amended soils. Water Air Soil Pollut. 12, 247−254.

Phung, H.T., Lund, I.J., Page, A.L., 1978. Potential use of fly ash as a liming material. In: Adriano, D.C., Brisbin, I.L. (Eds.), Proceedings of the Conference 'Environmental Chemistry and Cycling Processes'. US Department of Energy, p. 504.

Phung, H.T., Lund, L.J., Page, A.L., Bradford, G.R., 1979b. Trace elements in fly ash and their release in water and treated soils. J. Environ. Qual. 8, 171−175.

Phunshon, T., Adriano, D.C., Weber, J.T., 2002. Restoration of drastically eroded land using coal fly ash and poultry bio-solid. Sci. Total. Environ. 296, 209−225.

Pichtel, J.R., Hayes, J.M., 1990. Influence of fly ash on soil microbial activity and populations. J. Environ. Qual. 19, 593−597.

Pichtel, J.R., Dick, W.A., Sutton, P., 1994. Comparison of amendments and management practices for long-term reclamation of abandoned mine lands. J. Environ. Qual. 23, 766–772.

Pierzynski, G.M., Heitman, J.L., Kulakow, P.A., Kluitenberg, G.J., Carlson, J., 2004. Revegetation of waste fly ash landfills in a semiarid environment. J. Rangel. Ecol. Manag. 57, 312–319.

Prakash, M., Karmegam, N., 2010. Vermistabilization of pressmud using Perionyx ceylanensis Mich. Bioresour. Technol. 101 (21), 8464–8468.

Pramanik, P., Chung, Y.R., 2011. Changes in fungal population of fly ash and vinasse mixture during vermicomposting by Eudrilus eugeniae and Eisenia fetida: documentation of cellulose isozymes in vermicompost. Waste Manag. 3, 1169–1175.

Qadir, S.U., Raja, V., Siddiqui, W.A., 2016. Morphological and biochemical changes in Azadirachta indica from coal combustion fly ash dumping site from a thermal power plant in Delhi, India. Ecotoxicol. Environ. Saf. 129, 320–328.

Ram, L.C., Masto, R.E., 2010. Review: an appraisal of the potential use of fly ash for reclaiming coal mine spoil. J. Environ. Manag. 91, 603–617.

Ram, L.C., Masto, R.E., 2014. Fly ash for soil amelioration: a review on the influence of ash blending with inorganic and organic amendments. Earth Sci. Rev. 128, 52–74.

Ram, L.C., Tripathi, P.S.M., Mishra, S.P., 1995. Moessbauer spectroscopic studies on the transformations of Fe-bearing minerals during combustion of coal: correlation with fouling and slagging. Fuel Process. Technol. 42, 47–60.

Ram, L.C., Srivastava, N.K., Jha, S.K., Sinha, A.K., Masto, R.E., Selvi, V.A., 2007. Management of lignite fly ash through its bulk use via biological amendments for improving the fertility and crop productivity of soil. Environ. Manag. 40, 438–452.

Ram, L.C., Srivastava, N.K., Tripathi, R.C., Jha, S.K., Sinha, A.K., Singh, G., et al., 2006. Management of mine spoil for crop productivity with lignite fly ash and biological amendments. J. Environ. Manag. 79, 173–187.

Rama Rao, D., Nanda, S.K., Kareemulla, K. 2012. Technological Forecasting and Assessment of Future Fly Ash use in Agriculture in India, DST Project Report, National Academy of Agricultural Research Management, Hyderabad, India. pp 140.

Rani, K., Agarwal, S., Kalpana, S., 2012. Effect of sugarcane waste on soil properties of coal fly ash affected land in, cultivation of Lycopersicum esculentum. J. Chem. Pharm. Res. 4 (3), 1505–1510.

Ransome, L.S., Dowdy, R.H., 1987. Soybean growth and boron distribution in a sandy soil amended with scrubber sludge. J. Environ. Qual. 16, 171–175.

Rautaray, S.K., Ghosh, B.C., Mittra, B.N., 2003. Effect of fly ash, organic wastes and chemical fertilizers on yield, nutrient uptake, heavy metal content and residual fertility in a rice–mustard cropping sequence under acid lateritic soils. Bioresour. Technol. 90, 275–283.

Reynolds, K., Kruger, R., Rethman, N., Truter, W., 2002. The production of an artificial soil from sewage sludge and fly-ash and the subsequent evaluation of growth enhancement, heavy metal translocation and leaching potential. Water SA Suppl. 73–77.

Roy, G., Joy, V.C., 2011. Dose-related effect of fly ash on edaphic properties in laterite crop land soil. Ecotoxicol. Environ. Saf. 74, 769–775.

Sahin, U., Angin, I., Kiziloglu, F.M., 2008. Effect of freezing and thawing processes on some physical properties of saline–sodic soils mixed with sewage sludge or fly ash. Soil. Tillage Res. 99, 254–260.

Sahu, S.K., Bhangare, R.C., Ajmal, P.Y., Sharma, S., Pandit, G.G., Puranik, V.D., 2009. Characterization and quantification of persistent organic pollutants in fly ash from Coal fueled thermal power stations in India. Microchem. J. 92, 92–96.

Sajwan, K.S., Ornes, W.H., Youngblood, T., 1995. The effect of fly ashy sewage sludge mixtures and application rates on biomass production. J. Environ. Sci. Health 30, 1327−1337.

Sajwan, K.S., Paramasivam, S., Alva, A.K., Adriano, D.C., Hooda, P.S., 2003. Assessing the feasibility of land application of fly ash, sewage sludge and their mixtures. Adv. Environ. Res. 8, 77−91.

Salingar, Y., Sparks, D.L., Pesek, J.D., 1994. Kinetics of iron removal from an iron-rich industrial coproduct: III. Manganese and chromium. J. Environ. Qual. 23, 1201−1205.

Samaras, P., Papadimitriou, C.A., Haritou, I., Zouboulis, A.I., 2008. Investigation of sewage sludge stabilization potential by the addition of fly ash and lime. J. Hazard. Mater. 154, 1052−1059.

Sankari, S.A., Narayanasamy, P., 2007. Bio-efficacy of fly-ash based herbal pesticides against pests of rice and vegetables. Curr. Sci. 92, 811−816.

Sarangi, P.K., Mahakur, D., Mishra, P.C., 2001. Soil biochemical activity and growth responses of rice Oryza sativa in fly ash amended soil. Bioresour. Technol. 76, 199−205.

Sarkar, A., Rano, R., Mishra, K.K., Sinha, I.N., 2005. Particle size distribution profile of some Indian fly ash: a comparative study to assess their possible uses. Fuel Process. Technol. 86, 1221−1238.

Sarojini, S., Ananthakrishnasamy, S., Manimegala, G., Prakash, M., Gunasekaran, G., 2009. Effect of lignite fly ash on the growth and reproduction of earthworm Eisenia fetida. E-J. Chem. 6, 511−517.

Saxena, M., Chauhan, A., Asokan, P., 1998. Flyash vemicompost from non-friendly organic wastes. Pollut. Res. 17, 5−11.

Schnappinger Jr., M.G., Martens, D.C., Plank, C.O., 1975. Zinc availability as influenced by application of fly ash to soil. Environ. Sci. Technol. 9, 258−261.

Schumann, A.W., Sumner, M.E., 2004. Formulation of environmentally sound waste mixtures for land application. Water Air Soil. Pollut. 152, 195−217.

Schutter, M.E., Fuhrmann, J.J., 2001. Soil microbial community responses to fly ash amendment as revealed by analyses of whole soils and bacterial isolates. Soil Biol. Biochem. 33, 1947−1958.

Selvakumari, G., Jayanthi, D., 1999. Effect of fly ash and compost on yield and nutrient uptake of groundnut in laterite soil. Proceedings of the National Seminar on Developments in Soil Science. Tamil Nadu Agricultural University, Coimbatore, India.

Selvam, A., Mahadevan, A., 2002. Effect of ash pond soil and amendments on the growth and arbuscular mycorrhizal colonization of Allium cepa and germination of Arachis hypogaea, Lycopersicon esculentum, and Vigna mungo seeds. Soil Sed. Contam. 11, 673−686.

Sen, A., 1997. Microbial Population Dynamics in Fly Ash Amended Acid Lateritic Soil (B. Tech. thesis). Indian Institute of Technology, Kharagpur, India.

Seneviratne, S.I., Corte, T., Davin, E.L., Hirschi, M., Jaeger, E.B., Lehner, I., et al., 2010. Investigating soil moisture−climate interactions in a changing climate: a review. Earth Sci. Rev. 99, 125−161.

Seshadri, B., Bolan, N.S., Kunhikrishnan, A., 2013a. Effect of clean coal combustion products in reducing soluble phosphorus in soil I. Adsorption study. Water Air Soil Pollut. 224, 1524.

Seshadri, B., Bolan, N., Choppala, G., Naidu, R., 2013b. Differential effect of coal combustion products on the bioavailability of phosphorus between inorganic and organic nutrient sources. J. Hazard. Mater. 261, 817−825.

Shaheen, S.M., Tsadilas, C.D., 2010. Sorption of cadmium and lead by acidic Alfisols as influenced by fly ash and sewage sludge application. Pedosphere 20, 436–445.

Shaheen, S.M., Peter, S., Hooda, P.S., Tsadilas, C.D., 2014. Opportunities and challenges in the use of coal fly ash for soil improvements -a review. J. Environ. Manag. 145, 249–267.

Sharma, A.P., Tripathi, B.D., 2009. Biochemical responses in tree foliage exposed to coal-fired power plant emission in seasonally dry tropical environment. Environ. Monit. Assess 158, 197–212. Available from: https://doi.org/10.1007/s10661-008-0573-2.

Sheoran, H.S., Duhan, B.S., Kumar, A., 2014. Effect of fly ash application on soil properties: a review. J. Agr. Nat. Resour. Manag. 1, 98–103.

Shen, J.F., Li, S.R., Sun, D.S., Li, G.H., 2004. Study on desertificated soil rehabilitation by solid wastes — a case study in Baotou, Inn Mongolia, China. Soil Bull 35 (3), 267–270.

Sikka, R., Kansal, B.D., 1994. Characterization of thermal power plant fly ash for agronomic purposes and to identify pollution hazards. Bioresour. Technol. 50, 269–273.

Sikka, R., Kansal, B.D., 1995. Effect of fly ash application on yield and nutrient composition of rice, wheat and on pH and available nutrient status of soils. Bioresour. Technol. 51, 199–203.

Simpson, G.G., King, L.D., Carlile, B.L., Blickensderfer, P.S., 1983. Paper mill sludges, coal fly ash, and surplus lime mud as soil amendments in crop production. Tappi J. 66, 71–74.

Sims, J.T., Vasilas, B.L., Ghodrati, M., 1995. Evaluation of fly ash as soil amendment for the Atlantic coastal plain: II. Soil chemical properties and crop for growth. J. Water Soil Pollut. 81, 363–372.

Singh, L.P., Siddiqui, Z.A., 2003. Effects of fly ash and *Helminthosporium oryzae* on growth and yield of three cultivars of rice. Bioresour. Technol. 86, 73–78.

Singh, A., Agrawal, S.B., 2010. Response of mung bean cultivars to fly ash: growth and yield. Ecotox. Environ. Safe 73, 1950–1958.

Singh, J.S., Pandey, V.C., 2013. Fly ash application in nutrient poor agriculture soils: impact on methanotrophs population dynamics and paddy yields. Ecotoxicol. Environ. Saf. 89, 43–51.

Singh, S.N., Kulshreshtha, K., Ahmad, K.J., 1997. Impact of fly ash soil amendment on seed germination, seedling growth and metal composition of *Vicia faba* L. Ecol. Eng. 9, 203–208.

Singh, A., Sharma, R.K., Agrawal, S.B., 2008. Effects of fly ash incorporation on heavy metal accumulation, growth and yield responses of *Beta vulgaris* plants. Bioresour. Technol. 99, 7200–7207.

Singh, J.S., Pandey, V.C., Singh, D.P., 2011. Coal fly ash and farmyard manure amendments in dry-land paddy agriculture field: effect on N-dynamics and paddy productivity. Appl. Soil Ecol. 47, 133–140.

Singh, K., Pandey, V.C., Singh, B., Patra, D.D., Singh, R.P., 2016. Effect of fly ash on crop yield and physico-chemical, microbial and enzyme activities of sodic soils. Environ. Eng. Manag. J. 15 (11), 2433–2440. Available from: http://omicron.ch.tuiasi.ro/EEMJ/.

Singh, N., Yunus, M., 2000. Environmental impacts of fly ash. In: Iqbal, M., Srivastava, P.S., Siddiqui, T.O. (Eds.), Environmental Hazards – Plants and People. C.B.S. Publishers and Distributors, New Delhi, pp. 60–79.

Sinha, S., Rai, U.N., Bhatt, K., Pandey, K., Gupta, A.K., 2005. Fly-ash induced oxidative stress and tolerance in *Prosopis juliflora* L. grown on different amendments substrates. Environ. Monit. Assess. 102, 447–457.

Sinha, R.K., Herat, S., Bharambe, G., Brahambhatt, A., 2010. Vermistabilization of sewage sludge (biosolids) by earthworms: converting a potential biohazard destined for landfill disposal into a pathogen-free, nutritive and safe biofertilizer for farms. Waste Manag. Res. 28, 872–881.

Skousen, J., Ziemkiewicz, P., Yang, J.E., 2012. Use of coal combustion by-products in mine reclamation: review of case studies in the USA. Geosyst. Eng. 15, 71–83.

Srivastava, N.K., Ram, L.C., 2010. Reclamation of coal mine spoil dump through fly ash and biological amendments. Int. J. Ecol. Dev. 17 (F10), 17–33.

Stout, W.L., Sharpley, A.N., Gbyrek, W.J., Pionke, H.B., 1999. Reducing phosphorus export from croplands with FBC fly ash and FGD system. Fuel 78, 175–178.

Su, D.C., Wong, J.W.C., 2002. The growth of corn seedings in alkaline coal fly ash stabilized sewage sludge. Water Air Soil Pollut. 133, 1–13.

Su, D.C., Wong, J.W.C., 2004. Chemical speciation and phytoavailability of Zn, Cu, Niand Cd in soil amended with fly ash-stabilized sewage sludge. Environ. Int 29, 895–900.

Taylor Jr., E.M., Schuman, G.E., 1988. Fly ash and lime amendment of acidic coal spoil to aid revegetation. J. Environ. Qual. 17, 120–124.

Tiwari, K.N., Sharma, D.N., Sharma, V.K., Dingar, S.M., 1992. Evaluation of fly ash and pyrite for sodic soil rehabilitation in Uttar Pradesh, India. Arid. Soil Res. Rehabil. 6, 117–126.

Tripathi, R.C., Masto, R.E., Ram, L.C., 2009. Bulk use of pond ash for cultivation of wheat–maize–eggplant crops in sequence on a fallow land. Resour. Conserv. Recy. 54, 134–139.

Tripathi, R.C., Jha, S.K., Ram, L.C., 2010. Impact of fly ash application on tracemetal content in some root crops.. Energ. Source. Part A. 32, 576–589.

Truter, W.F., Rethman, N.F., Potgieter, C.E., Kruger, R.A., 2005. The international scenarios on the use of fly ash in agriculture: a synopsis. In: Proceedings of the Fly Ash India 2005, International Congress. Technical Session, XII. FAUP, TIFAC, DST, New Delhi, India, pp. 1.1–1.10.

Tsadilas, C.D., Shaheen, S.M., Samaras, V., 2009. Influence of coal fly ash application individually and mixing with sewage sludge on wheat growth and soil chemical properties under field conditions. In: 15th International Symposium on Environmental Pollution and its Impact on Life in the Mediterranean Region. October, 7–11. Bari, Italy, p. 145.

Ukwattage, N.L., Ranjith, P.G., Bouazza, M., 2013. The use of coal combustion fly ash as a soil amendment in agricultural lands (with comments on its potential to improve food security and sequester carbon). Fuel 109, 400–408.

Ulrichs, C., Dolgowski, D., Mucha, T., Reichmuth, C., Mewis, I., 2005. Insecticide and phytotoxic effects of hard coal fly ash. Gesunde Pflanz. 57, 110–116.

Urvashi, Masto, R.E., Selvi, V.A., Lal Ram, C., Srivastava, N.K., 2007. An international study: effect of farm yard manure on the release of P from fly ash. Remediation Autumn. 70–81.

Veeresh, S., Tripathy, D., Chaudhuri, B.R., Hart, Powell, M.A., 2003. Sorption and distribution of adsorbed metals in three soils of India. Appl. Geochem. 18, 1723–1731.

Verma, S.K., Singh, K., Gupta, A.K., Pandey, V.C., Trivedi, P., Verma, R.K., et al., 2014. Aromatic grasses for phytomanagement of coal fly ash hazards. Ecol. Eng. 73, 425–428.

Vincini, M., Carini, F., Silva, S., 1994. Use of alkaline fly ash as an amendment for swine manure. Bioresour. Technol. 49, 213–222.

Wang, O., Zhang, Y., Lian, J., Chao, J., Gao, Y., Yang, F., et al., 2013. Impact of fly ash and phosphatic rock on metal stabilization and bioavailability during sewage sludge vermicomposting. Bioresour. Technol. 136, 281–287.

Weinstein, L.H., Osmeloski, J.F., Rutzke, M., Beers, A.O., McCahan, J.B., Bache, C.A., et al., 1989. Elemental analysis of grasses and legumes growing on soil covering coal fly ash landfill sites. J. Food Saf. 9, 291−300.

Wilkinson, M.T., Richards, P.J., Humphreys, G.S., 2009. Breaking ground: pedological, geological, and ecological implications of soil bioturbation. Earth Sci. Rev. 97, 257−272.

Wong, J.W.C., 1995. The production of artificial soil mix from coal fly ash and sewage sludge. Environ. Technol. 16, 741−751.

Wong, J.W.C., Wong, M.H., 1987. Co-recycling of fly ash and poultry manure in nutrient deficient sandy soil. Res. Conser. 13, 291−304.

Wong, J.W.C., Wong, M.H., 1990. Effects of fly ash on yields and elemental composition of two vegetables, *Brassica parachinensis* and *B. chinensis*. Agri. Ecosys. Env. 30, 251−264.

Wong, J.W.C., Lai, K.M., 1996. Effect of an artificial soil mix from coal fly ash and sewage sludge on soil microbial activity. Biol. Fertil. Soil. 23, 420−424.

Wong, J.W.C., Su, D., 1997. The growth of *Agropyron elongatum* in an artificial soil mix from coal fly ash and sewage sludge. Biores. Technol. 59, 57−62.

Wong, J.W.C., Selvam, A., 2009. Growth and elemental accumulation of plants grown in acidic soil amended with coal fly ash−sewage sludge co-compost. Arch. Environ. Contam. Toxicol. 57, 515−523.

Wong, J.W.C., Jiang, R.F., Su, D.C., 1998. The accumulation of boron in *Agropyron elongatum* grown in coal fly ash and sewage sludge mixture. Water Air Soil Pollut. 106, 137−147.

Wong, J.W.C., Fang, M., Jiang, R., 2001. Persistency of bacterial indicators in biosolids stabilization with coal fly ash and lime. Water Environ. Res. 73, 607−611.

Woodard, M.A., Bearce, B.C., Cluskey, S., Townsend, E.C., 1993. Coal bottom ash and pinewood peelings as root substrate in circulating nutriculture. Hort. Sci. 28, 636−638.

Xenidis, A., Mylona, E., Paspaliaris, I., 2002. Potential use of lignite fly ash for the control of acid generation from sulphidic wastes. Waste Manag. 22, 631−641.

Xing, S., Zhao, Z., Zhou, B., Wu, X., 2001. Effect of fly ash-filtered mud mixture on soil properties and radish yield and quality. J. Appl. Ecol. 12, 121−125.

Xu, J.Q., Yu, R.-L., Dong, X.-Y., Hu, G.-R., Shang, X.-S., Wang, Q., et al., 2012. Effects of municipal sewage sludge stabilized by fly ash on the growth of Manila grass and transfer of heavy metals. J. Hazard. Mater. 217−218, 58−66.

Yeledhalli, N.A., Prakash, S.S., Gurumurthy, S.B., Patil, C.V., 2005. Long-term effect of fly ash/pond ash application on physico chemical and biochemical properties of soils. Fly Ash India 2005, Fly Ash Utilization Programme. Technical Session, XII. TIFAC, DST, New Delhi, India, pp. 5.1−15.

Yunusa, I.A.M., Eamus, D., De Silva, D.L., Murray, B.R., Burchett, M.D., Skilbeck, G. C., et al., 2006. Fly-ash: an exploitable resource for management of Australian agricultural soils. Fuel 85, 2337−2344.

Yunusa, I.A.M., Braun, M., Lawrie, R., 2009. Amendment of soil with coal fly ash modified the burrowing habits of two earthworm species. Appl. Soil. Ecol. 42, 63−68.

Zhang, H., Sun, L., Sun, T., 2008a. Solubility of ion and trace metals from stabilized sewage sludge by fly ash and alkaline mine tailing. J. Environ. Sci. 20, 710−716.

Zhang, H.L., Sun, L.N., Sun, T.H., 2008b. Heavy metals and trace elements of sewage sludge stabilized by coal fly ash. J. Liaoning Tech. Univer Nat. Sci. 27, 944−946.

# CHAPTER 4

# Opportunities and challenges in fly ash—aided paddy agriculture

## Contents

## 4.1 Introduction

The history of fly ash (FA) application in soil as a plant nutrient starts about 60 years ago (Rees and Sidrak, 1956), thereafter, a number of publications on the use of FA for soil amelioration have been published in different Journals. In addition, an extensive Library and Scopus-based survey revealed that FA has been used in many crops as a fertilizer, but paddy is one of them on which maximum researchers worked across the nations. However, there are very few studies on soil ecology in FA-treated soils. Rice consumption is highest in Asia, which consumes around 90% of the world's rice. There are many research papers on the application of FA in paddy field with respect to FA doses, paddy varieties, physicochemical parameters, growth parameters, enzymatic activities, nutritional value, toxicological assessment, etc.

Huge quantities of FA released from coal-based thermal power plants are a major problem worldwide. FA production is projected to hike 750 million tons $year^{-1}$ globally, but only less than 50% of FA is utilized (Izquierdo and Querol, 2012), whereas in India, the FA production is estimated to exceed to 300 million tons $year^{-1}$ by 2016 and 17 (Skousen et al., 2012). Generally, a huge amount of FA is disposed in FA basins, which create a number of pollutions in that area. However, FA phytoremediation has been advocated through naturally growing plants (Maiti and Jaiswal, 2008; Pandey and Singh, 2011, 2012; Pandey, 2012a, b, 2013; Pandey et al., 2012; Mitrović et al., 2012), but it is extremely limited in practice. FA disposal is an acute problem in India, where several organizations such as Fly Ash Mission (FAM), Technology Information Forecasting and Assessment Council, and the Department of Science and Technology have been involved to handle the menace. FAM started 15 technology demonstration projects at 55 sites all over India to evaluate the field application of FA as well as to assess the impact of FA application on nutritional quality and food safety issues associated with the products from agricultural produce for possible human consumption. Many crops were examined for the consumer acceptability/toxic elements, and they were recognized safe within the permissible limits and meet the food quality standard. FA application along with organic residue [press mud (PM)] in paddy cropping is helpful in increasing methanotroph numbers (Singh and Pandey, 2013).

Efforts are underway for the safe application of coal FA in soil—plant system for enhancing crop productivity and soil fertility. Coal FA has been listed as a Green List waste under the Organization for Economic Cooperation and Development, and is not considered as a waste under Basel Convention, 1992. FA amendment in soil system has been advocated due to the presence of micro- and macronutrients (Rees and Sidrak, 1956; Kim et al., 1994a, b; Ram et al., 2007; Pandey et al., 2009a,b, 2010; Pandey and Singh, 2010; Singh et al., 2012, 2016a,b). The commercialization of FA as a fertilizer in paddy fields may be established on the basis of soil—plant—microbe interaction effects over a long periods in many harvests. FA contains most of the essential plant nutrients, except

N, because it is oxidized into gaseous constituents during coal combustion (Adriano et al., 1980). Some additional amendments like farmyard manure (FYM), sewage sludge (SS), PM, biofertilizers, gypsum, etc. along with FA have been used with paddy crop to enhance soil fertility. Besides, these amendments are also helpful for soil pH buffering, enhancing soil organic matter content, chelation of toxic metals, microbial stimulation, reducing soil bulk density (BD), improving soil structure, increasing soil cation exchange capacity (CEC), more retention of water and nutrients, nutrient mineralization, and increasing the bioavailability of the nutrients. Thus the selection of appropriate amendments for a particular soil along with FA is a key factor for mitigating the FA disposal problem and tapping economic potentials. Though there are several review articles on "FA use for soil amelioration" (Iyer and Scott, 2001; Gupta et al., 2002; Kumar and Singh, 2003; Yunusa et al., 2006; Jala and Goyal, 2006; Pandey et al., 2009a; Basu et al., 2009; Pandey and Singh, 2010; Ahmaruzzaman, 2010; Ram and Masto, 2014; Shaheen et al., 2014), the role of FA application at ecosystem level with paddy crop is not fully addressed. Therefore this chapter addresses the critical issues related to the feasibility of FA application in rice fields to ensure soil amelioration and enhance rice production without compromising the grain quality.

## 4.2 Methods

This study focuses on using FA for paddy soil amelioration and addresses cases from pot-scale to field scale (Table 4.1). The methodological approach for the collection of literature review is briefly described. This review is based on different types of literature from the entire world. Many specific keywords were chosen for mining a large range of papers. The keywords selected were: fly ash agriculture; fly ash paddy crop; fly ash rice fields; fly ash methane mitigation; fly ash methanotroph; fly ash for soil amelioration; fly ash effect on soil enzyme; agronomic application of fly ash; and fly ash effect on microbes. An extensive survey was done for literature review mining. Literature from more than 1200 publications was screened, and then relevant papers on "using fly ash for paddy crop" were selected for this review. The 111 references were finally selected from various types of literature.

**Table 4.1** A suggestive list of fly ash amendment approach for the paddy soil amelioration and rice production.

| Amendments | FA dose range | Soil types | Remarks | Country | References |
|---|---|---|---|---|---|
| FA + soil | FA at 0%, 2%, 4%, and 8% (w/w) | Sand, sandy loam 1, loamy sand, loam, sandy loam 2, silty clay, and silty loam | 2%—4% had a beneficial effect on the dry-matter yield, while 8% had a significant depressing effect | India | Sikka and Kansal (1995) |
| FA + soil | FA at 17.5 and 20 t ha$^{-1}$ | — | Grain and straw yield increased by 21% and 18%, respectively, at 17.5-t ha$^{-1}$ FA amendment when compared with the control | India | Sarangi et al. (2001) |
| FA + acid lateritic soils | FA at 10 t ha$^{-1}$ | Sandy loam acid lateritic soil | Integrated use of FA, OW, and CF was beneficial in improving crop yield, soil pH, organic carbon, and available N, P, and K | Kharagpur, India | Rautaray et al. (2003) |
| FA + soil | FA at 20%, 40%, 60%, 80%, and 100% (v/v) | Loam | 40% FA caused a higher increase in growth and yield than did 20% | India | Singh and Siddiqui (2003) |
| Control, FA, CF, FA + CF, FYM + CF, FA + FYM + CF, L + FYM + CF, PFS + CF, FA + PFS + CF, L + PFS + CF, CR + CF, and FA + CR + CF | FA at 10 t ha$^{-1}$; organic sources (FYM, PFS, and CR) at 30 kg N ha$^{-1}$; lime at 2 t ha$^{-1}$ | Acid lateritic soil | 10 t ha$^{-1}$ of FA in combination with organic sources and CF increased the grain yield and nutrient uptake of rice yield compared with CF alone | Kharagpur, India | Mittra et al. (2005) |
| FA + soil | FA at 40, 80, and 120 Mg ha$^{-1}$ | Silt loam and loamy sand | Highest yield of rice at 90 Mg ha$^{-1}$ | Korea | Lee et al. (2006) |
| FA + soil | FA at 0, 1, 2.5, 5, 10, and 15 Mg ha$^{-1}$ | Sandy-soil and sandy loam | Improved physical properties of soil and growth and yield of rice at 10 Mg ha$^{-1}$ | India | Mishra et al. (2007) |
| FA + soil | FA at 0, 40, 80, and 120 Mg ha$^{-1}$ | Silt loam and loamy sand | Increase of available phosphorus by FA application in paddy soils | Korea | Lee et al. (2007a) |

| Treatments | Application rate | Soil type | Findings | Country | References |
|---|---|---|---|---|---|
| FA + soil | FA at 0, 20, 40, and 60 Mg ha$^{-1}$ | Siltly loam and loamy sand | Reducing phosphorus release from paddy soils by a fly ash–gypsum mixture | Korea | Lee et al. (2007b) |
| FA + soil | FA at 10%, 25%, 50%, 75%, and 100% | Garden soil | 10%–25% FA amended in agricultural soils gives better rice crop yields | India | Dwivedi et al. (2007) |
| FA + soil | FA at 0, 40, 80, and 120 Mg ha$^{-1}$ | Loamy fine sand | FA can be a good soil amendment for rice production without B toxicity | South Korea | Lee et al. (2008a) |
| FA$_{10}$ + soil + NF$_{90}$ + BGA$_{12.5}$ and FA$_{100}$ + soil + NF$_{120}$ + BGA$_{12.5}$ | FA at 10 and 100 t ha$^{-1}$, NF at 90 and 120 kg ha$^{-1}$, and BGA at 12.5 kg ha$^{-1}$ | Garden soil | Integrated use of FA, BGA, and NF for improved growth, yield, and mineral composition of rice plants besides reducing high demand of NF | India | Tripathi et al. (2008) |
| Control, CF, FA, FYM + CF, FYM + FA + CF, FYM + RHA + CF, FYM + L + CF, PFS + CF, PFS + FA + CF, PFS + RHA + CF, and PFS + L + CF | FA at 5 and 10 Mg ha$^{-1}$, RHS at 5 Mg ha$^{-1}$, and L at 2 Mg ha$^{-1}$ | Acid lateritic soil | The integrated use of FA or RHA, PFS, FYM, and CF improves growth, yield, and nutrient uptake of rice | India | Karmakar et al. (2010) |
| Control, FA, FYM, and FA + FYM | FA at 50 t ha$^{-1}$ and FYM at 50 t ha$^{-1}$ | Slightly alkaline and sandy loam | FA and FYM amendments enhance the rates of N-transformation processes, plant available N, and paddy productivity | India | Singh et al. (2011) |
| Control, PM + FA | PM at 10 t ha$^{-1}$ and FA at 10, 50, and 100 t ha$^{-1}$ | Nutrient poor agriculture soils | FA at lower doses with PM seems to offer the potential amendments to improving soil methanotroph population and paddy crop yields | India | Singh and Pandey (2013) |
| FA + soil | FA at 200 t ha$^{-1}$ | Field soil | Improve the physical and chemical properties of deficient soil and improves the soil fertility and crop yield | India | Patra et al. (2012) |

(Continued)

**Table 4.1** (Continued)

| Amendments | FA dose range | Soil types | Remarks | Country | References |
|---|---|---|---|---|---|
| FA at 0–30, RHA at 0–20, and BA 0–20 Mg ha$^{-1}$ | FA at 15 and 30 Mg ha$^{-1}$ | Alkaline loamy sand | Addition of RHA and BA significantly increased the grain yield of wheat and rice but FA caused small increases in crop yields | India | Thind et al. (2012) |
| FA + soil | FA at 0, 2, 10, and 20 Mg ha$^{-1}$ | Field acidic soil | FA is a potential soil ameliorant on reducing CH$_4$ emissions as well as increasing rice grain productivity. At 10 Mg ha$^{-1}$ FA level, the total CH4 emission was reduced by 20%, along with 17% yield increment over the control | South Korea | Ali et al. (2007) |
| FA + soil | FA at 5%, 10%, 20%, and 40% w/w for laboratory, while 50–400 t ha$^{-1}$ for field plots | Laterite cropland soil | This short-term study indicated that the use of up to 10% (100 t ha$^{-1}$) dose of fly ash is apparently safe to improve chemical and biological properties of soil in laterite croplands | India | Roy and Joy (2011) |
| FA + soil | FA at 0%, 5%, 10%, 15%, and 20% (w/w) | Rice field soil | At lower levels, FA might be beneficial to both rice field and rice crop and can be utilized as soil fertilizer or nutritional supplement, but, at higher levels, FA incorporation is potentially harmful for agricultural crops | India | Singh et al. (2012) |
| FA + soil + gypsum | Fly ash–gypsum mixture (50:50 w/w) | Poorly drained silt loam and loamy fine sand | A fly ash–gypsum mixture (50:50 w/w) could be used as an alternative soil amendment to improve soil properties and enhance rice productivities in Korean paddies | Korea | Lee et al. (2008b) |

| FA + soil | FA at 200 t ha$^{-1}$ | At two different geographical locations of India | Results indicated that there is no difference between rice samples grown in soils with or without FA. Furthermore, there were no adverse effects on hematological, biochemical, or histopathological parameters when rice was fed to rats for 6 months. This indicates that rice grown on FA-treated soils may be safe for human consumption | India | Bhaskarachary et al. (2012) |
|---|---|---|---|---|---|
| FA + mine spoil + PM | FA at 20 t ha$^{-1}$ and PM at 10 t ha$^{-1}$ | Mine spoil | 35.6% yield increase | Neyveli, India | Ram et al. (2006) |
| FA + PM + soil | FA at 2 t ha$^{-1}$ and PM at 10 t ha$^{-1}$ | Paddy soil | 9.2% yield increase | Anaimalai, India | Jeyabal et al. (2000) |
| FA + FYM + soil | FA at 3% and FYM at 10 t ha$^{-1}$ | Alkali soil | 175.8% yield increase | Puresndi, India | Chaudhary et al. (2001) |
| Acidic FA + gypsum + sodic soil | FA levels of 0.0%, 1.5%, 3.0%, 4.5%, 6.0%, and 7.5% used alone as well as in combination with 100%, 80%, 60%, 40%, 20%, and 10% gypsum requirement of the soil, respectively | Sodic soil | For reclaiming sodic soils of the southwest Punjab, gypsum was advocated to be substituted up to 40% of its requirement with 3.0% acidic FA | India | Kumar and Singh (2003) |
| FA + gypsum + soil | FA at 0, 20, 40, and 60 Mg ha$^{-1}$ | Silt loam paddy soil | Higher uptake of heavy metals by rice in submerged silt loam paddy soil and increased in the available B with increasing FA dose (both alone and in mixture with gypsum), but not beyond toxicity levels | Korea | Lee et al. (2002) |

(Continued)

**Table 4.1** (Continued)

| Amendments | FA dose range | Soil types | Remarks | Country | References |
|---|---|---|---|---|---|
| FA + sewage sludge + sand soil | FA at 20% FA (w/w) | Sand soil | Soil N, P, and K contents and their uptake by rice at all the growth stages on 20% FA (w/w) application were recorded | China | Wong and Lai (1996) |
| FA + soil | FA at 200 t ha$^{-1}$ | Soil | Heavy metal study in rice grain and straw | India | Vijayan et al. (1999) |
| FA + alkaline soil | FA at 100 t ha$^{-1}$ | Alkaline soil | 28.57% grain increase over control and 11.32% straw increase over control | India | Ram et al. (2011) |
| FA + saline soil | FA application up to 20% | Saline soil | Grain and straw yield increased with FA application up to 20% | India | Singh and Singh (1986) |
| FA + soil | | Soil | FA can at best be used as a supplement to chemical fertilizer | India | Kumar et al. (1998) |
| FA + soil | | Soil | The heavy metal contents (Cd, Pb, and Ni) in the rice crop from FA-treated plots were not significantly different in comparison with control | Korea | Kim et al. (1994a) |
| Control, FA, IF, FA + IF, FA + GM + IF, and FA + PS + IF | FA at 10 t ha$^{-1}$ | Acid lateritic soil | At the end of the cropping season, pH, organic carbon, and dehydrogenase activity were higher under the mixed treatment (FA + GM + IF) by 3%, 15%, and 154%, respectively, compared with IF alone | India | Rautaray (2005) |

| Treatments | Application | Soil type | Remarks | Country | Reference |
|---|---|---|---|---|---|
| No-amendment, soil amendment with FA–gypsum mixture | FA + gypsum [ratio of 75:25 (w/w)] mixture at 25 Mg ha$^{-1}$ and nitrogen at 0, 40, 80, 120, and 160 kg ha$^{-1}$ | Silty clay loam | Results suggest that FA–gypsum mixture may increase the nitrogen uptake of rice and decrease potential N losses from paddy soils. B toxicity in rice was not observed maybe due to the dilution and leaching effects in the submerged paddy soil | Korea | Lee et al. (2003) |
| FA + soil | FA + gypsum (50:50%) mixture at 0, 20, 40, and 60 Mg ha$^{-1}$ | Silt loam | Improving rice yield and soil quality by FA–phosphogypsum mixture application | Korea | Lee et al. (2005) |
| FA + soil + SS | FA at 20 and 40 g kg$^{-1}$ and SS (3 and 6 g kg$^{-1}$) | Paddy soil | Increased soil pH from 4.0 to 5.0–6.4, decreased the phytoavailability of heavy metals by at least 60%, and suppressed metal uptake by rice | China | Gu et al. (2011) |
| Control, FA, CF, FA + CF, FYM + CF, FA + FYM + CF, L + FYM + CF, PFS + CF, FA + PFS + CF, L + PFS + CF, CR + CF, and FA + CR + CF | FA at 10 t ha$^{-1}$, L at 2 t ha$^{-1}$, FYM, PFS, and CR was selected to provide 30 kg N ha$^{-1}$ | Sandy loam | It is proved that alkaline FA is a better amendment than lime for improving productivity of acid soil and enriching the soil with P and K | India | Swain et al. (2007) |
| FA + soil | FA at 25% | Garden soil | FA amended soil at 25% (low concentration) could be valuable in rice seedling as it improves growth and photosynthesis | India | Panda et al. (2019) |
| FA + soil | FA at 50% | Garden soil | 50% FA application declines the soil quality and rice production, and enhances the toxic metals in rice grains, which may pose a risk toward rice (grown on FA) consumption | India | Singh et al. (2015) |

(Continued)

**Table 4.1** (Continued)

| Amendments | FA dose range | Soil types | Remarks | Country | References |
|---|---|---|---|---|---|
| FA + Soil + $N_2$-fixing cyanobacteria | FA levels, 0, 0.5, 1.0, 2.0, 4.0, 8.0, and 10.0 $kg\,m^{-2}$ | Sandy loam | FA dose 4.0 $kg\,m^{-2}$, with 1.0-$kg\,m^{-2}$ cyanobacteria could be used ideal as it improves essential micronutrients and decreases toxic chemicals with usual $N_2$-fixation | India | Padhy et al. (2016) |
| FA + garden soil and MS + garden soil | FA at 50% and 100%, and MS at 50% and 100% | Garden soil | Indian wild rice has potential to tolerate 50% FA and mining soil, and can be noted as a good candidate | India | Bisoi et al. (2017) |
| FA + soil | FA dose (20, 40, and 60 t $ha^{-1}$) with and without FYM | Clay loam | The results showed that most of the toxic elements accumulated mainly in the paddy roots except Cd, which was reserved in the leaves | India | Kumar et al. (2018) |
| FA + soil + RDF | FA dose (0, 5, 10, and 15 t $ha^{-1}$) and RDF dose (50% and 100%) | Sandy clay loam | Mixture of FA, FYM, and RDF positively improved soil pH, organic C, N, P, and K. Therefore FA can be a component of integrated plant nutrition system in enhancing yield | India | Das et al. (2016) |
| FA + soil + N sources FA + soil | FA dose (0%, 5%, and 10% w/w), N sources (no input, urea, compost, and vetch), and FA dose (0, 5, 10, 20, 40, and 100 v/v) | Inceptisol (coarse loamy, mixed, mesic family of Fluvaquetic Endoaquepts) Aeric Endoaquept | FA is effective in augmenting soil carbon sequestration without decreasing paddy yield. FA application at lower doses (10–20 v/v) in soil enhanced micronutrients content, microbial activities, and crop yield | Korea India | Lim et al. (2017) and Nayak et al. (2015) |

BA, Bagasse ash; BGA, blue green algae; CF, chemical fertilizers; CR, crop residue; FA, fly ash; FYM, farmyard manure; GM, green manure; GS, garden soil; IF, inorganic fertilizer; L, lime; MS, mining soil; NF, nitrogen fertilizer; OW, organic waste; PFS, paper factory sludge; PM, press mud; PS, paddy straw; RDF, recommended dose of chemical fertilizer; RHA, rice husk ash; SS, steel slag.

## 4.3 Beneficial role of fly ash input in paddy fields

There are many beneficial role of the FA input in paddy crop agriculture. A suggestive list of FA amendment approach for the paddy crop production is given in Table 4.1. Furthermore, Fig. 4.1 summarizes the beneficial role of FA application in paddy fields. Here are discussed some benefits of FA application such as the improvement of the physicochemical properties of paddy soil, paddy soil biota, rice growth, paddy yield, and nutrient availability as well as methane mitigation from paddy fields. Natural resource conservation (i.e., water and soil) and saving of chemical fertilizer (CF) are other benefits of using FA in paddy fields (Lee et al., 2006; Ali et al., 2007; Pandey et al., 2009a; Pandey and Singh, 2010; Singh and Pandey, 2013). Here are presented some important FA characteristics to understand the beneficial role of FA input in paddy fields. The physicochemical and mineral properties of coal FA have many favorable factors for adequate nourishment to the rice crop. Generally, FA characteristics depend on combustion process, coal quality, coal source, particle size, extent of weathering, and age of the ash (Singh and Kolay, 2002, Sarkar et al., 2006, Pandey et al., 2009a; Ram and Masto, 2010). FA has high surface area, low BD, light texture, high moisture retention capacity, high electrical conductivity, and low CEC in comparison with normal soil. FA contains both micronutrients (Fe, Mn, Zn, Cu, Co, B, and Mo) and macronutrients (P, K, Ca, Mg, and S), which are essential to plants. The pH of FA ranges from 4.5 to 12.0 and depends on the sulfur content of the parent coal (Pandey and Singh, 2010). Chemically, FA mostly consists of $SiO_2$ (silica), $Al_2O_3$ (alumina), CaO (calcium oxide), $Fe_2O_3$ (iron oxide), MgO (magnesium oxide), $Na_2O$ (sodium oxide), $K_2O$ (potassium oxide), $SO_4^{2-}$ (sulfate), unburned carbon, and other metals in trace amount (Pandey and Singh, 2010; Ram, 1992; Pandey et al., 2009a). Pandey and Singh (2010) reviewed several beneficial effects of FA application in soil, but here, we are more specific to the paddy fields (Fig. 4.2).

## 4.3.1 Fly ash impact on the physicochemical properties of paddy soil

The physicochemical properties of soil plays an important role in the establishment of rice seedling and enhancement of crop growth and yield attributes. In this direction, several scientists have worked to modify the physicochemical properties of paddy soil to enhance high yield of paddy by the application of FA alone or coapplication with other organic and

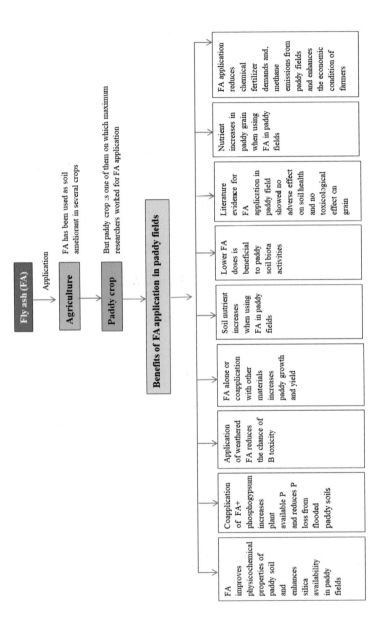

**Figure 4.1** Detailed beneficial role of FA application in paddy fields.

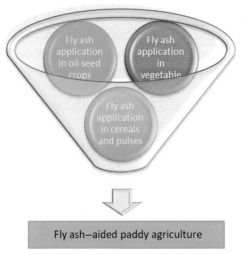

**Figure 4.2** This funnel-shaped diagram shows the clarified information on FA-aided paddy agriculture toward implication for safe utilization of FA for agricultural production.

inorganic materials. Generally, the impact of FA on the soil physical characteristics is related to the changes in soil texture. Modification in the soil texture is linked with BD, porosity, water holding capacity (WHC), and hydraulic conductivity, which have a direct impact on the plant growth, nutrient retention, and biological activity of the soil (Campbell et al., 1983; Fulekar and Dave, 1986; Biliski et al., 1995; Fang et al., 2001; Guest et al., 2001; Jayasinghe and Tokashiki, 2012). Singh et al. (2012) observed that FA increases the WHC of the paddy field due to reduced BD. Singh et al. (2011) reported the influence of coapplication of FA + FYM on WHC and BD of paddy field. They noticed the greatest value of WHC for FA + FYM-treated plot in comparison with control, while the situation was reverse for BD. FA application from 10% to 30% level to coarse-textured soils improved the WHC manifold depending on the FA dose and soil texture (Campbell et al., 1983). FA application alone or coapplication with organic material to soil at a lower rate has been reported to promote the growth of paddy crop through the improvement of soil conductivity, soil porosity, and WHC (Singh and Siddiqui, 2003; Mittra et al., 2005; Mishra et al., 2007; Singh et al., 2011). A wide range of BD values (0.81–1.16 Mg/m$^3$) of coal FAs have been noticed (Chatterjee et al., 1988; Sarkar et al., 2005). It was also observed that coal FA has a lower BD value than arable soil (Sikka and Kansal, 1994).

A marked decrease was noticed in the BD of the paddy soils after FA addition (Singh et al., 2011, 2012), and improvement in porosity and WHC is well documented (Singh and Siddiqui, 2003; Mishra et al., 2007). Soil moisture is a vital factor of the soil system, which has significant effects on the biogeochemical cycles. The FA supports to conserve soil moisture in the soil system (Seneviratne et al., 2010). In this regard, Singh et al. (2011) observed that FA and FYM, alone or coapplication, increased the soil moisture of the paddy field in comparison with control or other treatments. On the addition of FA + FYM at $50 \, t \, ha^{-1}$ each to dry-land paddy field, a significant improvement was observed in the soil moisture and WHC with the reduction in BD, which aid to crop production (Singh et al., 2011).

## 4.3.2 Fly ash and paddy soil biota

Lots of information exists on the role of FA application in rice cropping. Several scientists have advocated the use of FA application for improvement in paddy crop yield. However, the influence of FA on the soil biota during paddy cropping is less known. Only few researchers have worked on the soil biota of paddy fields. In this regard, Tripathi et al. (2008) studied the role of blue green algae (BGA) biofertilizer to accomplish the nitrogen (N) demand and mitigate FA stress to the growth and yield of paddy crops. BGA are the primitive algae (currently called cyanobacteria) or the diverse group of photosynthetic prokaryotes growing often in paddy fields, which helps to fix atmospheric N and to convert insoluble phosphorus into soluble form (Irisarri et al., 2001). Generally, coal FA is rich in B and deficient in N content. It has been recognized that B is essential for N-fixation by heterocyst-forming BGA strains like Anabaena (Blevins and Lukaszewski, 1998). Hence, BGA are suitable candidates for improving N status in FA amended paddy fields. It is well known that N and P are the major limiting factors to plant growth. Tripathi et al. (2008) studied the impact of different doses of FA with and without BGA on phytotoxic, tolerance responses, growth, and yield of paddy crop var. Saryu-52. Thus it seems that BGA are suitable biofertilizers, which will help for a safe use of FA in paddy cultivation. This is low-cost agrotechnology for the farmers to cultivate paddy crop in FA amended soils (Tripathi et al., 2008).

The impact of FA application on methanotroph population dynamics and paddy yields was studied in nutrient poor soils by Singh and Pandey (2013).

They concluded that FA at lower doses along with PM seems to be a suitable amendment for improving soil methanotroph population and paddy yields for the nutrient poor soils (Singh and Pandey, 2013). Rautaray (2005) studied the effects of FA, organic fertilizers, and inorganic fertilizers (IFs) on changes in dehydrogenase activity in an acid lateritic soil. Rautaray noticed the highest dehydrogenase activity (8.47-mg triphenyl formazon $g^{-1}$ 24 $h^{-1}$) under the mixed fertilization treatments (FA + IF + green manure). Earthworm is an eco-friend of the farmers in agriculture. Several endemic earthworm species accumulate heavy metals in their intestine by forming organometallic complexes, which needs research for refining biocompatibility of FA in agriculture (Bhattacharya et al., 2012). Furthermore, Pandey and Singh (2010) suggested for exploring the impact of FA addition in agriculture on the main features of earthworm burrows, for instance, their diameter, length, tortuosity, sinuosity, branching, volume, connectivity, and orientation. Thus the paddy-based agroecosystem is a thrust area of tremendous scope for mitigating the toxicity of FA application on soil biota and decomposer food web.

### 4.3.3 Fly ash-silica and rice growth

Rice (*Oryza sativa* L.) is the staple food crop in several countries all over the world including India, which feeds more than half of the global population. Rice is the sole cereal crop grown under waterlogged condition in irrigated as well as rain-fed fields in two seasons (wet and winter) in the Asian countries (Mishra et al., 2007). Silica plays an important role in the paddy productivity, because paddy needs high amounts of silica, which enhances the resistance of lodging and pathogens (Mengel and Kirkby, 1987; Deren et al., 1994) and induces the absorption of primary elements such as K, P, and N (Hu and Wang, 1995). Silica promotes healthy plants through the formation of a thick silicate epidermal layer to maintain erect leaves that enhance photosynthesis. The silicate layer serves as a barrier against fungal disease, chewing, and sucking insect pests and nematodes in rice crop. Silica is also helpful to enhance the resistance of toxic metals, which might be due to the stimulation of antioxidant systems, alleviation of photosynthetic inhibition, and complexation of heavy metals with silica (Neumann and Nieden, 2001; Shi et al., 2005; Liang et al., 2007). The mechanism involved in silica-mediated reduced metal accumulation and detoxification in paddy crop is still not clear (Gu et al., 2011).

High silica content is an important characteristic of FA. Plant's roots absorb Si from soil matrix as uncharged monosilicic acid in the form of $H_4SiO_4$ or $Si(OH)_4$. However, silica uptake differs significantly between plant species. Silica is not considered as an essential mineral element for all higher plants (Epstein, 1999), but there is an increasing evidence as essential element in large amount for high levels of rice yield and vigorous paddy growth (Ma and Yamaji, 2006). Therefore silica is a most important fertilizing element for rice crop. Silica accumulation is more in monocotyledons than in dicotyledons (Epstein, 1994). There is a dearth of available silica in the paddy soil of many countries like Korea. Therefore Korean government needs silicate fertilizer in the form of by-product slag from iron manufacturing. In fact, such country needs an alternative source of low-priced silica (Lee et al., 2006, 2008b). FA, which is a ferroaluminosilicate mineral, has a very high silica content (Pandey and Singh, 2010). In this regard, the field experiments were done on two soils (silty loam and loamy sand) by Lee et al. (2006) to evaluate rice productivity in FA amended soil (0, 40, 80, and 120 Mg ha$^{-1}$) and 2 Mg ha$^{-1}$ Si was used as a control. They found that FA increased the available silica contents of the both soils and could be a good alternative source of cheaper silicate fertilizer in paddy soil. In addition, in a paddy field experiment, Lee et al. (2007a) observed that the increased water-soluble and available silica contents by FA application increase available P contents in soil and reported a very strong positive correlation between the available P and water-soluble/available silica. The applied FA contained about 510 mg kg$^{-1}$ of available silica that is soluble in 1 M sodium acetate solution (pH 4.0). Approximately 20, 40, and 60 kg ha$^{-1}$ of available silica were added by the application of 40, 80, and 120 Mg ha$^{-1}$ FA, respectively. At the time of paddy harvesting stage, available silica was 23 and 22 mg kg$^{-1}$ at 0 FA level (control) and increased to 62 and 52 mg kg$^{-1}$ by the application of 120 Mg ha$^{-1}$ FA in Yehari and Daegok, respectively. In a previous study, Obihara and Russell (1972) also observed that the addition of water-soluble silica to a soil may increase the amount of P and reported a positive relationship between the water-soluble silica in soil and its P status.

## 4.3.4 Fly ash and paddy yield

FA incorporation in paddy fields as soil ameliorant is an emerging research area from the considerations of utilizing its huge amount and tapping economic potentials. The application of FA has significantly increased the growth and yield of paddy crops in a number of research studies. In this regard, numerous studies showed the good response of paddy crops to coapplication of FA with PM or SS or FYM. Singh et al. (2012) assessed the feasibilities of safe and effective utilization of FA as soil amendment in paddy field experiment and its impact on growth and yield of three rice cultivars. They observed that different yield attributes of all rice cultivars (the number of grain per plant, number of years per plant, harvest index, and weight of 1000 grains) showed significant increments under 5% and 10% FA level (w/w). The rice cultivar Sugandha-3 exhibited higher yield than cultivars Sambha and Saryu-52. The number of grains per plant increased in cultivar Sambha, Sugandha-3, and Saryu-52 by 21%, 33%, and 8% at 10% FA in comparison with control, respectively. Results indicated that up to a certain level, FA amendment can be utilized in Indian paddy field as a good ameliorant. In another study, the application of 20% and 40% FA with soil caused a significant increase in plant growth and yield of all the three Indian rice cultivars, that is, Pusa Basmati, Pant-4, and Pant-10 (Singh and Siddiqui, 2003). Lee et al. (2006) showed that an application of 90 Mg ha$^{-1}$ FA can significantly increase the grain yield in Korean rice cultivars. Dwivedi et al. (2007) noted that the grain weight was optimum with 25% FA application with three Indian rice cultivars viz., Saryu-52, Sabha-5204, and Pant-4. But Saryu-52 and Sabha-5204 were more tolerant to FA and have potential to be grown at a lower FA dose for good crop yields. FA application increased rice yields by 7%–13% following the rate of FA 40, 80, or 120 Mg ha$^{-1}$ (Kim et al., 1994c). In Odisha state of India, Patra et al. (2012) demonstrated that FA application at the rate of 200 t ha$^{-1}$ increased the rice grains by 40% and 13% over control values at Malud and Dhenkanal, respectively. In another experiment of Odisha state, Mishra et al. (2007) showed that the grain yield increased maximum at 10% FA-treated paddy field as compared with other treatments.

Karmakar et al. (2010) reported that the integrated use of FA or rice husk ash (RHA), paper factory sludge (PFS), or FYM and CFs improves growth and yield of paddy crop. Such integrated nutrient management increased the paddy production to a considerable extent. Singh et al. (2011) conducted an experiment in a dry-tropical nutrient poor soil to assess the effect of FA/FYM or in combination with paddy yield and N-dynamics. The used treatments in this experiment were control, 50 t ha$^{-1}$ FA, 50 t ha$^{-1}$ FYM, and FA + FYM (each 50 t ha$^{-1}$). The best treatment noticed for paddy grain yield was FA + FYM (92% increase over control). Lee et al. (2002) studied the effect of an FA and gypsum mixture (75:25 ratio w/w) on rice cultivation in Korean paddy soil. They reported the maximum yield of rice at an application rate of 40 Mg ha$^{-1}$ of the combination. In continuation, Lee et al. (2003) further demonstrated the evaluation of use of FA—gypsum mixture for rice production at different nitrogen rates. The application of FA—gypsum mixture in Korean paddy field increased the maximum rice yield by about 530 kg ha$^{-1}$ (from 6,507 kg ha$^{-1}$ in the no-amendment treatment to 7,035 kg ha$^{-1}$ in the FA—gypsum mixture treatments). This mixture could decrease about 50 kg ha$^{-1}$ of the N application in rice field to attain the maximum rice yield. Recently, Singh and Pandey (2013) demonstrated that the rice grain increased from 1658 to 3686 kg ha$^{-1}$ in all FA + PM amended plots compared with controls. Results indicated that the best treatment was the combination PM + FA at 10 t ha$^{-1}$ each, which increased 138% of rice grain yield relative to control. The acidic FA (pH, 5.89) alone or in blend with gypsum improved paddy crop yield and physicochemical properties of the sodic soil (pH, 9.07) (Kumar and Singh, 2003). Singh and Singh (1986) reported significant increase in grain and straw yield of rice at FA application rate up to 20%. However, FA addition higher than 20% decreased the rice yield.

## 4.3.5 Fly ash and phosphorus accumulation

In Korea, a high use of fertilizer has resulted in excess P accumulations in some arable soils. These P enriched runoff from rice fields can cause new environmental problems like eutrophication of water bodies (streams, lakes, and freshwater of estuaries). This high P content of arable soils has increased the amount of water-soluble P in the paddy soil, consequently increasing the potential of P export through surface runoff to water bodies (Sharpley, 1995; Sharpley et al., 1996). Lee et al. (2003) reported that around 35 kg ha$^{-1}$ year$^{-1}$ of $P_2O_5$ was lost from the surface layer through plant uptake in a long-term fertilized paddy soil with standard management practices. The physicochemical approach is an effective treatment process for controlling soil P loss through surface runoff by reducing its solubility through precipitation with Ca, Fe, or Al (Stout et al., 1998). Coal FA and phosphogypsum are the sources of elements and compounds that can form insoluble precipitates with P (Lee et al., 2007b). Gypsum is an acidic by-product from the phosphate fertilizer industry in many countries, which is easily available. Lee et al. (2005) prepared a mixture of FA and gypsum (50:50, wt wt$^{-1}$) to satisfy the Si and Ca requirements of paddy crop and found positive effects by improving soil fertility and stimulating rice growth. In another study, they applied this mixture at rates of 0, 20, 40, and 60 Mg ha$^{-1}$in two paddy soils of contrasting textures (silty loam in *Yehari* and loamy sand in *Daegok*). The FA—gypsum mixture significantly reduced water-soluble phosphate in the surface soils by shifting from water-soluble phosphate and iron-bound phosphate to calcium-bound phosphate and aluminum-bound phosphate during rice cultivation in both soils, thereby reducing the potential for P transport by surface runoff into lakes (Lee et al., 2007b) and increase soil fertility. The FA—gypsum mixture should be beneficial to the increment of plant available phosphorus and rice productivity as well as reducing P export from flooded Korean paddy soils, where P loss in surface runoff is of prime concern (Lee et al., 2007b).

Lee et al. (2007a) studied the role of FA application as a liming material in two Korean acidic soils (silt loam in *Yehari* and loamy sand in *Daegok*). The main objective of this experiment was to examine: (1) the effect of FA application rate on soil P levels; and (2) the characteristics of P fraction of two Korean paddy soils. In this experiment, they observed that available P contents in the surface soil increased significantly with the increase in FA doses because of high P content (total P 786 mg kg$^{-1}$) in applied FA. The recommended level of available P for paddy cultivation is 35—52 mg kg$^{-1}$ in Korean country (RDA, 1999). Available P, determined by Lancaster method, was 25 and 66 mg kg$^{-1}$ at 0 FA level and increased to 53 and 120 mg kg$^{-1}$ by the application of 120 Mg ha$^{-1}$ FA in *Yehari* and *Daegok*, respectively. There are several factors that may be responsible for the enhancement of available P by FA application. First, applied FA has high content of P (about 786 mg kg$^{-1}$), which may increase soil P concentration. In this experiment, about 31, 63, and 94 kg ha$^{-1}$ of P were supplied in paddy agricultural fields by the addition of 40, 80, and 120 Mg ha$^{-1}$ FA, respectively (Lee et al., 2007a). Second, alkaline FA increases the pH of acidic paddy soils. Lee et al. (2007a) noticed that when FA was applied at120 Mg ha$^{-1}$ level in rice fields, the pH changed from 5.2 to 6.0 and 5.7 to 6.7 at *Yehari* and *Daegok* sites, respectively. The presence of available P in soil mostly depends on the pH of the soil and shows a high availability in the range of neutral pH (Bohn et al., 1979). Therefore Lee et al. (2007a) observed that "the neutralizing effect of the alkaline FA on acidic soil is a very important factor for increased P availability during rice cultivation." Furthermore, the high

Si content of FA is another important factor to increase the available P concentration in paddy soils. It is reported that silicate ions help to increase the P solubility and availability in soil by displacing P ion from ligand exchange sites (Roy et al., 1971) and by inhibiting P ion sorption for the same specific anion exchange site (Hingston et al., 1972; Jepson et al., 1976).

Lee et al. (2007a) also observed the various P fractions in paddy soil by the application of FA and analyzed for various P fractions like iron-bound P (Fe-P), aluminum-bound P (Al-P), calcium-bound P (Ca-P), and water-soluble P (W-P) by sequential extraction procedures. Generally, the distribution of different P fractions mainly depends on the soil mineralogical characteristics, soil pH, the kinds of P fertilizer applied, and other soil factors. The Ca-P fraction is the major form of P in alkaline soils, while Al-P and Fe-P fractions are the main form of P in acidic soils (Dean, 1949). They analyzed the FA amended paddy soil for various P fractions by sequential extraction procedures using water (W-P), $25 \, g \, L^{-1}$ acetic acid, and 1-M ammonium chloride (Ca-P), 1-M ammonium fluoride with pH 7.0 (Al-P), and 0.1 M sodium hydroxide (Fe-P) following the method of Sekiya (1983) and Watanabe and Kato (1983). Several researchers reported that the conversion of W-P to less soluble forms with lime- or calcium-containing substances like coal combustion by-products can decrease the release of soil P to surface runoff in aerobic uplands (Stout et al., 1998, 1999, 2000). In contrast to aerobic upland soil, Lee et al. (2007a) found that FA application significantly increased W-P content in anaerobic paddy soils. At the time of paddy transplanting stage, W-P was approximately $6.7 \, mg \, kg^{-1}$ (1.1% of extractable P) in *Daegok* site (control) at 0 FA level, then it was increased to 8.1, 8.5, and $11.1 \, mg \, kg^{-1}$ by incorporating 40, 80, and $120 \, Mg \, ha^{-1}$ of FA, respectively. The same pattern trend was also seen in *Yehari* site (Lee et al., 2007a). Due to the flooded condition of paddy field, ferric iron ($Fe^{3+}$) having lower water solubility reduced to ferrous iron ($Fe^{2+}$) with a high solubility (Bohn et al., 1979). The Ca-P, Al-P, and extractable P contents enhanced significantly with the FA application, while the Fe-P fraction reduced noticeably with FA application (Lee et al., 2007a).

The availability of many nutrients in FA is very low. It may be increased through either vermicomposting or addition of FYM. In this regard, Bhattacharya and Chattopadhyay (2002) evaluated the possibility of enhancing P bioavailability in FA through vermicomposting in a field-based experiment. They mixed FA with organic matter (cow dung) at

1:3, 1:1, and 3:1 ratios. These mixtures were incubated with and without earthworm for 50 days. They reported manyfold increases in the concentration of phosphate-solubilizing bacteria in the earthworm-treated mixture (FA + organic matter) in comparison with untreated mixture. This technology transforms considerable amounts of insoluble P from FA into more soluble forms. They observed maximum P availability in a 1:1 ratio of FA and cow dung in comparison with other ratios (Bhattacharya and Chattopadhyay, 2002).

## 4.3.6 Fly ash and methane mitigation from paddy fields

There is a great challenge to sustain the rice productivity without methane emissions from flooded paddy fields. Flooded paddy fields are one of the major anthropogenic sources of methane emissions to the atmosphere that accounts for 20%—26% of the global methane emissions (Neue and Roger, 1993). There are two microbes, such as methanogens and methanotroph, that are present in paddy fields as source and sink of methane, respectively. Methane emission from paddy cropping results from the anaerobic decomposition of organic compounds, where $CO_2$ acts as an inorganic electron acceptor. Microbes that are capable of reducing more favorable electron acceptors, that is, $NO^{3-}$, $Mn^{4+}$, $Fe^{3+}$, and $SO_4^{2-}$, may outcompete methanogens using the less favorable electron acceptor such as $CO_2$ (Lovely et al., 2004). Thus methane emission from flooded paddy fields could be reduced by supplying alternative electron acceptors like $NO^{3-}$, $Mn^{4+}$, $Fe^{3+}$, and $SO_4^{2-}$ (Jakobsen et al., 1981; Achtnich et al., 1995).

FA has been recognized as a good soil ameliorant to neutralize acidic soil, because it contains high amount of silica, iron, and manganese oxides (Ko, 2002; Pandey et al., 2009a; Pandey and Singh, 2010). Ferric oxide and manganese oxide are good electron acceptors. Thus alkaline FA may be used to suppress methane emission in acidic paddy soil. In this regard, Ali et al. (2007) investigated the effects of FA applications on methane emissions during rice cultivation. They applied FA at the rate of 0, 2, 10, and 20 Mg ha$^{-1}$ into potted soils before rice transplanting. Methane emission rates gradually decreased with the increasing FA doses, but paddy yield significantly increased up to 10 Mg ha$^{-1}$ FA level. At this optimum dose, total methane emission was declined by 20% along with 17% paddy grain yield increment over the control. The suppression of methane emission from paddy fields with FA amendment may be attributed owing to high content of active and free iron and manganese oxides and sulfate ions in soil, which acted as oxidizing agents as well as electron acceptors (Ali et al., 2007). Furthermore, Singh and Pandey (2013) revealed that FA at lower doses with PM increases soil methanotroph population and paddy crop yield in nutrient poor soils of dry-tropical regions. As we know, methanotroph utilizes methane as a sole source of carbon and energy from paddy fields as well as atmosphere due to the presence of broad spectrum methane monooxygenase enzyme (Pandey et al., 2013). It means FA decreases methane emissions from paddy fields in two ways by using lower doses.

## 4.3.7 Coapplication of fly ash with other amendments

FA has been used as a coapplication with organic or inorganic amendments for the amelioration of nutrient poor paddy soil to increase rice production across the globe (Wong and Lai, 1996; Jeyabal et al., 2000; Kumar and Singh, 2003; Lee et al., 2003, 2005, 2008b; Mittra et al., 2005; Rautaray, 2005; Ram et al., 2006, 2007; Tripathi et al., 2008; Karmakar et al., 2010; Gu et al., 2011; Singh et al., 2011; Thind et al., 2012; Singh and Pandey, 2013). Some important examples of coapplication of FA with organic or inorganic amendments are discussed in this section. Recently, Singh et al. (2011) studied the coapplication of FA and FYM, mixture treatment (FA + FYM, each $50 \, t \, ha^{-1}$) on rice, which was found to be most effective to enhance (92%) the grain yield of paddy crops. Combined application of acidic FA (pH, 5.89) with gypsum as well as acidic FA (pH, 5.89) alone improved the physicochemical characteristics of sodic soil (pH, 9.07) and paddy crop yield (Kumar and Singh, 2003). Another coapplication of FA (1 and $2 \, t \, ha^{-1}$) and PM (10 and $20 \, t \, ha^{-1}$) increased panicle numbers, tiller numbers, dry-matter production, and straw and grain yields of paddy crops (Jeyabal et al., 2000). Lee et al. (2005) set a combination of FA and gypsum (50:50, wt/wt) to know the Si and Ca requirements of rice and noticed favorable effects by improving soil fertility and increasing rice growth. A mixture of FA—gypsum at $25 \, t \, ha^{-1}$ increased the rice grain yield (8%) and increased N and Si uptake of paddy. A pot-based experiment showed that the coapplication of FA (20 and $40 \, g \, kg^{-1}$) and steel slag (3 and $6 \, g \, kg^{-1}$) enhanced the pH of soil from 4.0 to 5.0—6.4, reduced the phytoavailability of heavy metals by at least 60%, and likewise suppressed metal uptake by paddy crops (Gu et al., 2011). Lee et al. (2008b) assessed the coapplication of FA and phosphogypsum for reducing dissolved phosphorus in arable soil, and reported that coapplication reduces P loss from paddy soil and improved paddy soil fertility. Thus coapplication of FA and gypsum is a feasible soil ameliorator to improve the nutrient balance in paddy soils and to decrease the N requirement rate of paddy crops (Lee et al., 2003). Some previous field studies (Ram et al., 2006, 2007) with mine spoil and red soils revealed positive effect of coapplication of FA and PM on soil quality and crop yield. Various FA dosages (0, 5, 10, 20, 50, 100, and 200 t ha$^{-1}$) were applied for mine spoil trials, with and without PM (10 t ha$^{-1}$ before cultivation of the first crop). The rice crops were used in rotation as the first, third, fifth, and sixth crops. The other crops like green gram (second) and sun hemp (fourth) were grown as green manure. The application of FA from 5 to 20 t ha$^{-1}$ (with and without PM) increased significantly the grain and straw yield (3.0%–42.0%) over the corresponding control.

Thus coapplication of FA with various organic and inorganic amendments can be used as a low-priced effective mixture to improve soil quality as well as crop production. A wide range of amendments like RHA, PFS, FYM, CFs, lime, BGA, nitrogen fertilizer, organic waste, crop residue, PM, bagasse ash, IF, paddy straw, green manure, steel slag, etc. have been used along with FA. The precise benefits linked with coapplication of FA with the above-discussed amendments are summarized in Table 4.1. Coapplication of FA with locally available sources of organic matter could be exploited well to reduce the cost of cultivation and soil amelioration.

## 4.4 Nutrient assessment

There is a growing concern about the application of FA as a plant nutrient in agricultural sector; however, there are limited attempts to understand the determinants of stability in agroecosystems. FA can enhance the availability of Ca, Si, Mg, Na, K, S, B, and other trace nutrients (Elseewi et al., 1981; Wong and Wong, 1989; Pandey and Singh, 2010). FA applications have supplemented the paddy crop nutritional deficiencies of P (Lee et al., 2008b); N and S (Sikka and Kansal, 1995); Mn, Mg, and Zn (Mittra et al., 2005); K (Lee et al., 2006); and Si (Lee et al., 2006). Nutrient deficiencies in paddy crop fields can be ameliorated by either direct application of FA in soil or coapplication of FA with organic and inorganic materials or regulating the soil pH corrections. Coapplication of FA with organic matter adds N and P to the paddy soils. Coutilization of

FA with suitable and locally available waste material can assist in meeting the nutritional requirements of paddy crop as well as minimize the adverse conditions of paddy fields. Lee et al. (2006) proved that FA could be a good supplement alike to other inorganic soil amendments to restore the nutrient balance in Korean paddy soils. Results showed a high nutrient content in rice crop under FA treatment and increased Si, P, and K uptake by paddy crops (Lee et al., 2006). Lee et al. (2008b) showed that mixture of FA and Gypsum (1:1) at 60 Mg ha$^{-1}$ increased the uptake of Si, P, and K by rice crop. Lee et al. (2002) used a combination of FA and gypsum to meet Si and Ca requirements of rice crop and found favorable results in improving soil quality and stimulating rice growth.

Mittra et al. (2005) noticed higher uptake of micronutrients (Zn, Mn, Fe, Co, and Cu) and macronutrients (Ca, Mg, P, K, and S) in test plants at higher FA doses, while nutrient contents were constant in rice grains, because soil nutrient contents were also increased by higher FA doses. Singh et al. (2012) reported with the help of field experiment that major nutrients (Ca, Na, Mg, K, and Fe) in the grain of all rice cultivars (Sambha, Sugandga-3, and Saryu-52) increased significantly with different FA doses, in comparison with their controls. The S content in paddy crop increased significantly with the application of FA, up to the 8% level, because FA consists of as high as 61 ppm S content (Sikka and Kansal, 1995). Elseewi et al. (1978) also observed the increased S content of plants after FA application. Singh et al. (2011) reported that FA and FYM amendments enhance the rates of N-transformation processes to release more plant available N in dry-tropical nutrient poor soils. They observed that the effect of all treatments was significant for soil inorganic-N ($NH_4^+$ and $NO_3^-$) concentration, N-mineralization, and nitrification rates over the period. Across all treatments and months, inorganic-N concentration, N-mineralization, and nitrification rates were greatest in FA + FYM-treated paddy field in the month of August. Wong and Lai (1996) recorded the N, P, and K contents of soil and their efficient uptake by paddy crop at all the growth stages on 20% FA (w/w) application. Bhattacharya and Chattopadhyay (2004) investigated the possibility of improving the N status in blend of FA and cow dung by using vermicomposting method. They used five combinations of FA and cow dung such as FA alone, cow dung alone, FA + cow dung (1:1), FA + cow dung (1:3), and FA + cow dung (3:1). These combinations were incubated with and without earthworms for 50 days. The bioavailable forms of N such as $NH_4^+$ and $NO_3^-$ tended to increase significantly in the combinations treated with earthworms. It may be due to increased microbial activity in the vermi-composted samples and also due to a significant rise in the concentration of nitrogen-fixing bacteria in this combination. In all the mixtures, FA + cow dung at 1:1 ratio looked to show the maximum availability of N content (Bhattacharya and Chattopadhyay, 2004).

## 4.5 Risk assessments

FA recycling in agriculture may pose health hazard, especially where the ash is exposed to water such as agricultural fertilizer, or a landfill or mine-fill material. These practices have some risk of water pollution through leaching into ground water or surface water. If FA is used in agriculture, then risk assessment is very essential because of the presence of heavy metals and other toxicants in FA, which may be absorbed in crops and may ultimately enter into the food chain. Biomonitoring and risk assessment of FA application in rice cropping is very necessary before the recommendation of commercial level is passed on. The risk assessment of FA application in agriculture may be divided in three categories, that is, agro-toxicological aspects of FA application in crops, boron toxicity on paddy crops, and DNA damage analysis.

### 4.5.1 Agrotoxicological aspects of fly ash application in paddy crops

The apprehensions regarding the quality assessment of FA grown paddy crop products and possible negative impacts due to heavy metals and radionuclides in FA have been addressed through in-depth studies. In Indian scenario, the comprehensive laboratory and clinical researches have been taken up at the National Institute of Nutrition, Indian Council of Medical Research, Ministry of Health, and GOI to evaluate the production of crop grown on FA-treated soil for toxicological impact. Rautaray et al. (2003) reported that heavy metal content such as Cd and Ni decreased in rice grain and straw under the FA application due to the increase in pH. Kim et al. (1994a) observed that the contents of Cd, Pb, and Ni did not increase in white rice at an FA application rate of 40, 80, or 120 Mg ha$^{-1}$. Furr et al. (1976) reported that the heavy metals did not increase in white rice with increasing application of FA and gypsum mixture, perhaps due to dilution and leaching effects, metal sulfide formation, etc. in the submerged paddy fields. In other evaluations too, the concentration of trace elements was found in the normal range and was below the critical levels in the tested crops; therefore the grains were safe for consumption (Patra et al., 2012). Bilski et al. (2012) reported that the transfer of heavy metals from FA to plants does not indicate the possibility of heavy metal transmission in the food chain, so there is no danger for human health. Bhaskarachary et al. (2012) demonstrated nutritional and toxicological aspects in rats by feeding rice grown in FA amended soils. Paddy crop was treated with FA ($200 \, t \, ha^{-1}$) and without FA at two different geographical regions of India. This study showed that no adverse effects were seen on hematological (red blood cell, white blood cell, and hemoglobin level), biochemical (serum alanine transaminase, serum aspartate transaminase, serum alkaline phosphatase, serum creatinine, and serum urea), or histopathological parameters (liver, spleen, kidneys, lungs, heart, testis, ovary, and brain) when rice was fed to rats for 6 months. Therefore it is anticipated that paddy grown on FA-treated soils may be safe for human consumption. Lee et al. (2002) prepared a mixture of FA and gypsum to decrease B toxicity in paddy soils and improving soil quality. Bhattacharya et al. (2012) showed that there is no difference in trace and heavy metals content in the rice samples, which was grown in FA-treated ($200 \, t \, ha^{-1}$) and control soils. However, sometimes heavy metals concentrate in rice crops and increase available B with increasing FA dose (alone or in combination with gypsum), within the permissible limits. These were attributed to the dilution and leaching effects under submerged growing conditions of paddy crops (Lee et al., 2002).

The heavy metal accumulation in rice grain cannot be generalized due to FA application in agricultural field, because it is governed by FA—soil—paddy crop interactions. The same FA may have different effects from soil to soil and plant to plant. Most of the rice studies revealed significantly enhanced the rice growth and yield under FA application without increasing the accumulation of heavy metals beyond toxicity levels. The main concern of FA application in agriculture is the presence of heavy metals, which directly depends on the nature of coal. Generally, foreign countries have low ash coal and metal-enriched FA, while Indian coal produces relatively high amount of FA with low heavy metal content. FA addition in soil forms iron oxyhydroxide (goethite) by ash—water interactions. The oxyhydroxide plays important role in scavenging trace metals, primarily through the formation of flocs, immobilizing them, and reducing their bioavailability (Leekie et al., 1980). A study has been done on the metal(loid) composition (As, Cd, and Ni) of various plant tissues (grain, seed husk, leaves, and roots) for securing health safety related to rice consumption (Tripathi et al., 2008). Presence of radioactive elements in FA is another health concern of animals and human being. Karmakar et al. (2009) observed that the radioactivity in treated rice crop and soil was either below the detection limit or remained under permissible limit. Tripathi et al. (2013) also studied the presence of natural radionuclides ($Ra^{226}$, $Th^{232}$, and $K^{40}$) in FA and their mobilization through soil amelioration with FA and crop production. The results revealed the variation in radionuclide contents in soil, FA, and crop production depending on soil and FA types. They also studied the mobilization of the radionuclides from FA amended soil to plant and interactions between radionuclides and soil. Finally, it was concluded that the content of the radionuclides in FA amended soil and crop production was within the permissible limits.

## 4.5.2 Boron toxicity and paddy crops

The use of FA as a commercial fertilizer in agriculture is unusual in most of the countries, because FA also contains toxic metal(loid)s (i.e., Cd, Cr, As, B, Se, etc.) that may harmfully affect soil quality and crop growth (Pandey and Singh, 2010; Pandey et al., 2011). But most researchers found that the effect of heavy metals of FA is insignificant in paddy crop at lower doses. But one of the most important issues for plants grown in FA amended soil is potential boron (B) toxicity due to the presence of significant level in FA. Although B is important to plant growth at μg concentrations, the difference between sufficiency and toxicity is the smallest among the micronutrient (Mengel and Kirkby, 1987). It causes toxicity in many plants when the level surpasses $1 \text{ mg kg}^{-1}$ (Ayers and Westcot, 1985). The fresh FA application may be responsible for B toxicity in some plants, but B toxicity was not found in plants grown on soils amended with weathered FA, because most of the plant available B is easily leached out of soil (Clark et al., 2001). Hodgson and Townsend (1973) suggested that crop selection is an important factor with respect to B toxicity of FA. Normally, 17%—64% of B in FA was water-leachable, but B toxicity was observed in many plants during 2—3 years after FA addition in upland soils ( James et al., 1982). Lee et al. (2006) observed that the amount of available B and B content of the rice plants increased to a maximum of 2.2—2.6 and $52—53 \text{ mg kg}^{-1}$, respectively, on the addition of $120 \text{ Mg ha}^{-1}$ FA. The paddy crops did not show B toxicity or an excessive uptake of heavy metals in the submerged paddy soil, perhaps due to dilution and leaching effects in the submerged paddy fields.

The maximum rice yields were recorded in around $90 \, Mg \, ha^{-1}$ FA-treated soil without showing toxicity effects (Lee et al., 2006). Boron is present in soils under various forms, that is, specifically adsorbed B (SPA-B), nonspecifically adsorbed B (NSA-B), B occluded in amorphous Fe and Al oxides (AMO-B), B occluded in crystalline Fe and Al oxides (CRO-B), B occluded in Mn oxyhydroxides (MOH-B), residual B (RES-B), and having a different availability to plant (Tsalidas et al., 1994). There are several factors for example organic matter, pH, Fe oxides, Al oxides and clay minerals, which are responsible for the transformations among the various forms of Boron (Xu et al., 2001). During B fractions, SPA-B, NSA-B, and MOH-B fractions are the most available to plants, while B fractions occluded in Fe and Al oxyhydroxides are relatively unavailable forms (Jin et al., 1987). Therefore the distribution of all B forms may be changed by the application of FA in paddy soil during rice cultivation. In this direction, Lee et al. (2008a) demonstrated the characteristics of B distribution in paddy field by FA application. B fraction in soils (hot water-extractable) increased with increasing FA doses and reached up to $55 \, mg \, kg^{-1}$ at $120 \, Mg \, ha^{-1}$ FA level, but B toxicity was not detected in rice (Lee et al., 2008a), because of the critical toxicity level of B in matured paddy crop is $100 \, mg \, kg^{-1}$ (FFTC, 2001). Most of the B from applied FA was in RES-B form, which contained >60% of the total soil B content, while labile B fractions (SPA-B, NSA-B, and MOH-B) contained <3% of total soil B content (Lee et al., 2008a). AMO-B fraction was not dependent on FA doses and comprised from 18.2% to 37.8%. MOH-B and CRO-B forms were not noticed. Thus FA appears as a potential soil ameliorant for raising the paddy crop productivity without B toxicity in the soil (Lee et al., 2008a).

### 4.5.3 Fly ash and DNA damage analysis

Recently, Markad et al. (2012) studied the effect of FA on DNA damage in earthworm (*Dichogaster curgensis*). They performed comet and neutral red retention assays on earthworm coelomocytes to assess genetic damages and lysosomal membrane stability. The comet assay reported the FA-induced DNA damage and DNA–protein crosslinks in earthworm coelomocytes. Likewise, Chakraborty and Mukherjee (2011) showed that there was no any damage in nuclear DNA in Vetiver (*Vetiveria zizanioides*) grown on FA, which suggests the long-term survival of the plant on the FA site. Comet assay may be used for initial, effective, and quick selection of plants for phytoremediation and restoration of contaminated sites. Chakraborty et al. (2009) studied on *Allium* bulbs that were grown to germinate directly in FA, and after five days, the germinating roots were processed for detecting DNA damage through comet assay. The results indicate that 100% FA level inhibits root growth and mitotic indices; and induces binucleated cells as a function of the proportion, but is not toxic at very low level. DNA strand breaks were observed only at higher concentrations by the comet assay. The *Allium* test can give a more reliable data in combination with the comet assay, which is simpler, faster, and efficient. When FA is used in agriculture for soil amelioration, it should be carefully used at very low levels in order to protect the agriculture system from any potential adverse effects (Chakraborty et al., 2009). FA verses DNA damage analysis in rice-based agroecosystem should be done in different microrelief for a perfect risk assessment and safe utilization of FA in agriculture.

## 4.6 Further research needs

There are some research gaps in the area of FA application in rice-based agroecosystem such as: (1) paddy—microbial interaction in FA amended soil systems; (2) mycorrhizal response in FA—paddy soil systems; (3) effect on microbial diversity and microbial DNA damage in FA amended paddy field; (4) investigating nutrient cycling and nutrient use efficiency in FA—paddy soil systems; (5) reducing the $CH_4$ flux in FA—paddy soil systems by manipulating the population of methanogen and methanotroph; and (6) assessment of FA-aided phytostabilization of arsenic contaminated paddy soils.

## 4.7 Conclusions

Several case studies have proved the beneficial role of FA application in paddy fields for increasing yield and soil fertility, as well as reducing methane emissions from paddy fields. The mixture of FA and phosphogypsum is more suitable for enhancing plant available phosphorus and rice productivity while decreasing P loss from flooded paddy fields. A number of studies of FA application, covering a wide range of soil types, paddy crop varieties, and agroclimatic conditions from pot to field scale, have recommended its slight use in agriculture. Long-term field trials with FA alone or coapplication in soil—paddy crop systems did not reveal any adverse effect rather beneficial effect on rice yield and soil quality. If FA were used continuously in paddy cropping, then monitoring of health-based risk assessment would be very necessary time to time in respect to heavy metals and radionuclides from safety point of view. The optimum dose of FA application for particular paddy soil and paddy crop variety should be established after repeated field studies. The information provided in this review may be used for developing a suitable agrobiotechnology for eco-friendly cultivation of rice by using FA as a future fertilizer for paddy soil. It will also help policy makers, researchers, and farmers to come up at one platform to make useful guidelines for exploiting FA potential for the amelioration of substandard soil sites to boost paddy production.

# References

Achtnich, C., Bak, F., Conrad, R., 1995. Competition for electron donors among nitrate reducers, ferric iron reducers, sulfate reducers and methanogens in anoxic paddy soil. Biol. Fertil. Soil. 19, 65−72.

Adriano, D.C., Page, A.L., Elseewi, A.A., Chang, A., Straughan, I.A., 1980. Utilization and disposal of fly ash and other coal residues in terrestrial ecosystem: a review. J. Environ. Qual. 9, 333−344.

Ahmaruzzaman, M., 2010. A review on the utilization of fly ash. Prog. Energ. Combust. Sci. 36, 327−363.

Ali, M.A., Oh, J.H., Kim, P.J., 2007. Suppression of methane emission from rice paddy soils with fly ash amendment. Korean J. Environ. Agric. 26 (2), 141−148.

Ayers, R.S., Westcot, D.W., 1985. Water quality for agriculture. FAO Irrigation Drainage Paper, No. 22−82.

Basu, M., Pande, M., Bhadoria, P.B.S., Mahapatra, S.C., 2009. Potential fly-ash utilization in agriculture: a global review. Prog. Nat. Sci. 19, 1173−1186.

Bhaskarachary, K., Ramulu, P., Udayasekhararao, P., Bapurao, S., Kamala, K., Syed, Q., et al., 2012. Chemical composition, nutritional and toxicological evaluation of rice (Oryza sativa) grown in fly ash amended soils. J. Sci. Food Agric. 92, 2721−2726.

Bhattacharya, S.S., Chattopadhyay, G.N., 2002. Increasing bioavailability of phosphorus from fly ash through vermicomposting. J. Environ. Qual. 31, 2116−2119.

Bhattacharya, S.S., Chattopadhyay, G.N., 2004. Transformation of nitrogen during vermicomposting in fly ash. Waste Manag. Res. 22, 488−491.

Bhattacharya, S.S., Iftikar, W., Sahariah, B., Chattopadhyay, G.N., 2012. Vermicomposting converts fly ash to enrich soil fertility and sustain crop growth in red and lateritic soils. Resour. Conserv. Recy. 65, 100−106.

Biliski, J.J., Alva, A.K., Sajwan, K.S., 1995. Fly ash. In: Rechcigl, J.E. (Ed.), Soil Amendments and Environmental Quality. Lewis Publishers, London, pp. 327−363.

Bilski, J., Jacob, D., Mclean, K., McLean, E., Mardee, S.F., 2012. Agro-toxicological aspects of coal fly ash (FA) phytoremediation by cereal crops: effects on plant germination, growth and trace elements accumulation. Adv. Biores. 3, 121−129.

Bisoi, S.S., Mishra, S.S., Barik, J., Panda, D., 2017. Effects of different treatments of fly ash and mining soil on growth and antioxidant protection of Indian wild rice. Int. J. Phytoremediat. 19 (5), 446−452.

Blevins, D.G., Lukaszewski, K.M., 1998. Boron in plant structure and function. Annu. Rev. Plant. Physiol. Plant Mol. Biol. 49, 481—500.

Bohn, M., McNeal, G., O'Connor, G., 1979. Soil chemistry. Wiley-Interscience, New York.

Campbell, D.J., Fox, W.E., Aitken, R.L., Bell, L.C., 1983. Physical characterization of sands amended with fly ash. Aust. J. Soil Res. 21, 147—154.

Chakraborty, R., Mukherjee, A., 2011. Technical note: vetiver can grow on coal fly ash without DNA damage. Int. J. Phytoremediat. 13, 206—214.

Chakraborty, R., Mukherjee, A.K., Mukherjee, A., 2009. Evaluation of genotoxicity of coal fly ash in Allium cepa root cells by combining comet assay with the Allium test. Environ. Monit. Assess. 153, 351—357.

Chatterjee, T., Mukhopadhya, M., Dutta, M., Gupta, S.K., 1988. Studies on some agro-chemical properties of fly ash. Clay Res. 7, 19—23.

Clark, R.B., Ritchey, K.D., Baligar, V.C., 2001. Benefits and constraints for use of FGD products on agriculture land. Fuel 80, 821—828.

Das, K.N., Patgiri, D.K., Basumatari, A., Das, K., Sharma, A., 2016. Direct and residual effects of fly ash based integrated fertilization on rice (Oryza Sativa L.) and rapeseed (Brassica Camprestris L.). Commun. Soil Sci. Plant Anal. 47 (6), 679—691.

Dean, L.A., 1949. Fixation of soil phosphorus. Adv. Agron. 1, 391—411.

Deren, C.W., Datnoff, L.E., Snyder, G.H., Mari, F.G., 1994. Silicon concentration, disease, and yield components of rice genotypes grown on flooded organic histosols. Crop. Sci. 34, 733—737.

Dwivedi, S., Tripathi, R.D., Srivastava, S., Mishra, S., Shukla, M.K., Tiwari, K.K., et al., 2007. Growth performance and biochemical responses of three rice (Oryza sativa L.) cultivars grown in fly-ash amendment soil. Chemosphere 67, 140—151.

Elseewi, A.A., Straughan, I.R., Page, A.L., 1978. Sequential cropping of fly ash-amended soils: effects on soil chemical properties and yield and elemental composition of plants. Sci. Total Environ. 20, 785—790.

Elseewi, A.A., Grimm, S.R., Page, A.L., Straughan, I.R., 1981. Boron enrichment of plants and soils treated with coal ash. J. Plant Nutr. 3, 409—427.

Epstein, E., 1994. The anomaly of silicon in plant biology. In: Proceedings of the National Academy of Sciences, USA, 91, pp. 11—17.

Epstein, E., 1999. Silicon. Annu. Rev. Plant Physiol. Plant Mol. Biol. United States National Academy of Sciences (United States), Washington, D.C., 50, pp. 641—664.

Fang, M., Wong, M.H., Wong, J.W.C., 2001. Digestion activity of thermophilic bacteria isolated from ash amended sewage sludge compost. Water Air Soil Pollut. 126, 1—12.

Fulekar, M.H., Dave, J.M., 1986. Disposal of fly ash-an environmental problem. Int. J. Environ. Stud. 26, 191—215.

Furr, A.K., Kelly, W.C., Bache, C.A., Gutenmann, W.H., Lisk, Dl, 1976. Multielement uptake by vegetables to millet growth in pots on fly ash amended soil. J. Agric. Food Chem. 24, 885—888.

Gu, H.H., Qiu, H., Tian, T., Zhan, S.S., Deng, T.H.B., Chaney, R.L., et al., 2011. Mitigation effects of silicon rich amendments on heavy metal accumulation in rice (Oryza sativa L.) planted on multi-metal contaminated acidic soil. Chemosphere 83, 1234—1240.

Guest, C.A., Johnston, C.T., King, J.J., Alleman, J.E., Tishmack, J.K., Norton, L.D., 2001. Chemical characterization of synthetic soil from composting coal combustion and pharmaceutical by-product. J. Environ. Qual. 30, 246—25.

Gupta, D.K., Rai, U.N., Tripathi, R.D., Inouhe, M., 2002. Impacts of fly ash on soil and plant responses. J. Plant Res. 115, 401—409.

Hingston, F.J., Posner, A.M., Quirk, J.P., 1972. Anion adsorption by goethite and gibbsite, I: the role of the proton in determining adsorption envelops. J. Soil Sci. 23, 177—192.

Hodgson, D.R., Townsend, W.N., 1973. The amelioration and revegetation of pulverized fuel ash. In: Chadwick, M.J., Goodman, G.T. (Eds.), Ecology and Reclamation of Devastated Land, Vol. 2. Gordon and Breach, London, pp. 247–270.

Hu, Dl, Wang, F.H., 1995. Silica nutrition of rice. Agric. Sci. Hubei 5, 33–36.

Irisarri, P., Gonnet, S., Monza, J., 2001. Cyanobacteria in Uruguayan rice fields: diversity, nitrogen fixing ability and tolerance to herbicides and combined nitrogen. J. Biotechnol. 91, 95–103.

Iyer, R.S., Scott, J.A., 2001. Power station fly ash—a review of value-added utilization outside of the construction industry. Resour. Conserv. Recy. 31, 217–228.

Izquierdo, M., Querol, X., 2012. Leaching behavior of elements from coal composition fly ash: an overview. Int. J. Coal Geol. 94, 54–66.

Jakobsen, P., Patrick Jr., W.H., Williams, B.G., 1981. Sulfide and methane formation in soils and sediments. Soil Sci. 132, 279–287.

Jala, S., Goyal, D., 2006. Fly ash as a soil ameliorant for improving crop production: a review. Bioresour. Technol. 97, 1136–1147.

James, W.D., Granham, C.C., Glascock, M.D., Hanna, A.S.G., 1982. Water-leachable boron from coal ashes. Environ. Sci. Technol. 16, 195–197.

Jayasinghe, G.Y., Tokashiki, Y., 2012. Influence of coal fly ash pellet aggregates on the growth and nutrient composition of *Brassica campestris* and physicochemical properties of greysoils in Okinawa. J. Plant Nutr. 35, 453–47.

Jepson, W.B., Jeffs, D.G., Ferris, A.P., 1976. The adsorption of silica on gibbsite and its relevance to the kaolinite surface. J. Colloid Interface Sci. 5, 454–461.

Jeyabal, A., Arivazhagan, K., Thanunathan, K., 2000. Utilisation of lignite fly ash as source of plant nutrient for rice. Fertiliser N. 45, 55–58.

Jin, J., Martens, D.C., Zelazny, L.W., 1987. Distribution and plant availability of soil boron fractions. Soil Sci. Soc. Am. J. 51, 1228–1231.

Karmakar, S., Mittra, B.N., Ghosh, B.C., 2009. Influence of industrial solid wastes on soil-plant interaction in rice under acid lateritic soil. World of Coal Ash (WOCA) Conference, May 4 – 7, 2009, Lexington, KY.

Karmakar, S., Mittra, B.N., Ghosh, B.C., 2010. Enriched coal ash utilization for augmenting production of rice under acid lateritic soil. Coal Combustion and Gasification. Products 2, 45–50.

Kim, B.Y., Lim, S.U., Park, J.H., 1994a. Influence of fly ash application on content of heavy metal in the soil. I. Content change by the application rate. J. Korean Soc. Soil Sci. Fert. 27, 65–71.

Kim, B.Y., Jung, G.B., Lim, S.U., Park, J.H., 1994b. Influence of fly ash application on content of heavy metals in the soil. II. Content change by the successive application. J. Korean Soc. Soil Sci. Fertilizer 27, 72–77.

Kim, B.Y., Jung, G.B., Lim, S.U., Park, J.H., 1994c. Influence of fly ash application on content of heavy metals in the soil. III. Content change in the rice and soybean by the application rate. J. Korean Soc. Soil Sci. Fertilizer 27, 220–225.

Ko, B.G., 2002. Effects of Fly Ash and Gypsum Application on Soil Improvement and Rice Cultivation (Ph.D. thesis). GNU, Jinju.

Kumar, D., Singh, B., 2003. The use of coal fly ash in sodic soil reclamation. Land. Degrad. Dev. 14, 285–299.

Kumar, A., Sarkar, A.K., Singh, R.P., Sharma, V.N., 1998. Yield and trace metal levels in rice (*Oryza sativa*) as influenced by fly ash, fertilizer and farmyard manure application. Indian J. Agric. Sci. 68, 590–592.

Kumar, T., Tedia, K., Samadhiya, V., Kumar, R., Kumar, H., Singh, A.K., 2018. Effect of different dose of fly ash with and without FYM on heavy metals status in soil and plant parts of rice. J. Pharmacogn. Phytochem. 7 (2), 1872–1876.

Lee, Y.B., Ha, H.S., Park, B.K., Cho, J.S., Kim, P.J., 2002. Effect of a fly ash and gypsum mixture on rice cultivation. Soil Sci. Plant Nutr. 48, 171–178.

Lee, Y.B., Ha, H.S., Lee, K.D., Park, K.D., Cho, Y.S., Kim, P.J., 2003. Evaluation of use of fly ash-gypsum mixture for rice production at different nitrogen rates. Soil Sci. Plant Nutr. 49, 69—76.

Lee, Y.B., Ha, H.S., Lee, C.H., Lee, H., Ha, B.H., Kim, P.J., 2005. Improving rice productivity and soil quality by coal ash-phosphogypsum mixture application. Korean J. Soil Sci. Fertilizer 38, 45—50.

Lee, H., Ha, H.S., Lee, C.H., Lee, Y.B., Kim, P.J., 2006. Fly-ash effect on improving soil properties and rice productivity in Korean paddy soils. Bioresour. Technol. 97, 1490—1497.

Lee, C.H., Lee, H., Lee, Y.B., Chang, H.H., Ali, M.A., Min, W., et al., 2007a. Increase of available phosphorus by fly-ash application in paddy soils. Commun. Soil Sci. Plant Anal. 38, 1551—1562.

Lee, C.H., Lee, Y.B., Lee, H., Kim, P.J., 2007b. Reducing phosphorus release from paddy soils by a fly ash-gypsum mixture. Bioresour. Technol. 98, 1980—1984.

Lee, S.B., Lee, Y.B., Lee, C.H., Hong, C.O., Kim, P.J., Yu, C., 2008a. Characteristics of boron accumulation by fly ash application in paddy soil. Bioresour. Technol. 99, 5928—5932.

Lee, Y.B., Ha, H.S., Lee, C.H., Kim, P.J., 2008b. Coal fly ash and phospho-gypsum mixture as an amendment to improve rice paddy soil fertility. Commun. Soil Sci. Plant Anal. 39, 1041—1055.

Leekie, J.O., Benjamin, M.M., Hayes, K.A., Altman, S., 1980. Adsorption/co-precipitation of trace elements from water with iron oxyhydroxide. Electric Power Research Institute, Palo Alto, CA. Final Report, EPRI RP-910.

Liang, Y.C., Sun, W.C., Zhu, Y.G., Christie, P., 2007. Mechanisms of silicon-mediated alleviation of abiotic stresses in higher plants: a review. Environ. Pollut. 147, 422—428.

Lim, S.S., Choi, W.J., Chang, S.X., Arshad, M.A., Yoon, K.S., Kim, H.Y., 2017. Soil carbon changes in paddy fields amended with fly ash. Agric. Ecosyst. Environ. 245, 11—21.

Lovely, D.R., Holmes, D.E., Nevin, K.P., 2004. Dissimilarity Fe (III) and Mn (IV) reduction. Adv. Microb. Physiol. 49, 219—286.

Ma, J.M., Yamaji, N., 2006. Silicon uptake and accumulation in higher plants. Trends Plant. Sci. 11 (8), 392—397.

Maiti, S.K., Jaiswal, S., 2008. Bioaccumulation and translocation of metals in the natural vegetation growing on fly ash lagoons: a field study from Santaldih thermal power plant, West Bengal, India. Environ. Monit. Assess. 136, 355—370.

Markad, V.L., Kodam, K.M., Ghole, V.S., 2012. Effect of fly ash on biochemical responses and DNA damage in earthworm, *Dichogaster curgensis*. J. Hazard. Mater. 215—216, 191—198.

Mengel, K., Kirkby, E.A., 1987. Further elements of importance. In: Mengel, K., Kirkby, E.A. (Eds.), Principles of Plant Nutrition, fourth ed. International Potash Institute, Bern, Switzerland, pp. 577—582.

Mishra, M., Sahu, R.K., Padhy, R.N., 2007. Growth, yield and elemental status of rice (*Oryza sativa*) grown in fly ash amended soils. Ecotoxicology 16, 271—278.

Mitrović, M., Jarić, S., Kostić, O., Gajić, G., Karadžić, B., Djurdjević, L., et al., 2012. Photosynthetic efficiency of four woody species growing on fly ash deposits of a Serbian 'Nikola Tesla-A' thermoelectric plant. Pol. J. Environ. Stud. 21, 1339—1347.

Mittra, B.N., Karmakar, S., Swain, D.K., Ghosh, B.C., 2005. Fly-ash a potential source of soil amendment and a component of integrated plant nutrient supply system. Fuel 84, 1447—1451.

Nayak, A.K., Raja, R., Rao, K.S., Shukla, A.K., Mohanty, S., Shahid, M., et al., 2015. Effect of fly ash application on soil microbial response and heavy metal accumulation in soil and rice plant. Ecotox. Environ. Safe 114, 257—262.

Neue, H.U., Roger, P.A., 1993. Rice agriculture: factors controlling emission. In: Khalil, M.A.K., Shearer, M. (Eds.), *Global* Atmospheric *Methane*. Springer-Verlag Berlin Heidelberg, NATO ASI/ARW series.

Neumann, D., Nieden, U., 2001. Silicon and heavy metal tolerance of higher plants. Phytochemistry 56, 685−692.

Obihara, C.H., Russell, E.W., 1972. Specific adsorption of silicate and phosphate by soils. J. Soil Sci. 23, 105−117.

Padhy, R.N., Nayak, N., Dash-Mohini, R.R., Rath, S., Sahu, R.K., 2016. Growth, metabolism and yield of rice cultivated in soils amended with fly ash and cyanobacteria and metal loads in plant parts. Rice Sci. 23 (1), 22−32.

Panda, D., Mandal, L., Barik, J., Mishra, S.S., Padhan, B., 2019. Improvement of growth, photosynthesis and antioxidant defense in rice (*Oryza sativa* L.) grown in fly ash-amended soil. Proc. Natl. Acad. Sci. India Sect. B Biol. Sci 89 (3), 853−860.

Pandey, V.C., 2012a. Invasive species based efficient green technology for phytoremediation of fly ash deposits. J. Geochem. Explor. 123, 13−18.

Pandey, V.C., 2012b. Phytoremediation of heavy metals from fly ash pond by *Azolla caroliniana*. Ecotoxicol. Environ. Saf. 82, 8−12.

Pandey, V.C., 2013. Suitability of *Ricinus communis* L. cultivation for phytoremediation of fly ash disposal sites. Ecol. Eng. 57, 336−341.

Pandey, V.C., Singh, N., 2010. Impact of fly ash incorporation in soil systems. Agric. Ecosyst. Environ. 136, 16−27.

Pandey, V.C., Singh, K., 2011. Is *Vigna radiata* suitable for the revegetation of fly ash landfills? Ecol. Eng. 37, 2105−2106.

Pandey, V.C., Singh, B., 2012. Rehabilitation of coal fly ash basins: current need to use ecological engineering. Ecol. Eng. 49, 190−192.

Pandey, V.C., Abhilash, P.C., Singh, N., 2009a. The Indian perspective of utilizing fly ash in phytoremediation, phytomanagement and biomass production. J. Environ. Manag. 90, 2943−2958.

Pandey, V.C., Abhilash, P.C., Upadhyay, R.N., Tewari, D.D., 2009b. Application of fly ash on the growth performance and translocation of toxic, heavy metals within *Cajanus cajan* L.: Implication for safe utilization of fly ash for agricultural production. J. Hazard. Mater. 166, 255−259.

Pandey, V.C., Singh, J.S., Kumar, A., Tewari, D.D., 2010. Accumulation of heavy metals by chickpea grown in fly ash treated soil: effect on antioxidants. Clean − Soil Air Water 38, 1116−1123.

Pandey, V.C., Singh, J.S., Singh, R.P., Singh, N., Yunus, M., 2011. Arsenic hazards in coal fly ash and its fate in Indian scenario. Resour. Conserv. Recy. 55, 819−835.

Pandey, V.C., Singh, K., Singh, R.P., Singh, B., 2012. Naturally growing *Saccharum munja* on the fly ash lagoons: a potential ecological engineer for the revegetation and stabilization. Ecol. Eng. 40, 95−99.

Pandey, V.C., Singh, J.S., Singh, D.P., Singh, R.P., 2013. Methanotrophs: promising bacteria for environmental remediation. Int. J. Environ. Sci. Technol. 11, 241−250.

Patra, K.C., Rautray, T.R., Nayak, P., 2012. Analysis of grains grown on fly ash treated soils. Appl. Radiat. Isotopes 70, 1797−1802.

Ram, L.C., 1992. Moessbauer spectroscopic and gamma radiolytic studies of some Indian coals. PhD thesis (unpublished), Banaras Hindu University, Varanasi, India.

Ram, L.C., Masto, R.E., 2010. Review: an appraisal of the potential use of fly ash for reclaiming coal mine spoil. J. Environ. Manag. 91, 603−617.

Ram, L.C., Masto, R.E., 2014. Fly ash for soil amelioration: a review on the influence of ash blending with inorganic and organic amendments. Earth Sci. Rev. 128, 52−74.

Ram, L.C., Srivastava, N.K., Tripathi, R.C., Jha, S.K., Sinha, A.K., Singh, G., et al., 2006. Management of mine spoil for crop productivity with lignite fly ash and biological amendments. J. Environ. Manag. 79, 173−187.

Ram, L.C., Srivastava, N.K., Jha, S.K., Sinha, A.K., Masto, R.E., Selvi, V.A., 2007. Management of lignite fly ash through its bulk use via biological amendments for improving the fertility and crop productivity of soil. Environ. Manag. 40, 438—452.

Ram, L.C., Masto, R.E., Singh, S., Tripathi, R.C., Jha, S.K., Srivastava, N.K., et al., 2011. An appraisal of coal fly ash soil amendment technology (FASAT) of central institute of mining and fuel research (CIMFR). World Academy of Science. Eng. Technol. 52, 703—714.

Rautaray, S.K., 2005. Nutrient dynamics, dehydrogenase activity, and response of the rice plant to fertilization sources in an acid lateritic soil. Acta Agric. Scand. Sect. B Soil Plant Sci. 55, 162—169.

Rautaray, S.K., Ghosh, B.C., Mittra, B.N., 2003. Effect of fly ash, organic wastes and chemical fertilizers on yield, nutrient uptake, heavy metal content and residual fertility in a rice-mustard cropping sequence under acid lateritic soils. Bioresour. Technol. 90, 275—283.

Rees, W.J., Sidrak, G.H., 1956. Plant nutrition on fly ash. Plant Soil 8, 141—159.

RDA, 1999. Fertilization standard of crop plants. NIAST, RDA: Suwon, Korea.

Roy, G., Joy, V.C., 2011. Dose-related effect of fly ash on edaphic properties in laterite crop land soil. Ecotoxicol. Environ. Saf. 74, 769—775.

Roy, A.C., Ali, M.Y., Fox, R.Y., Silva, J.A., 1971. Influence of calcium silicate on phosphate solubility and availability in Hawaiian latosols, In: Proceedings of International Symposium on Soil Fertility Evaluation; New Delhi, India, 757—765.

Sarangi, P.K., Mahakur, D., Mishra, P.C., 2001. Soil biochemical activity and growth response of rice *Oryza sativa* in fly ash amended soil. Bioresour. Technol. 76, 199—205.

Sarkar, A., Rano, R., Mishra, K.K., Sinha, I.N., 2005. Particle size distribution profile of some Indian fly ash: a comparative study to assess their possible uses. Fuel Proc. Technol. 86, 1221—1238.

Sarkar, A., Rano, R., Udaibhanu, G., Basu, A.K., 2006. A comprehensive characterization of fly ash from a thermal power plant in Eastern India. Fuel Process. Technol. 68, 259—277.

Sekiya, K., 1983. Phosphorus. Methods of Soil Analysis (Dojou Youbu Bunsekihou). Agriculture, Forestry and Fishery, Youkendou, Tokyo, pp. 225—257.

Seneviratne, S.I., Corte, T., Davin, E.L., Hirschi, M., Jaeger, E.B., Lehner, I., et al., 2010. Investigating soil moisture—climate interactions in a changing climate: a review. Earth Sci. Rev. 99, 125—161.

Shaheen, S.M., Hooda, P.S., Tsadilas, C.D., 2014. Opportunities and challenges in the use of coal fly ash for soil improvements-a review. J. Environ. Manag. 145, 249—267.

Sharpley, A.N., 1995. Dependence of runoff phosphorus on extractable soil phosphorus. J. Environ. Qual. 24, 920—926.

Sharpley, A.X., Meisinger, J.J., Breeuwsma, A., Sims, T., Daniel, T.C., Schepers, J.S., 1996. Impacts of animal manure management on ground and surface water quality. In: Hatfield, J. (Ed.), Effective management of animal waste as a soil resource. Lewis Publishers, Boca Raton, FL, pp. 1—50.

Shi, Q.H., Bao, Z.Y., Zhu, Z.J., He, Y., Qian, Q.Q., Yu, J.Q., 2005. Silicon-mediated alleviation of Mn toxicity in *Cucumis sativus* in relation to activities of superoxide dismutase and ascorbate peroxidase. Phytochemistry 66, 1551—1559.

Sikka, R., Kansal, B.D., 1994. Characterization of thermal power plant fly ash for agronomic purposes and to identify pollution hazards. Bio. Tech. 50, 269—273.

Sikka, R., Kansal, B.D., 1995. Effect of fly ash application on yield and nutrient composition of rice, wheat and on pH and available nutrient status of soil. Bioresour. Technol. 51, 199—203.

Singh, N.B., Singh, M., 1986. Effect of fly ash application on saline soil and on yield components, yield and uptake of NPK of rice and wheat at varying fertility levels. Ann. Agric. Res. 7, 245—257.

Singh, D.N., Kolay, P.K., 2002. Simulation of ash-water interaction and its influence on ash characterization. J. Prog. Energ. Combust. Sci. 28, 267—299.

Singh, L.P., Siddiqui, Z.A., 2003. Effects of fly ash and *Helminthosporium oryzae* on growth and yield of three cultivars of rice. Bioresour. Technol. 86, 73—78.

Singh, J.S., Pandey, V.C., 2013. Fly ash application in nutrient poor agriculture soils: impact on methanotrophs population dynamics and paddy yields. Ecotoxicol. Environ. Saf. 89, 43—51.

Singh, J.S., Pandey, V.C., Singh, D.P., 2011. Coal FA and farmyard manure amendments in dry-land paddy agriculture field: effect on N-dynamics and paddy productivity. Appl. Soil Ecol. 47, 133—140.

Singh, A., Sarkar, A., Agrawal, S.B., 2012. Assessing the potential impact of fly ash amendments on Indian paddy field with special emphasis on growth, yield, and grain quality of three rice cultivars. Environ. Monit. Assess. 184, 4799—4814.

Singh, P.K., Tripathi, P., Dwivedi, S., Awasthi, S., Shri, M., Chakrabarty, D., et al., 2015. Fly-ash augmented soil enhances heavy metal accumulation and phytotoxicity in rice (Oryza sativa L.); A concern for fly-ash amendments in agriculture sector. Plant. Growth Regul. Available from: https://doi.org/10.1007/s10725-015-0070-x.

Singh, K., Pandey, V.C., Singh, B., Patra, D.D., Singh, R.P., 2016a. Effect of fly ash on crop yield and physico-chemical, microbial and enzyme activities of sodic soils. Environ. Eng. Manag. J. 15 (11), 2433—2440.

Singh, P.K., Tripathi, P., Dwivedi, S., Awasthi, S., Shri, M., Chakrabarty, D., et al., 2016b. Fly-ash augmented soil enhances heavy metal accumulation and phytotoxicity in rice (Oryza sativa L.); a concern for fly-ash amendments in agriculture sector. Plant Growth Regul. 78 (1), 21—30.

Skousen, J., Ziemkiewicz, P., Yang, J.E., 2012. Use of coal combustion by-products in mine reclamation: review of case studies in the USA. Geosystem Eng. 15, 71—83.

Stout, W.L., Sharpley, A.N., Pionke, H.B., 1998. Reducing soil phosphorus solubility with coal combustion by-products. J. Environ. Qual. 27, 111—118.

Stout, W.L., Sharpley, A.N., Gburek, W.J., Pionke, H.B., 1999. Reducing phosphorus export from croplands with FBC fly ash and FGD. Fuel 78, 175—178.

Stout, W.L., Sharpley, A.N., Landa, J., 2000. Effectiveness of coal combustion by-products in controlling phosphorus export from soils. J. Environ. Qual. 29, 1239—1244.

Swain, D.K., Rautaray, S.K., Ghosh, B.C., 2007. Alkaline coal fly ash amendments are recommended for improving rice-peanut crops. Acta Agric. Scand. Sect. B Soil Plant Sci. 57 (3), 201—211.

Thind, H.S., Singh, Y., Singh, B., Singh, V., Sharma, S., Vashistha, M., et al., 2012. Land application of rice husk ash, bagasse ash, and coal fly ash: effects on crop productivity and nutrient uptake in rice—wheat system on alkaline loamy sand. Field Crop. Res. 155, 137—144.

Tripathi, R.D., Dwivedi, S., Shukla, M.K., Mishra, S., Srivastava, S., Singh, R., et al., 2008. Role of blue green algae biofertilizer in ameliorating the nitrogen demand and fly-ash stress to the growth and yield of rice (Oryza sativa L.) plants. Chemosphere 70, 1919—1929.

Tripathi, R.C., Jha, S.K., Ram, L.C., Vijayan, V., 2013. Fate of radionuclides present in Indian fly ashes on its application as soil ameliorant. Radiat. Prot. Dosimetry 156 (2), 198—206.

Tsalidas, C.D., Yassoglou, N., Kosmas, C.S., Kallianou, C.H., 1994. The availability of soil boron fractions to olive trees and barley and their relationships to soil properties. Plant Soil 162, 211—217.

Vijayan, V., Behera, S.N., Tripathy, P.S., Singh, G., 1999. Study of heavy metals in wheat and paddy grains. Int. J. PIXE 9, 365—371.

Watanabe, M., Kato, N., 1983. Research on the behavior of applied phosphorus fertilized in soil. Nat. Inst. Agric. Sci. Serv. 251, 1−31.

Wong, M.H., Wong, J., 1989. Germination and seedling growth of vegetable crops in fly ash-amended soils. Agric. Ecosyst. Environ. 26 (1), 23−35.

Wong, J.W.C., Lai, K.M., 1996. Effect of an artificial soil mix from coal fly ash and sewage sludge on soil microbial activity. Biol. Fertil. Soil. 23, 420−424.

Xu, J.M., Wang, K., Bell, R.W., Yang, Y.A., Huang, L.B., 2001. Soil boron fractions and their relationship to soil properties. Soil Sci. Am. J. 65, 133−138.

Yunusa, I.A.M., Eamus, D., De Silva, D.L., Murray, B.R., Burchett, M.D., Skilbeck, G. C., et al., 2006. Fly-ash: an exploitable resource for management of Australian agricultural soils. Fuel 85, 2337−2344.

# CHAPTER 5

# Microbial responses to fly ash—aided soil

## Contents

## 5.1 Overview to fly ash

Coal-based thermal power stations generate fly ash (FA) in huge quantities, which needs thousands of hectare land for its dumping. FA consists of very fine and glass-like particles ranging in size from 0.01 to 100 μm (Davison et al., 1974). FA physical properties depend on the type of coal, its ash content, method used for combustion, type of boiler, and collector setup. Some of the noticeable properties of FA are light texture, low bulk density, and high surface area (Pandey and Singh, 2010). The pH of FA varies from acidic to alkaline, which depends especially on the S content of parent coal. Importantly, majority of the FAs across the globe are alkaline (Ram and Masto, 2014). Incorporation of FA into soil has a substantial effect on soil pH, which is controlled by factors such as soil and FA pH, neutralizing capacity of FA, buffering capacity of soil, Ca or S content (molar ratio) in FA, and the content of other minor alkalis or alkaline earth cations (Izquierdo and Querol, 2012; Ram and Masto, 2014).

*Phytomanagement of Fly Ash*
ISBN: 978-0-12-818544-5
DOI: https://doi.org/10.1016/B978-0-12-818544-5.00005-5
**141**

FA contains micro- and macronutrients that are required by flora and fauna (Pandey and Singh, 2010). The macronutrients of FA include P, Ca, K, S, and Mg, whereas micronutrients of FA contain Fe, Mn, Cu, Zn, B, Mo, and Co. In addition, some toxic elements such as Cr, Cd, Pb, As, Ni, V, and Hg are also present in FA (Pandey and Singh, 2010; Pandey et al., 2011). The level of elements either nutrient or toxic in FA depends generally on some factors such as the coal type and the combustion technique (Singh et al., 2011a; 2016). FA can be used for the correction of sulfur and boron deficiency in acidic soils (Ram and Masto, 2014). Acidic components of soil are neutralized by the lime present in FA and in turn release nutrients, such as Mo, S, and B, in plant available forms. Several studies revealed to the potential use of FA to improve the fertility of barren soil for agriculture or revegetation (Pandey and Singh, 2010; Pandey et al., 2010; Singh et al., 2011c, 2016; Singh and Pandey, 2013; Srivastava et al., 2014). Moreover, some review articles also suggested that the safe utilization of FA as a soil amender is a potential area for solving sustainability issues and tapping economic possibilities (Pandey and Singh, 2010; Pandey et al., 2009a; Ram and Masto, 2014). A significant enhancement in soil fertility and nutrient uptake of oil seed crops (Selvakumari and Jayanthi, 1999), nutrient content (Ca, Mg, B, K, Zn, and S) of *Brassica* crops (Menon et al., 1993), and plant root biomass (Taylor and Schuman, 1988) were observed upon the addition of FA into soil. FA had also been used for decreasing P solubility in soil (Stout et al., 1999).

There are numerous works related to the application of FA as a soil amendment that was found to benefit a wide range of crops with respect to their growth and yield (Pandey et al., 2009a, 2010; Ram et al., 2006a, b; 2007a,b; Ram and Masto, 2010, 2014; Singh et al., 2011c, 2016; Singh and Pandey, 2013). In particular, the addition of sewage sludge−FA combination on agricultural soils was also recommended as a promising study that considered the safety, environmental protection, and agronomic production (Basu et al., 2009). Most of the research work done on FA application is primarily focused on its use as a soil conditioner in agricultural and soil science (Pandey and Singh, 2010), but limited studies are based on the activities of microbial communities in FA-aided soil (Ram and Masto, 2010). Soil microbes are vital to agroecosystem functions. They are involved in some important functions such as soil aggregation, humus formation, nutrient cycling, decomposition of various compounds, and other transformations (Wu et al., 2011). Managing the dose of FA application in soil is an urgent need to replace chemical fertilizers toward better agricultural practices. This chapter will describe some themes such as FA-aided soil and microbial sustainability, microbial activities in FA-aided soil (carbon mineralization, nitrogen mineralization, and phosphorus mineralization), enzyme activities in FA-aided soil, impact of FA-aided soil on methanotroph, and microbial monitoring in FA-aided soil in detail.

## 5.2 Fly ash–aided soil and microbial sustainability

Soil degradation and increasing chemical fertilizer cost are two important things that make marginal farmers unaffordable. Apart from this, the soil degradation due to the release of different kinds of manmade pollutants such as heavy metals, etc. through chemical fertilizer utilization in agriculture. Likewise, FA amendments can improve the physicochemical and biological characteristics of the soil; however, the negative aspect of FA is the presence of toxicants, which poses a serious threat to soil and human health. Lower doses of FA is beneficial to the quality of the soil and thereby several ecosystem services offered by the soil system toward a good quality of life and human well-being. Growing concern about agricultural sustainability and increasing chemical fertilizer cost to sustainable development are some of the other reliable reasons for the promotion of FA as a soil ameliorator. FA has micro- and macronutrients that are essential for plant growth and metabolism. It is also evident that crops use nutrients in larger amounts from soil in the current cultivation and need replenishment. Under such conditions, microbes offer good alternative tools to replenish plant nutrients. In agroecosystem, microbes play a significant role in the nutrient recycling, fixing, solubilizing, or mobilizing. These potential microbes present in fertile soil, but their numbers are often scanty in degraded soil. Lower amount of FA application in soil has been recommended to increase the microbial populations and to improve the soil quality (Pandey and Singh, 2010; Ram and Masto, 2014). There are a number of studies regarding the positive aspects of FA amendment in soil on microbial activities such as improving tyrosinase, peroxidase, and phenol oxidase (Amonette et al., 2003a; Masto et al., 2014), nitrogen mineralization and nitrification processes (Cervelli et al., 1986), microbial biomass carbon up to 2.5% of FA amendment (Nayak et al., 2014), and the bioavailability of P (Pathan et al., 2002; Urvashi et al., 2007).

Agriculture offers a foremost role to national income, employment to people, food security, and export earnings in the developing countries. However, in the current scenario, increasing land degradation due to over exploitation of land and ever-increasing use of chemical fertilizers is a major concern worldwide. This is the reason for reduced microbial populations and decreased nutrient contents. In this regard, FA addition in soil plays a major role in the improvement of the physicochemical and biological properties of the soil. However, FA dose depends on crop to crop and soil to soil (Pandey and Singh, 2010). Most of the researches on FA in the agriculture sector recommended its use as a fertilizer. "Sustainable agriculture involves the successful management of agricultural resources to assure human needs without harming environmental quality and conserving natural resources for future" (Pandey et al., 2017). In contrast with the traditional agricultural practices, sustainable agriculture has a wide-ranging concept and involves advanced approaches for the management of agricultural technology and practices. Currently, soil amendment technologies have been applied to remediate the problematic soils of agriculture, mainly nutrient poor soils. The recent attention of scientific community is to use potential soil ameliorators for the sustainable practices of agriculture. The differentiation between negative and positive aspects of soil ameliorators like FA for agricultural practices is based on how they affect the soil quality (mainly nutrient contents and microbial populations), and thereby the soil system improves crop growth and yield. The rhizospheric soil system has different groups of potential microbes that have positive effects on crop growth and their yield. In this regard, Kumar and Patra (2013) studied the effect of FA amended soil on plant growth and metal uptake by *Zea mays* in the presence and absence of plant growth promoting bacteria (PGPB). The bacterial strains used for doing the above study were *Bacillus* sp. BC29 and *Pseudomonas* sp. PS5 and PS14, which were isolated from FA polluted sites. Other properties like phosphate-solubilizing ability of the strains, including indole acetic acid, siderophores, and hydrogen cyanide acid producing potential, were also evaluated. The bacterial strains significantly improved the crop growth, while the highest growth and biomass were recorded with BC29 and PS14 strains at 25% of FA dose. In addition, the selected PGPB strains also protected the plants from the toxicity of FA. FA tolerant PGPB, FA tolerant fungi, and FA toxicant degrading microbes have been considered as helpful microbes for sustainable remediation and agricultural practices (Tiwari et al., 2008; Kumar and Patra, 2013; Tastan, 2017).

Microbial processes and their populations affect the soil fertility and structure in a different method. Each of which has an improving effect on the key soil-based limitations to yield, for instance.

- PGPB increase, the ability of nutrient acquisition by crops.
- Soil microbial community that mineralizes, decomposes, and makes nutrients' availability to plants.
- Microbial community that influences water-holding capacity, cation exchange capacity, soil acidity, soil toxicity, and the soil N, P, and S reserves.
- Bacteria and fungi that influence the soil structure and water regime.

## 5.3 Microbial processes in fly ash—aided soil

The impacts of FA amendments on physicochemical and biological properties of the soil have been broadly studied. FA has been used in agriculture, forestry, and degraded land reclamation due to its nutrient supplying and soil ameliorating capacities (Pandey and Singh, 2010). Being an excellent ameliorant of soil, FA has received a great attention over the decades (Basu et al., 2009; Jala and Goyal, 2006; Juwarkar and Jambhulkar 2008; Kalra et al., 2000; Krzaklewski et al., 2012; Kumari et al., 2013; Pandey et al., 2009a; Ram et al., 2006b, 2007b; Ram and Masto, 2014; Sajwan et al., 2003; Tripathi et al., 2009, 2010). A lot of information is available on the impacts of FA on the physicochemical properties, but there is not much work regarding the impacts of FA on the microbial community and its activities in FA-aided soil (Ram and Masto, 2010). In this direction, Rippon and Wood (1975) reported heightened microbial counts in soil amended with FA. However, Arthur et al. (1984) reported a modest microbial activity under lower doses of FA in soil, whereas the activity was inhibited under higher doses of FA. Wong and Wong (1986) reported increased microbial respiration in sandy soil unlike sandy loam soil after the application of FA. An improvement was reported in microbial community numbers in degraded subsoil amended with FA, which might have benefitted gram-negative bacteria along with fungi compared with the other microbes in soil (Schutter and Fuhrmann, 2001). Gaind and Gaur (2004) found that *Azospirillum brasilense*, *Azotobacter chroococcum*, and *Bacillus circulans* are tolerant to the toxicities of raw FA, whereas *Pseudomonas striata* survives mostly in FA amended with soil at 1:1 (w/w) ratio. Some studies also recommended lower rates of FA as a beneficial dose for augmenting microbial activities, but higher rates of FA were noted as a lethal dose that inhibited microorganisms (Ram et al., 2006b, 2007b; Ram and Masto, 2010, 2014). The toxicity of higher rates of FA application is mainly due to the excessive amounts of toxic elements (like B and As) that are added into the soil from FA, which adversely affects the normal soil microbial community (Lim and Choi, 2014). FA application also impacts soil aggregation, which, in addition to the positive effect of plant growth on soil microbial diversity, may favor soil revitalization and increased plant growth. García-Sánchez et al. (2015) revealed that FA did not provoke significant changes in soil microbial and enzymatic activities when applied to the soil. Moreover, the qPCR and phospholipid fatty acid analysis (PLFA) quantification methods also showed that FA application did not modify soil microbial biomass and community structure. Therefore FA may be proposed as a fertilizer, since its impact on soil was not relevant. Bioindicators are biological species of living soil that can be used to monitor the health of ecosystem. Such bioindicators usually respond more rapidly to change in the soil environment compared with physicochemical indicators (Singh et al., 2011b). Microbes in soil ecosystem incur organic matter mineralization and increase their availability for other soil microbes. Mineralization is an important biological process, which involves the conversion of organic substances into inorganic substances by soil microorganisms, whereas decomposition could be due to physicochemical or biological processes. Such a soil microbial activity favors the formation of soil aggregates, cycling of nutrients, stabilization of heavy metals, and stimulation of the activity of soil enzymes, namely dehydrogenase, urease, phosphatases, and so on (Rippon and Wood, 1975; Pati and Sahu, 2004). The microbial activities in FA amended soil can be defined in detail into three major parts such as carbon mineralization, nitrogen mineralization, and phosphorus mineralization.

## 5.3.1 Carbon mineralization

It is well known that FA addition initiates an improvement in degraded lands by various mechanisms. Furthermore, it also plays an important role in the stabilization of the soil carbon. Carbon mineralization by the addition of FA in soil has been studied in detail (Masto et al., 2014; Nayak et al., 2014). The study of humification and litter decomposition is used to understand the carbon sequestration potential of FA. Tyrosinase is a soil enzyme responsible for carbon sequestration. Some different alkaline FAs enhanced humification (Amonette et al., 2003a) and stabilized polyphenol oxidase enzymes (i.e., tyrosinase, peroxidase, and phenol oxidase) (Amonette et al., 2003a; Masto et al., 2014). Furthermore, they also notified that FA plays as a catalyst for soil humification mainly through tyrosinase stabilization, monomers oxidation, and promotion of oxidation and condensation reactions by the alkaline pH (Amonette et al., 2003b). During the carbon mineralization, higher activities of carbon stabilization enzymes, like tyrosinase and peroxidase, were observed in soil amended with FA compared with the soil alone. Microbial biomass carbon (MBC) was higher in FA amended soil, whereas dissolved organic carbon was very low due to safe carbon sorption into FA. Importantly, Indian FAs can effectively stabilize organic carbon within the soil (Masto et al., 2014). The humification of organic compounds in soil and FA matrices has been shown in Fig. 5.1 (Masto et al., 2014).

Carbon mineralization is a sensitive and reliable method for the study of soil microbial activity, which shows the rate of carbon fixation along with the pollutant toxicity (Torstensson and Stenström, 1986). The different rates of $CO_2$ evolution from soil denote changes, especially in the carbon cycle during the microbial degradation of carbonaceous compounds. Carbon cycling also helps in mineralization of plant nutrients such as N, P, K, and S. Microbial biomass constitutes 1%—5% of the total soil

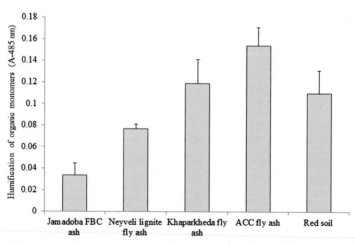

**Figure 5.1** Carbon mineralization (mg $CO_2$—C $kg^{-1}$ soil) by different doses of FA application. *Source: Data from Nayak, A.K., Kumar, A., Raja, R., Rao, K.S., Mohanty, S., Shahid, M., et al., 2014. Fly ash addition affects microbial biomass and carbon mineralization in agricultural soils. Bull. Environ. Contam. Toxicol. 92 (2), pp. 160—164. <http:// doi.org/10.1007/s00128-013-1182-5>.*

organic matter and is the most active fraction (Nsabimana et al., 2004). It serves as a potential indicator of changes in soil organic carbon (Cookson et al., 2007; Huang and Song, 2010). The alteration in MBC can provide valuable information on the impact of FA on the soil quality. MBC is also a part of soil organic matter that demonstrates the soil quality (Doran and Parkin, 1994) and soil microbiological status (Nannipieri et al., 1990) and predicts the pollutant degradation capacity (Voos and Groffman, 1997). MBC is used as a tool in field management practices (Perrott et al., 1992), nutrient applications (Nayak et al., 2012), pesticide applications (Kumar et al., 2012), and pollution monitoring (Powlson, 1994).

Some potential indexes of microbial activity, such as basal respiration rate ($CO_2$ evolution) and carbon mineralization, also serve as indicators for assessing the soil biological activity and quality due to their sensitivity to soil stress. This stress may be the FA addition in the soil system. For the consideration of FA as a soil ameliorant, its positive and negative impacts on soil microbial processes need to be assessed to know its use toward agricultural sustainability in terms of soil fertility and crop productivity. Microbial activity is an important bioindicator for the soil quality (Gregorich et al., 1994), which in turn is measured by carbon mineralization (Gray, 1990). MBC and carbon mineralization can be used for assessing the side effects of stress like FA in the soil system. In this direction, Nayak et al. (2014) studied the effect of 0% − 20% of FA application on soil MBC and carbon mineralization in laboratory conditions for 4 months at 60% of soil water-holding capacity and 25°C temperature. The FA application of 2.5% did not reduce soil respiration and microbial activities, which was significantly observed at 10% − 20% of FA addition compared with the control. The later also decreased the soil MBC with time. Thus the FA application of 2.5% can be safe for agricultural practices, as it can improve nutrient cycling in soils. Carbon mineralization depends on not only the doses of FA but also the period of exposure to the microbes. Fig. 5.2 shows the carbon mineralization at different days of incubation under different FA stress conditions (Nayak et al., 2014). It indicates that the FA application from 0.5% to 2.5% is not adequate to suppress the activities of soil microbes. Ogut et al. (2009) observed that FA with low salinity and neutral pH does not affect soil carbon mineralization. Soil carbon mineralization also significantly declined in a poorly buffered soil amended with 5% of unweathered FA of pH 12 (Pichtel, 1990; Pichtel and Hayes, 1990). The addition of unweathered acidic FA at a rate of 100 t ha$^{-1}$ did not affect $CO_2$−C production, which declined at the dose

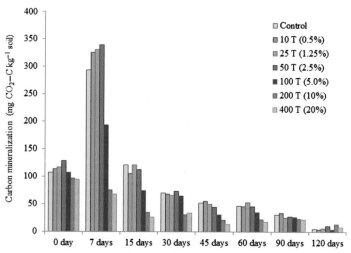

**Figure 5.2** Humification process under soil and different FA matrices. *Source: Data from Masto, R.E., Sengupta, T., George, J., Ram, L.C., Sunar, K.K., Selvi, V.A., et al., 2014. The Impact of Fly Ash Amendment on Soil Carbon. Energ. Source. Part A, 36, pp. 554—562. http://dx.doi.org/10.1080/15567036.2010.544004.*

of 400 and 700 t ha$^{-1}$ (Arthur et al., 1984). Pati and Sahu (2004) also observed less impact of FA addition (up to 2.5%) on $CO_2$ evolution, after which it significantly reduced $CO_2$ evolution and MBC. Inhibition of carbon mineralization mainly occurs due to amorphous aluminosilicates and unburnt carbon present in FA (Arbestain et al., 2009).

## 5.3.2 Nitrogen mineralization

Nitrogen mineralization is an important process and vital part of soil fertility. It is the process by which organic nitrogen is converted to plant available inorganic forms. It is regarded as a potential indicator to comprehend the soil's response to biological change (Stamatiadis et al., 1999). Studies regarding soil nitrogen mineralization with respect to FA application are less and limited to laboratory conditions. Therefore the FA impact on soil nitrogen cycling is very important, because FA has been identified as a potential soil modifier to increase soil fertility. A small study revealed that the amendment of soils with FA can enhance nitrogen mineralization and the nitrification processes (Cervelli et al., 1986). Figs. 5.3 and 5.4 represent the $NH_4^+-N$ and $NO_3-N$ concentrations in the Lamporecchio soils amended with different doses of Poland FA along with farmyard manure (FYM; Cervelli et al., 1987), and it is observed that mineralization

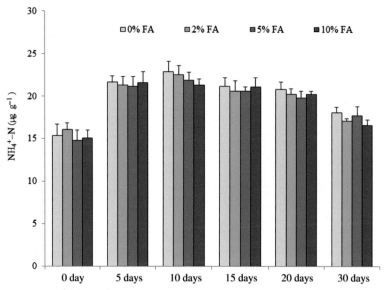

**Figure 5.3** $NH_4^+-N$ concentrations in the soils amended with different FA doses. *Source: Data from Cervelli, S., Petruzzelli, G., Perna, A., 1987. Fly ashes as an amendment in cultivated soils I. Effect on mineralization and nitrification. Water Air Soil Pollut. 33, pp. 331–338.*

of organic N contents of FYM led to higher value of inorganic N (Savant and De Datta, 1982; Ram et al., 2006b, 2007b; Singh et al., 2011b). Higher doses of FA application bring an adverse effect on soil microbes that perform nitrogen fixation and nodule formation (Cheung et al., 2000). Likely, the FA application rate of 40% was not found to bring any significant change in N mineralization compared with the control, and moreover, at 100% FA application, the mineralization rate was significantly reduced (Mishra et al., 2007). However, a positive trend in the nitrification potential of soil was found at an FA application rate of 40%, but no corresponding increase in nitrifier population together with a decreasing trend in denitrifying enzyme activity was reported. Some studies have reported a decrease in mineralization, nitrification, and denitrification potentials of nitrogen in soil after the addition of FA up to $50\ t\ ha^{-1}$. Laboratory studies also indicated a reduced nitrification potential in FA amended soil (Garau et al., 1991).

## 5.3.3 Phosphorus mineralization

Phosphorus is essential for plant nutrition followed by nitrogen. The key metabolic processes guided by it are signal transduction, photosynthesis,

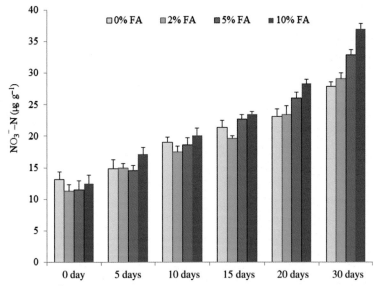

**Figure 5.4** $NO_3$—N concentrations in the Lamporecchio soils altered different FA doses. *Source: Data from Cervelli, S., Petruzzelli, G., Perna, A., 1987. Fly ashes as an amendment in cultivated soils I. Effect on mineralization and nitrification. Water Air Soil Pollut. 33, pp. 331—338.*

respiration, energy transfer, nitrogen fixation, and macromolecular biosynthesis (Saber et al., 2005). Usually, a substantial amount of phosphorus is present in soil, ranging from 400 to 1200 mg kg$^{-1}$, which is either in organic forms, that is, inositol phosphate (soil phytate), phosphomonoesters, phosphodiesters, and phosphotriesters, or in mineral forms such as apatite, hydroxyapatite, and oxyapatite (Ahemad et al., 2009), yet it becomes a limiting factor for plant growth due to its unavailability to the plant roots. The soluble form of phosphorus is usually present at a maximum limit of 1 mg kg$^{-1}$ of soil (Goldstein, 1994), while the inorganic form has a limited availability due to high reactivity of the phosphate ions in soil (Rengel and Marschner, 2005).

To avoid this scarcity, phosphate fertilizers are applied to soils. However, the forms of phosphorus in the fertilizers are in a precipitated state and hence become less available for root absorption, which adversely impacts soil health and also degrades both terrestrial and aquatic resources. Phosphate-solubilizing bacteria are eco-friendly and economical tools toward the exploitation of their phosphorus solubilization and mineralization traits to provide a significant quantity of soluble phosphorus for plant growth. Microbial-mediated phosphorus management is a cost-effective and eco-friendly approach for sustainable agriculture. Phosphorus solubilization potential of microorganisms was reported by Pikovskaya (1948), which was significantly increased by the studies of Rodríguez and Fraga (1999), Misra et al. (2012), and Oves et al. (2013). Various strains of bacteria and fungi have been identified on the basis of their phosphate-solubilizing potential (Glick, 1995); to name a few are bacterial species of *Pseudomonas* and *Bacillus* (Illmer and Schinner, 1992; Tiwari et al., 2008; Kumari and Singh, 2011) and fungal species of *Aspergillus* and *Penicillium* (Wakelin et al., 2004). In soil, phosphorus-solubilizing bacteria constitute 1%−50% of the total bacterial population (Zaidi et al., 2009). In addition to *Pseudomonas* and *Bacillus*, some other important bacteria have been noted as phosphorus solubilizers such as *Xanthomonas* (De Freitas et al., 1997), *Azotobacter* (Kumar et al., 2001), *Vibrio proteolyticus, Xanthobacter agilis* (Vazquez et al., 2000), *Pantoea, Enterobacter, Klebsiella* (Chung et al., 2005), *Arthrobacter, Rhodococcus, Chryseobacterium, Serratia, Gordonia, Delftia* sp., and *Phyllobacterium* (Wani et al., 2005; Chen et al., 2006). Some symbiotic nitrogenous *rhizobia* have also revealed phosphate-solubilization activities (Zaidi et al., 2009). For instance, *Rhizobium* isolated from *Crotalaria* (Sridevi et al., 2007) and *Rhizobium leguminosarum* bv trifolii (Abril et al., 2007) was found to enhance plant phosphorus nutrition by mobilizing the inorganic and organic forms. A variety of phosphorus solubilizers have also been identified from stressed habitats, for example, *Kushneria sinocarni*, which are halophilic bacteria obtained from the eastern coastal area of China, and can be used in other salt-affected areas (Zhu et al., 2011). The percentage of phosphorus-solubilizing fungi in soil is approximately 0.1%−0.5% out of the total fungal population. They can traverse long distances unlike bacteria and are more significant for phosphorus solubilization in soils (Kucey, 1983).

In addition, the phosphate-solubilizing organisms possess multifarious plant growth promoting characteristics along with the detoxifying ability for a toxic metal that can assist plants for phytoremediation by promoting their growth and health under hazardous conditions as present in coal FA due to the presence of toxic levels of metals that often decline plant growth (Kumar et al., 2008; Rajkumar and Freitas, 2008; Kumari and Singh, 2011). These microorganisms remediate metal-contaminated soils mainly by facilitating phytoextraction (metal accumulation in plant tissues) or phytostabilization (decreasing metal toxicity by transforming them into immobile forms). Secretion of siderophores, organic acid, 1-aminocyclopropane-1-carboxylic acid (ACC) deaminase, and indole 3 acetic acid (IAA) by the phosphate-solubilizing microorganisms enhances the phytoremediation potential of plants (Ahemad, 2015).

Furthermore, there is an urgent need to assess certain amendments that can mineralize the organic forms of phosphorus and orthophosphates and facilitate their release from the bounded forms to be available to the plant roots for absorption. All the applied phosphorus in the field is unavailable to the roots instantly due to their adsorption to soil particles or loss through runoff, and leaching or binding in organic forms. A spike in soil pH increases the activity of phosphatase and the subsequent mineralization of organic forms of phosphorus (Fuentes et al., 2006). Thus the pH increase is the most important characteristic of FA for the mineralization of organic phosphorus and the immobilization of inorganic phosphorus (Seshadri et al., 2013). The alkaline by-product FA can potentially decrease the loss of soil phosphorus and increase its utilization by the plants (Stout et al., 2000). Thus FA as an amendment can be used on soil, which will not only be cost effective but also reduce the hazard to the environment due to its storage. This aspect of improvement in the bio-availability of phosphorus in soil due to FA amendments is being steadily explored by the scientific community (Pathan et al., 2002; Urvashi et al., 2007). Pathan et al. (2002) reported an increase in bioavailable phosphorus from 18 to 43 $\mu g \, g^{-1}$ in soil due to FA amendments under laboratory conditions.

## 5.4 Enzyme activities in fly ash–aided soil

The soil enzymes are the main factors for the measurement of biological properties that play a significant biochemical role in decomposition of organic matter (Burns, 1982; Sinsabaugh et al., 1991). They catalyze many essential reactions related to the life processes of soil microbes. The soil enzymatic activities include nutrient cycling, stabilizing the soil structure, organic matter formation, and decomposition of organic wastes (Dick et al., 1994). The function of soil enzymes is a continuous process in which they are synthesized, accumulated, and decomposed to play a vital role in agriculture (Tabatabai, 1994; Dick, 1997). A group of enzymes present in all types of soils does metabolic processes, depending on the physicochemical, biological, and biochemical properties of the soils (McLaren, 1975). Soil enzymes are helpful to know the changes in soil management practices, because they respond quickly and are easy to measure. An understanding about their functions in the soil system potentially

**Table 5.1** Soil enzymes as indicators of soil quality and health.

| Soil enzyme | Enzyme reaction | Indicator of soil function | Significance |
|---|---|---|---|
| Dehydrogenase | Electron transport system | C–cycling | Pollution in soils |
| β–Glucosidase | Cellobiose hydrolysis | C–cycling | Energy for microorganisms |
| Urease | Urea hydrolysis | N–cycling | Plant available $NH_4$ |
| Amylase | N mineralization | N–cycling | Plant available $NH_4$ |
| Phosphatase | Release of $PO_4^-$ | P–cycling | Plant available P |
| Arylsulphatase | Release of $SO_4$ | S–cycling | Plant available S |

helps in evaluating the main biological status of soils (Bandick and Dick, 1999). The soil enzymes also affect soil management strategies along with the growth and nutrient uptake of plants. The enzyme activity is mainly essential for soil fertility, which can be measured on a regular basis to obtain this biological index, which indicates the biological processes occurring in the soil. They also reveal the effect on soil due to agricultural activities and fertilizer applications (Skujins, 1978). Therefore soil enzymes can be used as indicators to know the status of the soil quality. Soil enzymes as indicators of the soil quality and health are presented in Table 5.1.

FA incorporation into soil stimulates enzyme activities, that is, urease, phosphatases, and dehydrogenase (Pati and Sahu, 2004). Amendment of soils with FA incorporates elements, such as Ca, Mg, K, Mn, Cu, and Zn, into the soil system that alters the physicochemical properties (Yeledhalli et al., 2007). FA amendment in soil up to 10 t ha$^{-1}$ significantly increases the $CO_2$ evolution and soil enzyme (protease and dehydrogenase) activities that were found favorable for microbial activities (Sarangi et al., 2001). However, Pichtel and Hayes (1990) reported that increasing the dose of FA into the soil decreased the enzyme activities of sulfatase, phosphatase, invertase, and dehydrogenase. Yet FA addition into soil may benefit the gram-negative bacteria, fungi along with other soil microbes (Schutter and Fuhrmann, 2001). The highest level of dehydrogenase activity in soil amended with FA was reported at a dose of 10% of FA in the soil on w/w basis, because this amendment rate adds sufficient nutrients into the soil, which helps microbes for carrying out many metabolic activities (Wong and Wong, 1986; Saffigna et al., 1989). Higher dosage of FA tends to reduce substrate availability due to the accumulation of lignite-derived organic carbon compounds (Rumpel et al., 1998). Soil enzyme activities in FA-aided soil are presented in Table 5.2.

## 5.5 Impact of fly ash—aided soil on methanotroph

Methanotroph is a ubiquitous and unique prokaryote that uses atmospheric methane as the only carbon and energy source. They can be exploited for the remediation of a wide range of pollutants, including heavy metals and the organic pollutants, by virtue of their unique enzyme system, especially methane monooxygenases. They have the potential to exist in wide-ranging habitats from aerobically to anaerobically conditions owing to its physiologically versatile nature and need single-carbon compounds to survive. Therefore they are the only recognized important biological sinks for atmospheric methane and play a key role in decreasing about 15% of methane load to the total global methane destruction. Methane is a potential greenhouse gas that causes global warming (Pandey et al., 2014). Singh and Pandey (2013) showed that low doses of FA with press mud enhance the population of methanotroph in the soil and paddy productivity in nutrient poor soils. Methanotrophs metabolize methane for their carbon and energy with the help of a broad-spectrum methane monooxygenase enzyme (Pandey et al., 2014). It was determined that lower doses of FA decline methane emissions from the paddy plots by increasing the population of methanotroph. This short-term field trial shows an ample scope for the FA use with organic manures to enhance methanotrophs in dry-tropical zones.

## 5.6 Monitoring of microbial communities at genetic level in fly ash—aided soil

A traditional approach for bacterial identification is often not sufficient for monitoring specific microbes in a mixed microbial community, which are importantly responsible for significant biochemical reactions. However, numerous physiological (community-level physiological profiling), biochemical PLFA, and molecular (microbial DNA fingerprinting) techniques (culture-independent techniques) are currently used to study the soil microbial diversity. In this context, the monitoring study of microbial

**Table 5.2** Soil enzymes in fly ash–aided soil.

| S. no. | Soil enzyme | Remark | Reference |
|---|---|---|---|
| 1. | Dehydrogenase | Soil microbial biomass and dehydrogenase activity were reported to be highest at an FA amendment rate of 10% (w/w) | Wong and Wong (1986); Saffigna et al. (1989) |
| 2. | Dehydrogenase Phosphatase Sulfatase | Activities of such enzymes were inhibited, while the FA treatment level was increased | Pichtel and Hayes (1990) |
| 3. | Urease Cellulase | Fly ash added into soil at 16% (w/w) increases urease and cellulase enzyme activities | Lal et al. (1996) |
| 4. | Dehydrogenase Urease Phosphatase | Dehydrogenase and phosphatase activities were the highest at 5% of ash–sludge mixture amended soil than other treatments. Soil receiving 10% of ash–sludge mixture had the highest urease activity | Lai et al. (1999) |
| 5. | Dehydrogenase Protease Amylase | Little or no inhibition of soil enzyme activities up to 2.5% of FA amendment. With further addition of FA, all these activities were significantly decreased | Pati and Sahu (2004) |
| 6. | Dehydrogenase Urease Alkaline phosphatase | Soil enzyme activities were decreased with an increase in the FA dose application | Yeledhalli et al. (2007) |
| 7. | Dehydrogenase Alkaline phosphatase β-glucosidase | The impact of amendments (FA and press mud) was significant for enzymatic activities and paddy yields. Studies revealed a positive effect on soil enzyme activities of soil when amended with FA at 20 t ha$^{-1}$ | Singh and Pandey (2013) |
| 8. | Alkaline phosphatase Acid phosphatase | Activities of these enzymes were decreased with the application of FA | Nayak et al. (2015) |

communities at a genetic level in FA-aided soil is very insufficient and requires a detailed study during the large application of FA in the agricultural use. The application of FA at the rate of $40 \, t \, ha^{-1}$ along with the phosphate solubilizer *P. striata* can improve bean crop production and the subsequent phosphorous uptake in its grains, and ensure no negative effect on the *P. striata* population in soil (Gaind and Gaur, 2004). In a study, the behavior of transformed phosphate-solubilizing bacteria ($S_2$) pMMB277 was monitored in pure soil and FA amended soils (0%—30% of FA on w/ w basis) over a period of 28 days. The study showed that the FA application rate of 10% along with the above phosphate-solubilizing bacteria can have the best impact on their population and also provide a favorable condition for the growth of native bacteria (Kumar, 2004).

Plant growth promoting bacterial strains (NBRI K24 and NBRI K3) isolated from FA-contaminated soils can reduce Ni and Cr toxicities in *Brassica juncea* and promote its growth as observed in laboratory scale experiments. Characterization of the strains on the basis of their 16 S rDNA sequences revealed them to be *Enterobacter aerogenes* and *Rahnella aquatilis*, respectively. They were capable of producing siderophores to enhance metal phytoextraction by the above plant. Concurrently, the production of ACC deaminase and indole acetic acid, and the solubilization of phosphate by the bacteria were responsible for promoting plant growth (Kumar et al., 2009). Recently, Patel and Thakare (2015) isolated and identified the metal resistant bacterial species from FA-contaminated soil. The resistance strains showed the presence of multiple plasmids and identified species using 16SrDNA, which belongs to the genera *Pseudomonas*, *Bacillus*, *Achromobacter*, *Aeromonas*, *Klebsiella*, and *Staphylococcus*. Therefore it is clear that there are very limited works regarding microbial monitoring at a genetic level in FA amended soil and have wide scope in future.

## 5.7 Conclusion

In conclusion, FA addition in soil affects not only the soil physicochemical properties but also the microbial and enzyme activities. Usually, FA amendment in soil in lower doses have positive impacts on microbial communities (structure, function, and diversity), while higher doses of FA have harmful effects on the health of soil system. The microbial biomass, population, and values of potential soil enzyme activities mostly reduce with increasing the FA doses in soil.

# References

Abril, A., Zurdo-Pineiro, J.L., Peix, A., Rivas, R., Velazquez, E., 2007. Solubilization of phosphate by a strain of *Rhizobium leguminosarum* bv. Trifolii isolated from Phaseolus vulgaris in El Chaco Arido soil (Argentina). In: Velazquez, E., Rodriguez-Berrueco, C. (Eds.), Developments in Plant and Soil Sciences. Springer, The Netherlands, pp. 135–138.

Ahemad, M., 2015. Phosphate-solubilizing bacteria-assisted phytoremediation of metalliferous soils: a review. 3 Biotech. 5, 111–121.

Ahemad, M., Zaidi, A., Khan, M.S., Oves, M., 2009. Biological importance of phosphorus and phosphate solubilizing microbes. In: Khan, M.S., Zaidi, A. (Eds.), Phosphate Solubilizing Microbes for Crop Improvement. Nova Science Publishers Inc, New York, pp. 1–14.

Amonette, J.E., Kim, J., Russell, C.K., Palumbo, A.V., Daneils, W.L., 2003a. Enhancement of soil carbon sequestration by amendment with fly ash. Paper No.47.2003 International Ash Utilization Symposium, Lexington, KY, October 20–22.

Amonette, J.E., Kim, J., Russell, C.K., Palumbo, A.V., Daneils, W.L., 2003b. Fly ash catalyzes carbon sequestration. In: Proceedings of the Second Annual Conference on Carbon Sequestration, May 5–8, 2003, Alexandria, VA.

Arbestain, M., Camps, M.L., Ibargoitia, Z., Madinabeitia, M.V., Gil, S., Virgel, A., et al., 2009. Laboratory appraisal of organic carbon changes in mixtures made with different inorganic wastes. Waste Manage. 29, 2931–2938.

Arthur, M.F., Zwick, T.C., Tolle, D.A., Van Voris, P., 1984. Effects of fly ash on microbial $CO_2$ evolution from an agricultural soil. Water Air Soil Pollut. 22 (2), 209–211. Available from: https://doi.org/10.1007/bf00163101.

Bandick, A.K., Dick, R.P., 1999. Field management effects on soil enzyme activities. Soil Biol. Biochem. 31, 1471–1479.

Basu, M., Pande, M., Bhadoria, P.B.S., Mahapatra, S.C., 2009. Potential fly-ash utilization in agriculture: a global review. Prog. Nat. Sci. 19 (10), 1173–1186. Available from: https://doi.org/10.1016/j.pnsc.2008.12.006.

Burns, R.G., 1982. Enzyme activity in soil: location and possible role in microbial ecology. Soil Biol. Biochem. 14, 423–427.

Cervelli, S., Petruzzelli, G., Perna, A., Menicagli, R., 1986. Soil nitrogen and fly ash utilization: a laboratory investigation. Agrochemica 30, 27–33.

Cervelli, S., Petruzzelli, G., Perna, A., 1987. Fly ashes as an amendment in cultivated soils I. Effect on mineralization and nitr ification. Water Air Soil Pollut. 33, 331–338.

Chen, Y.P., Rekha, P.D., Arun, A.B., Shen, F.T., Lai, W.A., Young, C.C., 2006. Phosphate solubilizing bacteria from subtropical soil and their tricalcium phosphate solubilizing abilities. Appl. Soil Ecol. 34, 33–41.

Cheung, K., Wong, J.P., Zhang, Z., Wong, M., 2000. Revegetation of lagoon ash using the legume species *Acacia auriculiformis* and *Leucaena leucocephala*. Environ. Pollut. 109 (1), 75–82. Available from: https://doi.org/10.1016/s0269-7491(99)00235-3.

Chung, H., Park, M., Madhaiyan, M., Seshadri, S., Song, J., Cho, H., et al., 2005. Isolation and characterization of phosphate solubilizing bacteria from the rhizosphere of crop plants of Korea. Soil Biol. Biochem. 37, 1979-1974.

Cookson, W.R., Osman, M., Marschner, P., Abaye, D.A., Clark, I., Murphy, D.V., et al., 2007. Controls on soil nitrogen cycling and microbial community composition across land use and incubation temperature. Soil Biol. Biochem. 39 (3), 744–756. Available from: https://doi.org/10.1016/j.soilbio.2006.09.022.

Davison, R.L., Natusch, D.F.S., Wallace, J.R., Evans, C.A., 1974. Trace elements in fly ash. Dependence of concentration on particle size. Environ. Sci. Technol. 8 (13), 1107–1113. Available from: https://doi.org/10.1021/es60098a003.

De Freitas, J.R., Banerjee, M.R., Germida, J.J., 1997. Phosphate-solubilizing rhizobacteria enhance the growth and yield but not phosphorus uptake of canola (*Brassica napus*). Biol. Fertil. Soil. 36, 842—855.

Dick, R.P., 1997. Soil enzyme activities as integrative indicators of soil health. In: Pankhurst, C.E., Doube, B.M., Gupta, V.V.S.R. (Eds.), Biological Indicators of Soil Health. CABI, Wellingford,VA, pp. 121—156.

Dick, R.P., Sandor, J.A., Eash, N.S., 1994. Soil enzyme activities after 1500 years of terrace agriculture in the Colca Valley. Peru Agric. Ecosyst. Env. 50, 123—131.

Doran, J.W., Parkin, T.B., 1994. Defining and assessing soil quality. P.3-21. In: J.W. Doren et al. (Ed.) Defining Soil Quality for a Sustainable Environment. SSSA Special Publication 35. SSSA and ASA, Madison, WI.

Fuentes, B., Bolan, N.S., Naidu, R., Mora, M.L., 2006. Phosphorus in organic waste-soil systems. J. Soil Sci. Plant Nutr. 6, 64—83.

Gaind, S., Gaur, A.C., 2004. Evaluation of fly ash as a carrier for diazotrophs and phosphobacteria. Bioresour. Technol. 95, 187—190.

Garau, M.A., Dalmau, J.L., Felip, M.T., 1991. Nitrogen mineralization in soil amended with sewage sludge and fly ash. Biol. Fertil. Soil. 12 (3), 199—201. Available from: https://doi.org/10.1007/bf00337202.

García-Sánchez, M., Siles, J.A., Cajthaml, T., García-Romera, I., Tlustoš, P., Száková, J., 2015. Effect of digestate and fly ash applications on soil functional properties and microbial communities. Eur. J. Soil Biol. 71, 1—12. Available from: https://doi.org/10.1016/j.ejsobi.2015.08.004.

Glick, B.R., 1995. The enhancement of plant growth by free-living bacteria. Can. J. Microbiol. 41 (2), 109—117. Available from: https://doi.org/10.1139/m95-015.

Goldstein, A.H., 1994. Involvement of the quinoprotein glucose dehydrogenises in the solubilization of exogenous phosphates by gram-negative bacteria. In: Torriani Gorini, A., Yagil, E., Silver, S. (Eds.), Phosphate in Microorganisms: Cellular and Molecular Biology. ASM Press, Washington, DC, pp. 197—203.

Gray, T.R.G., 1990. Methods for studying the microbial ecology of soil. In: Grigorovia, R., Norris, J.R. (Eds.), Methods in Microbiology, Vol 22. Academic Press, London, pp. 309—342.

Gregorich, E.G., Carter, M.R., Angers, D.A., Monreal, C.M., Ellert, B.H., 1994. Towards a minimum data set to assess soil organic matterquality in agricultural soils. Can. J. Soil Sci. 74, 367—385.

Huang, J., Song, C., 2010. Effects of land use on soil water soluble organic C and microbial biomass C concentrations in the Sanjiang Plain in northeast China. Acta Agric. Scand. Sect. B Soil Plant Sci. 60 (2), 182—188. Available from: http://doi.org/10.1080/09064710802680387.

Illmer, P., Schinner, F., 1992. Solubilization of hardly-soluble $AlPO_4$ with P-solubilizing microorganisms. Soil Biol. Biochem. 24, 389—395.

Izquierdo, M., Querol, X., 2012. Leaching behaviour of elements from coal combustion fly ash: an overview. Int. J. Coal Geol. 94, 54—66. Available from: https://doi.org/10.1016/j.coal.2011.10.006.

Jala, S., Goyal, D., 2006. Fly ash as a soil ameliorant for improving crop production—a review. Bioresour. Technol. 97 (9), 1136—1147. Available from: https://doi.org/10.1016/j.biortech.2004.09.004.

Juwarkar, A.A., Jambhulkar, H.P., 2008. Restoration of fly ash dump through biological interventions. Environ. Monit. Assess. 139 (1-3), 355—365. Available from: https://doi.org/10.1007/s10661-007-9842-8.

Kalra, N., Harit, R., Sharma, S., 2000. Effect of flyash incorporation on soil properties of texturally variant soils. Bioresour. Technol. 75 (1), 91—93. Available from: https://doi.org/10.1016/s0960-8524(00)00036-5.

Krzaklewski, W., Pietrzykowski, M., Woś, B., 2012. Survival and growth of alders (*Alnus glutinosa* (L.) Gaertn. and *Alnus incana* (L.) Moench) on fly ash technosols at different substrate improvement. Ecol. Eng. 49, 35—40. Available from: https://doi.org/10.1016/j.ecoleng.2012.08.026.

Kucey, R.M.N., 1983. Phosphate solubilizing bacteria and fungi in various cultivated and virgin Alberta soils. Can. J. Soil Sci. 63, 671—678.

Kumar, R., 2004. Genetic monitoring of microbes in soils amended with fly ash. Thesis, Department of Biotechnology & Environmental Sciences, Thapar Institute of Engineering and Technology (Deemed University), PATI ALA-147 004.

Kumar, K.V., Patra, D.D., 2013. Effect of metal tolerant plant growth promoting bacteria on growth and metal accumulation in *Zea mays* plants grown in fly ash amended soil. Int. J. Phytoremediat. 15 (8), 743—755.

Kumar, V., Behl, R.K., Narula, N., 2001. Establishment of phosphate- solubilizing strains of *Azotobacter chroococcum* in the rhizosphere and their effect on wheat cultivars under greenhouse conditions. Microbiol. Res. 156, 87—93.

Kumar, K.V., Singh, N., Behl, H.M., Srivastava, S., 2008. Influence of plant growth promoting bacteria and its mutant on heavy metal toxicity in *Brassica juncea* grown in fly ash amended soil. Chemosphere 72, 678—683.

Kumar, K.V., Srivastava, S., Singh, N., Behl, H.M., 2009. Role of metal resistant plant growth promoting bacteria in ameliorating fly ash to the growth of *Brassica juncea*. J. Hazard. Mater. 170 (1), 51—57. Available from: https://doi.org/10.1016/j.jhazmat.2009.04.132.

Kumar, A., Nayak, A.K., Shukla, A.K., Panda, B.B., Raja, R., Shahid, M., et al., 2012. Microbial biomass and carbon mineralization in agricultural soils as affected by pesticide addition. Bull. Environ. Contam. Toxic. 88, 538—542.

Kumari, B., Singh, S.N., 2011. Phytoremediation of metals from fly ash through bacterial augmentation. Ecotoxicology 20 (1), 166—176.

Kumari, A., Pandey, V.C., Rai, U.N., 2013. Feasibility of fern *Thelypteris dentata* for revegetation of coal fly ash landfills. J. Geochem. Explor. 128, 147—152. Available from: https://doi.org/10.1016/j.gexplo.2013.02.005.

Lai, K.M., Ye, D.Y., Wong, J.W.C., 1999. Enzyme activities in a sandy soil amended with sewage sludge and coal fly ash. Water Air Soil Pollut. 113, 261—272.

Lal, J.K., Mishra, B., Sarkar, A.K., 1996. Effect of fly ash on soil microbial and enzymatic activity. J. Ind. Soc. Soil Sci. 44, 77—80.

Lim, S.S., Choi, W.J., 2014. Changes in microbial biomass, CH4 and CO2 emissions, and soil carbon content by fly ash co-applied with organic inputs with contrasting substrate quality under changing water regimes. Soil Biol. Biochem. 68, 494—502.

Masto, R.E., Sengupta, T., George, J., Ram, L.C., Sunar, K.K., Selvi, V.A., et al., 2014. The impact of fly ash amendment on soil carbon. Energ. Source. Part A 36, 554—562. Available from: https://doi.org/10.1080/15567036.2010.544004.

McLaren, A.D., 1975. Soil as a system of humus and clay immobilised enzymes. Chem. Scr. 8, 97—99.

Menon, M.P., Sajwan, K.S., Ghuman, G.S., James, J., Chandra, K., 1993. Fly ash-amended compost as a manure for agricultural crops. J. Environ. Sci. Health A Environ. Sci. Eng. Tox. 28 (9), 2167—2182. Available from: https://doi.org/10.1080/1093452930937599.

Mishra, M., Sahu, R.K., Padhy, R.N., 2007. Growth, yield, and elemental status of rice (Oryza sativa) grown in fly ash amended soil. Ecotoxicology 16, 271—278.

Misra, N., Gupta, G., Jha, P.N., 2012. Assessment of mineral phosphatesolubilizing properties and molecular characterization of zinc tolerant bacteria. J. Basic. Microbiol. 52, 549—558.

Nannipieri, P., Grego, S., Ceccanti, B., 1990. Ecological significance of the biological activity in soil. In: Bollag, J.-M., Stotzky, G. (Eds.), Soil Biochemistry, Vol. 6. Marcel Dekker, New York, pp. 293–355.

Nayak, A.K., Gangwar, B., Shukla, A.K., Mazumdar, S.P., Kumar, A., Raja, R., et al., 2012. Long-term effect of different integrated nutrient management on soil organic carbon and its fractions and sustainability of rice–wheat system in Indo Gangetic Plains of India. Field Crop. Res. 127, 129–139.

Nayak, A.K., Kumar, A., Raja, R., Rao, K.S., Mohanty, S., Shahid, M., et al., 2014. Fly ash addition affects microbial biomass and carbon mineralization in agricultural soils. Bull. Environ. Contam. Toxicol. 92 (2), 160–164. Available from: https://doi.org/10.1007/s00128-013-1182-5.

Nayak, A.K., Raja, R., Rao, K.S., et al., 2015. Effect of fly ash application on soil microbial response and heavy metal accumulation in soil and rice plant. Ecotoxicol. Environ. Saf. 114, 257–262.

Nsabimana, D., Haynes, R., Wallis, F., 2004. Size, activity and catabolic diversity of the soil microbial biomass as affected by land use. Appl. Soil Ecol. 26 (2), 81–92. Available from: https://doi.org/10.1016/j.apsoil.2003.12.005.

Ogut, M., Coyne, M., Thom, W.O., 2009. Weathered fly ash does not affect soil and biosolid carbon mineralization. Commun. Soil Sci. Plant Anal. 40, 1790–1801.

Oves, M., Khan, M.S., Zaidi, A., 2013. Chromium reducing and plant growth promoting novel strain *Pseudomonas aeruginosa* OSG41 enhance chickpea growth in chromium amended soils. Eur. J. Soil Biol. 56, 72–83.

Pandey, V.C., Singh, N., 2010. Impact of fly ash incorporation in soil systems. Agric. Ecosyst. Environ. 136, 16–27.

Pandey, V.C., Abhilash, P.C., Singh, N., 2009a. The Indian perspective of utilizing fly ash in phytoremediation, phytomanagement and biomass production. J. Environ. Manage 90, 2943–2958.

Pandey, V.C., Singh, J.S., Kumar, A., Tewari, D.D., 2010. Accumulation of heavy metals by chickpea grown in fly ash treated soil: effect on antioxidants. Clean. 38 (12), 1116–1123. Available from: https://doi.org/10.1002/clen.201000178.

Pandey, V.C., Singh, J.S., Singh, R.P., Singh, N., Yunus, M., 2011. Arsenic hazards in coal fly ash and its fate in Indian scenario. Resour. Conserv. Recy. 55, 819–835.

Pandey, V.C., Singh, J.S., Singh, D.P., Singh, R.P., 2014. Methanotrophs: promising bacteria for environmental Remediation. Int. J. Environ. Sci. Technol. 11, 241–250.

Pandey, V.C., Patel, D., Thakare, P.V., Ram, L.C., Singh, N., 2017. Microbial and enzymatic activities of fly ash amended soil. In: Singh, J.S., Singh, D.P. (Eds.), Microbes and Environmental Management. Studium Press LLC, USA, pp. 142–167. ISBN: 9789380012834.

Patel, D., Thakare, P.V., 2015. Occurrence of multiple plasmids in metal resistant bacterial population isolated from coal fly ash contaminated soil. J. Pure Appl. Microbiol. 9 (2), 1189–1196.

Pathan, S.M., Aylmore, L.A.G., Colmer, T.D., 2002. Reduced leaching of nitrate, ammonium and phosphorus in a sandy soil by fly ash amendment. Aust. J. Soil Res. 40, 1201–1211.

Pati, S.S., Sahu, S.K., 2004. $CO_2$ evolution and enzyme activities (dehydrogenase, protease and amylase) of fly ash amended soil in the presence and absence of earthworms (Drawida willsi Michaelsen) under laboratory conditions. Geoderma. 118 (3-4), 289–301. Available from: https://doi.org/10.1016/s0016-7061(03)00213-1.

Perrott, K.W., Sarathchandra, S.U., Dow, B.W., 1992. Seasonal and fertilizer effects on the organic cycle and microbial biomass in a hill country soil under pasture. Aust. J. Soil Res. 30, 383–394.

Pichtel, J., 1990. Microbial respiration in fly ash/sewage sludge-amended soils. Environ. Pollut. 63, 225–237.

Pichtel, J., Hayes, J., 1990. Influence of fly ash on soil microbial activity and populations. J. Environ. Qual. 19, 593–597.

Pikovskaya, R.I., 1948. Mobilization of phosphorus in soil connection with the vital activity of some microbial species. Microbiologiya 17, 362–370.

Powlson, D.S., 1994. The soil microbial biomass: before, beyond and back. In: Dighton, J., Giller, K.E., Ritz, K. (Eds.), Beyond the Biomass. Wiley, Chichester.

Rajkumar, M., Freitas, H., 2008. Effects of inoculation of plant-growth promoting bacteria on Ni uptake by Indian mustard. Bioresour. Technol. 99 (9), 3491–3498. Available from: https://doi.org/10.1016/j.biortech.2007.07.046.

Ram, L.C., Masto, R.E., 2010. An appraisal of the potential use of fly ash for reclaiming coal mine spoil. J. Environ. Manage 91, 603–617.

Ram, L.C., Masto, R.E., 2014. Fly ash for soil amelioration: a review on the influence of ash blending with inorganic and organic amendments. Earth Sci. Rev. 128, 52–74. Available from: https://doi.org/10.1016/j.earscirev.2013.10.003.

Ram, L.C., Srivastava, N.K., Tripathi, R.C., Jha, S.K., Sinha, A.K., Singh, G., et al., 2006a. Management of mine spoil for crop productivity with lignite fly ash and biological amendments. J. Environ. Manag. 79, 173–187.

Ram, L.C., Srivastava, N.K., Jha, S.K., Sinha, A.K., 2006b. Eco-friendly reclamation of mine spoil for agro-forestry through fly ash and biological amendments. Proceedings of the 23rd Annual International Pittsburgh Coal Conference, Session, 52, pp. 1–25.

Ram, L.C., Masto, R.E., Jha, S.K., Selvi, V.A., Srivastava, N.K., Sinha, A.K., 2007a. Reclamation of coal mine spoil using fly ash: a synoptic review. Proceedings of the 1st International Conference on Managing the Social and Environmental Consequences of Coal Mining in India (MSECCMI), pp. 771–801.

Ram, L.C., Srivastava, N.K., Jha, S.K., Sinha, A.K., Masto, R.E., Selvi, V.A., 2007b. Management of lignite fly ash through its bulk use via biological amendments for improving the fertility and crop productivity of soil. Environ. Manag. 40, 438–452.

Rengel, Z., Marschner, P., 2005. Nutrient availability and management in the rhizosphere: exploiting genotypic differences. N. Phytologist. 168, 305–312.

Rippon, J.E., Wood, M.J., 1975. Microbiological aspects of pulverized fuel ash. In: Chadwick, M.J., Goodman, G.T. (Eds.), The Ecology of Resource Degradation and Renewal. Wiley, New York, pp. 331–349.

Rodríguez, H., Fraga, R., 1999. Phosphate solubilizing bacteria and their role in plant growth promotion. Biotechnol. Adv. 17 (4-5), 319–339. Available from: https://doi.org/10.1016/s0734-9750(99)00014-2.

Rumpel, C., Knicker, H., Kogel-Knaber, I., Skjiemstad, J.O., Huuetti, R.F., 1998. Types and chemical composition of organic matter in reforested lignite-rich mine soils. Geoderma. 86, 123–142.

Saber, K., Nahla, L.D., Chedly, A., 2005. Effect of P on nodule formation and N fixation in bean. Agron. Sustain. Dev. 25, 389–393.

Saffigna, P.G., Powlson, D.S., Brookes, P.C., Thomas, G.A., 1989. Influence of sorghum residues and tillage on soil organic matter and soil microbial biomass in an Australian vertisol. Soil Biol. Biochem. 21, 759–765.

Sajwan, K., Paramasivam, S., Alva, A., Adriano, D., Hooda, P., 2003. Assessing the feasibility of land application of fly ash, sewage sludge and their mixtures. Adv. Environ. Res. 8 (1), 77–91. Available from: https://doi.org/10.1016/s1093-0191(02)00137-5.

Sarangi, P.K., Mahakur, D., Mishra, P.C., 2001. Soil biochemical activity and growth response of rice Oryza sativa in fly ash amended soil. Bioresour. Technol. 76 (3), 199–205. Available from: https://doi.org/10.1016/s0960-8524(00)00127-9.

Savant, N.K., De Datta, S.K., 1982. Nitrogen transformations in wetland rice soils, Advances in Agronomy, Vol. 35. Elsevier BV, pp. 241–302. Available from: http://doi.org/10.1016/s0065-2113(08)60327-2.

Schutter, M.E., Fuhrmann, J.J., 2001. Soil microbial community responses to fly ash amendment as revealed by analyses of whole soils and bacterial isolates. Soil Biol. Biochem. 33, 1947–1958.

Selvakumari, G., Jayanthi, D., 1999. Effect of fly ash and compost on yield and nutrient uptake of groundnut in laterite soil. In: Proceedings of the National Seminar on Developments in Soil Science. Tamilnadu Agricultural University, Coimbatore, India.

Seshadri, B., Bolan, N.S., Choppala, G., Naidu, R., 2013. Differential effect of coal combustion products on the bioavailability of phosphorus between inorganic and organic nutrient sources. J. Hazard. Mater. Available from: https://doi.org/10.1016/j.jbbr.2011.03.031.

Singh, J.S., Pandey, V.C., 2013. Fly ash application in nutrient poor agriculture soils: impact on methanotrophs population dynamics and paddy yields. Ecotoxicol. Environ. Saf. 89, 43–51.

Singh, S., Ram, L.C., Masto, R.E., Verma, S.K., 2011a. A comparative evaluation of minerals and trace elements in the ashes from lignite, coal refuse, and biomass fired power plants. Int. J. Coal Geol. 87, 112–120.

Singh, J.S., Pandey, V.C., Singh, D.P., 2011b. Efficient soil microorganisms: a new dimension for sustainable agriculture and environmental development. Agric. Ecosyst. Env. 140, 339–353.

Singh, J.S., Pandey, V.C., Singh, D.P., 2011c. Coal fly ash and farmyard manure amendments in dry-land paddy agriculture field: effect on N-dynamics and paddy productivity. Appl. Soil Ecol. 47, 133–140.

Singh, K., Pandey, V.C., Singh, B., Patra, D.D., Singh, R.P., 2016. Effect of fly ash on crop yield and physico-chemical, microbial and enzyme activities of sodic soils. Environ. Eng. Manag. J. 15 (11), 2433–2440.

Sinsabaugh, R.L., Antibus, R.K., Linkins, A.E., 1991. An enzymic approach to the analysis of microbial activity during plant litter decomposition. Agric. Ecosyst. Environ. 34, 43–54.

Skujins, J., 1978. Soil enzymology and fertility index-a fallacy? History of abiotic soil enzyme research. In: Burns, R.G. (Ed.), Soil Enzymes. Academic, London, pp. 1–49.

Sridevi, M., Mallaiah, K.V., Yadav, N.C.S., 2007. Phosphate solubilization by *Rhizobium* isolates from *Crotalaria* species. J. Plant Sci. 2, 635–639.

Srivastava, N.K., Ram, L.C., Masto, R.E., 2014. Reclamation of overburden and lowland in coal mining area with fly ash and selective plantation: a sustainable ecological approach. Ecol. Eng. 2014 (71), 479–489.

Stamatiadis, S., Werner, M., Buchanan, M., 1999. Field assessment of soil quality as affected by compost and fertilizer application in a broccoli field (San Benito County, California). Appl. Soil Ecol. 12 (3), 217–225. Available from: https://doi.org/10.1016/s0929-1393(99)00013-x.

Stout, W.L., Sharpley, A.N., Gburek, W.J., Pionke, H.B., 1999. Reducing phosphorus export from croplands with FBC fly ash and FGD gypsum. Fuel. 78 (2), 175–178. Available from: https://doi.org/10.1016/s0016-2361(98)00141-0.

Stout, W.L., Sharpley, A.N., Landa, J., 2000. Effectiveness of coal combustion by-products in controlling phosphorus export from soils. J. Env. Qual. 29, 1239–1244.

Tabatabai, M.A., 1994. Soil enzymes. In: Weaver, R.W., Angle, J.S., Bottomley, P.S. (Eds.), Methods of Soil Analysis, Part 2. Microbiological and Biochemical Properties. SSSA Book Series No. 5. Soil Science Society of America, Madison, WI, pp. 775–833.

Tastan, B.E., 2017. Clean up fly ash from coal burning plants by new isolated fungi *Fusarium oxysporum* and *Penicillium glabrum*. J. Environ. Manag. 200, 46–52.

Taylor, E.M., Schuman, G.E., 1988. Fly ash and lime amendment of acidic coal spoil to aid revegetation. J. Environ. Qual. 17 (1), 120. Available from: https://doi.org/10.2134/jeq1988.00472425001700010018x.

Tiwari, S., Kumari, B., Singh, S.N., 2008. Evaluation of metal mobility/immobility in fly ash induced by bacterial strains isolated from the rhizospheric zone of *Typha latifolia* growing on fly ash dumps. Bioresour. Technol. 99 (5), 1305–1310. Available from: https://doi.org/10.1016/j.biortech.2007.02.010.

Torstensson, L., Stenström, J., 1986. "Basic" respiration as a tool for prediction of pesticide persistence in soil. Toxic. Ass 1, 57–72.

Tripathi, R.C., Masto, R.E., Ram, L.C., 2009. Bulk use of pond ash for cultivation of wheat–maize–eggplant crops in sequence on a fallow land. Resour. Conserv. Recy. 54, 134–139.

Tripathi, R.C., Jha, S.K., Ram, L.C., 2010. Impact of fly ash application on trace metal content in some root crops. Energ. Source. Part A: Recovery, Utilization, Environ. Eff. 32, 576–589.

Urvashi, Masto, R.E., Selvi, V.A., Ram, L.C., Srivastava, N.K., 2007. An international study: effect of farm yard manure on the release of P from fly ash. Remediation. 70–81.

Vazquez, P., Holguin, G., Puente, M., Lopez-cortes, A., Bashan, Y., 2000. Phosphate solubilizing microorganisms associated with the rhizosphere of mangroves in a semi-arid coastal lagoon. Biol. Fertil. Soil. 30, 460–468.

Voos, G., Groffman, P.M., 1997. Relationships between microbial biomass and dissipation of 2,4-D and dicamba in soil. Biol. Fertil. Soil. 24, 106–110.

Wakelin, S.A., Warren, R.A., Harvey, P.R., Ryder, M.H., 2004. Phosphate solubilization by *Penicillium* sp. closely associated with wheat roots. Biol. Fertil. Soil. 40, 36–43.

Wani, P.A., Zaidi, A., Khan, A.A., Khan, M.S., 2005. Effect of phorate on phosphate solubilization and indole acetic acid (IAA) releasing potentials of rhizospheric microorganisms. Ann. Plant Protect. Sci. 13, 139–144.

Wong, M.H., Wong, J.W.C., 1986. Effects of fly ash on soil microbial activity. Environ. Pollut. Ser. A Ecol. Biol. 40 (2), 127–144. Available from: https://doi.org/10.1016/0143-1471(86)90080-2.

Wu, F., Dong, M., Liu, Y., Ma, X., An, L., Young, J.P.W., et al., 2011. Effects of long-term fertilization on AM fungal community structure and Glomalin-related soil protein in the Loess Plateau of China. Plant. Soil 342, 233–247.

Yeledhalli, N.A., Prakash, S.S., Gurumurthy, S.B., Ravi, M.V., 2007. Coal fly ash as modifi er of physicochemical and biological properties of soil. Karnataka. J. Agric. Sci. 20 (3), 531–534.

Zaidi, A., Khan, M.S., Ahemad, M., Oves, M., 2009. Plant growth promotion by phosphate solubilizing bacteria. Acta Microbiol. Immunol. Hung. 56, 263–284.

Zhu, F., Qu, L., Hong, X., Sun, X., 2011. Isolation and characterization of a phosphate-solubilizing halophilic bacterium Kushneria sp. YCWA18 from Daqiao Saltern on the coast of Yellow Sea of China. Evid. Base. Compl. Alternative Med. 615032:6.

# Further reading

Pandey, V.C., Abhilash, P.C., Upadhyay, R.N., Tewari, D.D., 2009b. Application of fly ash on the growth performance, translocation of toxic heavy metals within Cajanus cajan L.: implication for safe utilization of fly ash for agricultural production. J. Hazard. Mater. 166, 255–259.

Ram, L.C., Srivastava, N.K., Tripathi, R.C., Thakur, S.K., Sinha, A.K., Jha, S.K., et al., 2007c. Leaching behavior of lignite fly ash with shake and column tests. Environ. Geol. 51, 1119—1132.

Verma, S.K., Masto, R.E., Gautam, S., Choudhury, D.P., Ram, L.C., Maiti, S.K., et al., 2015. Investigations on PAHs and trace elements in coal and its combustion residues from a power plant. Fuel 162, 138—147.

# CHAPTER 6

# Fly ash application in reclamation of degraded land: opportunities and challenges

## Contents

## 6.1 Introduction

Worldwide fly ash (FA) production has been assessed up to 750 million tons year$^{-1}$ (Blissett and Rowson, 2012), but around 50% of global FA production is used (Izquierdo and Querol, 2012). In India, FA production is projected to hike up to 300 million tons year$^{-1}$ by 2016–17 (Singh et al., 2011a; Skousen et al., 2012). The fast depletion of good quality coal is a current concern to coal-based thermal power stations, because most of the power stations use high-ash coal with low calorific value, and their combustion would result in a high increase in FA production. Huge quantities of FA production from thermal power stations are being dumped in ash ponds with the help of FA–water slurry through an iron pipe. FA has the potential to be utilized in multiple applications, processes, as well as a substitute to other industrial resources. These areas are cement production, concrete products, building blocks, bricks, building tiles,

*Phytomanagement of Fly Ash*
ISBN: 978-0-12-818544-5
DOI: https://doi.org/10.1016/B978-0-12-818544-5.00006-7
**167**

structural fills, highways, cover materials, lightweight aggregates, coal mine fillings, degraded soil improvement, and source of crop nutrients in agriculture including forestry (Pandey and Singh, 2010; Pandey et al., 2009a). However, a major proportion of the global FA production is unutilized and needs urgent action to find out potential avenues. The other potential areas of FA utilization are wastewater treatment, metal extraction, and creation of cenospheres (Asokan et al., 2005). In Indian scenario, the extreme use of FA to the limit of 48.50% has been in the cement segment, followed by low lying areas (12.73%), embankments and roads (11.65%), coal mine filling (8.26%), tiles and bricks (6.3%), agriculture (1.74%), and others (10.82%). Even after the application in such areas, only 55.79% of the overall FA produced is utilized (CEA Central Electricity Authority, 2011). Therefore there is a broad opportunity and a pressing need to increase the quantity of FA use in every area, but agriculture sector has ample opportunity to the use of huge quantity of FA as a source of crop nutrients as well as soil improvement (Pandey and Singh, 2010; Pandey et al., 2009a). Increase in degraded soil throughout the world is a critical concern. It should be reclaimed with cheap, suitable, and easily available amendments. In this regard, FA has been identified as a potential and effective soil additive throughout the world. Furthermore, thousands of hectares of land engaged for the dumping of FA globally could be reclaimed through potential grass species including woody tree species (Pandey and Singh, 2012).

On the bases of UNCCD, $1.9 \times 10^9$ hectares of lands is degraded. Approximately 1.5 billion hectares of land are fit for agroforestry and agriculture purposes after restoration (UNCCD United Nations Convention to Combat Dessertification, 2012), mostly to meet the needs of the fastest growing population. The present world population is 7.2 billion, which is expected to rise up to 8.1 and 9.6 billion in 2025 and 2050, respectively (UNWPP United Nations World Population Prospects, 2013). Almost 99.9% of the food supply for human population is obtained from land through agriculture and agroforestry systems use. All over the world, cultivated land per individual has regularly declined from 0.44 hectare to less than 0.25 hectare in the last 50 years (FAO Food and Agriculture Organization of the United Nations, 2013). Therefore the reclamation of all types of degraded lands (i.e., wastelands, nonfertile lands, etc.) to make them fertile for food production is the current topic of attention (Ram and Masto, 2010). FA application in the degraded lands as a soil modifier is a current and potential sector from the emphasis of using its massive quantity, tackling of its environmental surmises, and tapping commercial prospects. A lot of studies demonstrate to the broader prospective of FAs to enhance soil fertility and ameliorate degraded land for cultivation and/ or revegetation (Ram et al., 2007, 2006; Pandey et al., 2009a, b, 2010; Singh and Pandey, 2013; Singh et al., 2011b, 2016; Srivastava and Ram, 2010; Ukwattage et al., 2013). The soil amelioration potential of FA depends on the favorable physicochemical properties of FA and appreci- able amount of plant micro- and macronutrients. But the history of FA application in soil as a plant nutrient starts to about 60 years ago (Rees and Sidrak, 1956). FA contains most of the essential plant nutrients, except nitrogen and carbon, which is attributable to the oxidation of nitrogen and carbon in coal combustion. The deficient nitrogen and car- bon have to be amended laterally with FA incorporation to soils. Adding other amendments along with FA application will boost to deliver nitro- gen and carbon, besides augmenting the FA effectiveness. A wide range of amendments are used in soil amelioration along with FA incorporation. From economical point of view, there are several locally available soil conditioners, such as press mud, agricultural wastes, sewage sludge, farm- yard manure, and green manures, that can facilitate enhancement in the soil systems through increasing the crop yield and solving the wastes dis- posal issues (TIFAC, 2001). The use of FA as an agronomic amendment can be boosted by mixing it with sewage sludge, cattle manure, and poul- try manure, which are acidic in nature and rich in nitrogen and phospho- rus (Adriano et al., 1980). Moreover, land application of such a combination can lessen the mobility of sewage sludge-borne potential toxic metals, which is known to increase the following cessation of sew- age sludge application (Sajwan et al., 2003; Hooda, 2010). Likewise, some studies shows that alkaline FA application with or without amendments can remediate at least marginally metal polluted sites by immobilization of mobile metal forms (Pandey and Singh, 2010).

The other important by-products are press mud, red mud, lime, gyp- sum, biochar, biofertlizers, and biosolids have been used as amendment with FA application. Incorporation of acid-forming organic by-products into the substrate of FA deposits could enhance the success of restoration programs (Haynes, 2009; Belyaeva and Haynes, 2012). There are limited studies on the restoration of FA deposits through plant species. In this regard, Pandey et al. (2015) evaluated the naturally colonized plant species for FA phytorestoration and identified some species for the reclamation of FA deposits. Pandey (2013) reported that *Ricinus communis*, castor oil plant (a crop of commercial importance), has potential to remediate FA dumps. Besides, castor can be employed on large scale for contributing toward FA dumps' utilization. Furthermore, *Thelypteris dentata* fern was found appropriate for phytostabilization/revegetation of FA lagoons (Kumari et al., 2013). Most of the environmental problems of FA could be effectively mitigated by phytomanagement (Pandey et al., 2009a) along with the selection of suitable plants (Pandey and Bajpai, 2019) and amendments (Ram and Masto, 2014). There are several important reviews on the FA utilization in the amelioration of different types of soil (Gupta et al., 2002; Yunusa et al., 2006; Jala and Goyal, 2006; Pandey et al., 2009a; Basu et al., 2009; Pandey and Singh, 2010; Ram and Masto, 2014; Shaheen et al., 2014). The present chapter focuses particularly on the FA application in the reclamation of different types of degraded lands throughout the world.

## 6.2 What makes fly ash a suitable amendment for degraded soils?

FA being a coal combustion residue contains essential plant micronutrients including Mn, Fe, Zn, Co, Cu, Mo, and B as well as macronutrients like Ca, K, Mg, P, and S (Pandey and Singh, 2010). The Si, Al, and Fe are the major elements in FA followed by Ca, K, Na, and Ti (Pandey and Singh, 2010). The sequence of dominant cation of FA is Ca, Mg, Na, and K (Matti et al., 1990). Mostly, Al is bound in insoluble aluminosilicate structures in FA, and this significantly limits its biological toxicity. The FA pH ranges from 4.5 to 12.0 and depends mainly on the coal's sulfur content (Pandey and Singh, 2010). If Ca/S ratio is less than 2.5, then FA will be acidic in nature, whereas a higher value of Ca/S ratio than 2.5 will produce alkaline FA (Anisworth and Rai, 1987). The FA color can differ as tan, gray, and black, which mostly depends on the unburned carbon content in coal ash (Kassim and Williams, 2005). FA has a low bulk density (BD), light texture, and high surface area (Jala and Goyal, 2006). The particle size of FA ranges from 0.01 to 100 μm (Jala and Goyal, 2006). The shape of the FA particle is mostly spherical, either solid or hallow and commonly glossy (amorphous) in nature. FA consists of primarily amorphous glass and a few crystalline phases. The crystalline phases of FA are aluminosilicate glass, gypsum, quartz, mullite, anhydrite, magnetite, hematite, ettringite, chlorite, lime, spinel, and feldspars, depending on the mineralogy of the parent coal, which are known as minerals (Moreno et al., 2005; Kutchko and Kim, 2006). A wide variation has been observed in FA mineralogical and physicochemical properties, depending on the origin of coal, combustion conditions, type of emission control devices, storage, and handling techniques (Jala and Goyal, 2006).

The FA impact on the soil physical characteristics is mainly due to the changes in the soil texture. Modification in the soil texture is associated with porosity, BD, hydraulic conductivity, water holding capacity (WHC), and void ratio, which have a direct impact on the plant growth, the nutrient retention, and the soil biological activity (Fulekar and Dave, 1986; Bilski et al., 1995; Pandey and Singh, 2010; Ram and Masto, 2014; Shaheen et al., 2014). The use of FA in degraded soil has been suggested to improve micro- and macronutrients (Pandey and Singh, 2010). FA should be used in sodic soil reclamation, because FA has the properties of gypsum and lime (Kumar and Singh, 2003). The alkaline FA incorporation to mine spoils and wasteland enhances the pH and WHC as well as reduces soil BD and compaction (Capp, 1978; Jastrow et al., 1981; Fail, 1987). Aged (weathered) FA from old ash dumping sites has a lower concentration of toxic metals, compared with fresh (unweathered) FA from newly ash dumping sites, due to leaching of toxic metals. Therefore weathered FA is safer in the use of the improvement of degraded soils (Page et al., 1979; Phung et al., 1978).

FA offers sites for the formation of organomineral complexes owing to the large specific surface area of the silt-sized particles (Jala and Goyal, 2006). Moreover, FA is able to fix $CO_2$ produced from soil respiration through carbonation of calcium (Ca) and magnesium (Mg), ($Ca^{2+}$ or $Mg^{2+}$) + $CO_3{}^{2-}$ → $CaCO_3$ or $MgCO_3$, which are present in FA, thereby reducing $CO_2$ emission while enhancing soil carbon content (Lim et al., 2012; Lim and Choi, 2014). Recently, it is also reported that FA application is effective in increasing soil carbon sequestration without reducing paddy yield in agricultural fields (Lim et al., 2017).

## 6.3 Fly ash use in amelioration of problematic soils

There are a number of studies regarding the use of FA in the amelioration of problematic soils. Several studies revealed the potential of FA application for improving a wide range of degraded soils such as sodic soil, acidic soil, mine spoil, and nutrient poor soil. These improvements are mostly related to the physicochemical and microbial properties of degraded soil. The following sections will explore the use of FA in the amelioration of

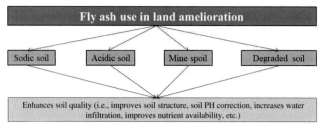

**Figure 6.1** Enhancement of the soil quality of problematic soils via FA application.

degraded soils in detail. FA enhances the soil quality of problematic soils via soil structure improvement, soil pH correction, enhancement of water infiltration, improvement of nutrient availability, etc. (Fig. 6.1). Furthermore, Table 6.1 describes the potential of FA for improving a range of problematic soil systems.

## 6.3.1 Sodic soil reclamation

Safe application of FA in the reclamation of sodic soils is a significant environmental issue. Studies are going on to find out more feasible approaches for using low-priced materials for the reclamation of widespread sodic areas. Gypsum ($CaSO_4.2H_2O$) and phosphogypsum have been used broadly to reclaim sodic soils as they release $Ca^{2+}$ cations (divalent calcium), which replace $Na^+$ cations (monovalent sodium) from exchange complex and make the clay particles free to flocculate and form macro aggregates (Singh et al., 2016). Numerous studies have also been showed that the application of organic amendments improves sodic soil properties more than gypsum including crop yield (Gill et al., 2008; Sahin and Anapali, 2005; Tejada et al., 2006). There is a pressing need to explore the feasibility of using FA as a soil additive for sodic soil reclamation. Normally, FA nature is highly alkaline (pH 10−12) due to the presence of hydroxides and carbonate salts of Ca and Mg, but sometimes, it may be extremely acidic (pH 3−4) in nature. Moreover, the presence of gypsum characteristics in FA play important role in the reclamation of sodic soil. Thus the application of FA with organic amendment seems to be a more effective ameliorant than gypsum or phosphogypsum. It will be better if weathered FA is used in association with organic amendment such as press mud that makes favorable changes in the physicochemical, microbial, and biochemical parameters of sodic soils (Pandey and Singh, 2010; Ram and Masto, 2014).

**Table 6.1** FA application for improving problematic soil system.

| Problematic soil | Amendments | FA doses | Concluding remarks | Reference |
|---|---|---|---|---|
| Calcareous soil and acidic soil | FA at 8% by wt. and sequential cropping of native desert sp. followed by barley | FA at 8% | Increased pH, EC, Ca, Mg, Na, B, and $SO_4$, shoots dry wt. of desert sp. and barley yield grain significantly improved, and availability of P, Zn, Fe, and Mn diminished but low Cu:Mo ratios in harvested plants. | Elseewi et al. (1980) |
| Sandy loam soil | FA + sandy loam soil | FA at 3%, 6%, and 12% (w/w) | FA amendment at the rate of 3% on sandy loam soil increased decomposition rate of soil organic carbon than control. | Wong and Wong (1986) |
| Nutrient deficient sandy soils | FA and PM in different ratios | | PM alone: increases yield; PM + FA: response depends on crop (*Brassica*) species; with increase in FA, tissue concentrations of Zn and Mn decreased, while Mo increased. | Wong and Wong (1987) |
| Acidic coal spoils | FA + lime + acidic coal spoil | FA at 0, 10, 20, 30, and 40 g kg$^{-1}$; lime at 0, 10, 20, 40, and 80 g kg$^{-1}$ | All doses of lime tested and FA doses at or above 20 g kg$^{-1}$ enhanced the soil pH, root biomass, and aboveground plant biomass. So FA is a possible substitute to lime for treating acidic coal spoils. | Taylor and Schumann (1988) |

(Continued)

**Table 6.1** (Continued)

| Problematic soil | Amendments | FA doses | Concluding remarks | Reference |
|---|---|---|---|---|
| Sandy soil, sandy loam | FA + soil | FA at 0%, 3%, 6%, 12%, and 30% (on a dry weight basis) | The EC and pH of both recipient soils were raised, but more so for the sandy soil. The increase in EC may limit the availability of soil water, because the enhanced pH and the high osmotic pressure would alter the availability of micronutrients to plants. | Wong and Wong (1989) |
| Silt loam soil | FA + silt loam soil and FA + silt loam soil + sludge (5%) | FA at 0%, 5%, 10%, and 20% (w/w); sewage sludge at 5% | High doses of FA to soil may inhibit nutrient cycling processes and normal decomposition. | Pichtel and Hayes (1990) |
| Sandy soil, sandy loam | FA + soil | FA at 0%, 3%, 6%, and 12% | The highest rate of FA increased the pH of sandy loam and sandy soil from 6.7 and 7.3 to 8.6 and 9.7, respectively. EC also increased from 135 and 56 to 341 and 2035 $\mu$mhos cm$^{-1}$ for sandy loam and sandy soil, respectively. Therefore sandy loam had a higher buffering capacity for receiving the FA amendment than sandy soil. Both accumulation and reduction of metals in plant tissue were significantly correlated with the pH of FA amended soils. | Wong and Wong (1990) |

| Soil | Treatment | FA rate | Observations | Reference |
|---|---|---|---|---|
| Acidic clay textured soils | FA + acidic soil | FA at 167 t ha⁻¹ | pH and concentration of B increased, B-tolerant alfalfa grown well. beneficial liming effect up to 110 t ha⁻¹ FA; uptake of As and Se is not studied | Warren (1992) |
| Acidic soil | FA + soil | FA at 0%, 2%, 5%, 10%, 15%, and 20% | Helped to decrease metal solubility and availability to plants. | Shende et al. (1994) |
| Acid alfisol | FA + soil, FA + soil + FYM at 5 g kg⁻¹ | FA at 80 and 160 g kg⁻¹ | Higher microbial activities in soil amended with up to 8% FA and combined application of FA and FYM showed to be useful in enhancing proliferation and activity of microbes in acid soil. | Lal et al. (1996) |
| Acidic soils (pH:4.5–5.8) | FA + acidic soil | Alkaline FA (pH:8.5–8.9) at 5, 20, and 50 g kg⁻¹ soil | FA increased pH and EC, FA increased dry biomass yield of ryegrass (L. perenne), increased uptake of B and P; alkaline FA (low B concentration) at 20 g kg⁻¹ recommended | Matsi and Keramidas (1999) |
| Poorly buffered acidic soils | FA + soil; soil + (Zn, Cu, Ni, and Cd); FA + soil + (Zn, Cu, Ni, and Cd) | 3% FA | Alkalizing effects of FA can be utilized to reduce plant accumulation of potentially toxic elements, particularly in poorly buffered acidic soils. | Scotti et al. (1999) |

*(Continued)*

**Table 6.1** (Continued)

| Problematic soil | Amendments | FA doses | Concluding remarks | Reference |
|---|---|---|---|---|
| Clayey, sandy–clay–loam, sandy, sandy loam | FA + soil | FA at 0%, 10%, 20%, 30%, and 40% by weight basis | FA addition in texturally different soils modifies the soil physicochemical characteristics, which in turn may influence the yields of crop. | Kalra et al. (2000) |
| Sandy loam | FA + sandy loam soil | FA at 0, 10, 12.5, 15, 17.5, and 20 t ha$^{-1}$ | Amylase, invertase, protease, and dehydrogenase activity increased with increasing FA application up to 10 t ha$^{-1}$, but decreased with higher doses of FA application. | Sarangi et al. (2001) |
| Soil | FA and soil | FA at 10%, 25%, and 50% | Improved soil physicochemical properties, plant growth, net primary productivity, leaf area, and photosynthetic pigments. | Ajaz and Tiyagi (2003) |
| Calcareous heavy loam soil (ustarents) | FA + soil | FA depths—5, 7, and 10 cm in bottom of pots; Soil depths—10, 8, and 5 cm; total depths of growth substrate—15-cm AM fungi treatments | AM fungi can make a significant contribution to successful crop establishment in soils overlying areas of FA. | Bi et al. (2003) |
| Water-logged sodic soils (pH: 9.07, EC: 3.87 dS/m, ESP:26.0) | Acidic FA (pH:5.9) alone at 0%–7.5%, and/or in combination with gypsum (10%–100%) and soil (10%), respectively, for the effect on paddy–wheat rotation crops | Acidic FA (pH:5.9) alone at 0%–7.5% | FA decreased pH and EC; 60% gypsum + 3% acidic FA recommended, FA application up to 4.5% increased the straw and grain yield of paddy and wheat | Kumar and Singh (2003) |

| Soil type | Treatment | Application | Results | Reference |
|---|---|---|---|---|
| Sandy loam acid lateritic soil | FA + sandy loam acid lateritic soil | | crops. In addition, FA increased N, K, Ca, Mg, S, Fe, Mn, B, Mo, Al, Pb, Ni, and Co, but decreased Na, P, and Zn in the seeds and straw of paddy and wheat crops. | Rautaray et al. (2003) |
| | | | Combined use of FA, chemical fertilizers, and organic wastes was beneficial to the improvement of crop yield, soil pH, organic carbon, and available N, P, and K in sandy loam acid lateritic soil. | |
| Acidic soil and duplex soil | FA + soil | FA to the top 0.15-m coarse textured (sandy) soil | Reduce hydraulic conductivity by 25% and improve water holding capacity. | Yunusa et al. (2006) |
| Sandy soils | Alkaline FA (pH:8.5) + compost amendments | FA at 12.5% | FA at 12.5% enhanced tomato plant growth but inhibited turf grass; Fe and Mn in tomato leaves were within permissible range | Chou et al. (2005) |
| Acid lateritic soil | Control, FA, CF, FA + CF, FYM + CF, FA + FYM + CF, L + FYM + CF, PFS + CF, FA + PFS + CF, L + PFS + CF, CR + CF, FA + CR + CF | FA at 10 t ha$^{-1}$; organic sources (FYM, PFS, and CR) at 30 kg N ha$^{-1}$; lime at 2 t ha$^{-1}$ | FA application at 10 t ha$^{-1}$ in combination with organic sources and chemical fertilizer enhanced the grain yield and nutrient uptake of rice and pod yield of peanut compared with chemical fertilizers alone. | Mittra et al. (2005) |

(Continued)

**Table 6.1** (Continued)

| Problematic soil | Amendments | FA doses | Concluding remarks | Reference |
|---|---|---|---|---|
| Two paddy soils of contrasting textures— silt loam and loamy sand | FA + PGY + paddy soil | FA + PGY (50:50 w/w) mixture at 0, 20, 40, and 60 Mg ha$^{-1}$ | Mixtures of FA + PGY should reduce P loss from rice paddy soils due to the high Ca content in this mixture, which might convert water-soluble P to less-soluble forms by precipitation process and enhance soil fertility. | Lee et al. (2007) |
| Soil | FA + soil | FA at 0,1, 2.5, 5, 10, and 15 Mg ha$^{-1}$ | Improved soil physical properties and rice growth and yield at 10 Mg ha$^{-1}$. | Mishra et al. (2007) |
| Loamy fine sand typic Hapludalfs | FA + soil | FA at 0, 40, 80, and 120 Mg ha$^{-1}$ | FA can be a good soil ameliorate for rice production without B toxicity. | Lee et al. (2008) |
| Sodic soils + press mud 10 t ha$^{-1}$ = Control | Control + FA at 10 t ha$^{-1}$, Control + FA at 20 t ha$^{-1}$, Control + PGY at 10 t ha$^{-1}$, and Control + PGY at 20 t ha$^{-1}$ | Press mud at 10 t ha$^{-1}$, FA at 10 and 20 t ha$^{-1}$, and PGY at 10 and 20 t ha$^{-1}$ | FA had beneficial impact on physico-chemical, microbial, and enzymatic activities as well crop growth. Thus the mixture of FA and press mud can be used for an efficient reclamation of sodic soils. | Singh et al. (2016) |

*AM*, Arbuscular mycorrhizal; *CF*, chemical fertilizer; *CR*, crop residue; *EC*, electrical conductivity; *FA*, fly ash; *FYM*, farmyard manure; *L*, lime; *PFS*, paper factory sludge; *PGY*, phosphogypsum; *PM*, poultry manure.
*Source*: Modified from Pandey, V.C., Singh, N., 2010. Impact of fly ash incorporation in soil systems. Agric. Ecosyst. Environ. 136, 16–27.

In the case of alkaline FA application, Singh et al. (2016) reported the effect of FA on the press mud-aided sodic soil and its resultant effects on the growth of *Vigna radiata*. Press mud blends with sodic soil at the rate of 10 tons hectare$^{-1}$ and named this mixture control sodic soil. The growth parameters of *V. radiata*, such as survival, root length, shoot length, number of nodules, number of pods, and biomass production, were significantly higher in the control sodic soil treated with FA at 20 tons hectare$^{-1}$. All the soil properties (i.e., physicochemical, biological, and biochemical) were analyzed at the crop harvest. The results revealed that FA improves the physical properties of the control sodic soil to a greater range than phosphogypsum. In addition, soil organic carbon, nitrogen, microbial biomass, bacterial population, and earthworms were highest in the control sodic soil amended with FA at 20 tons hectare$^{-1}$. The variations in the enzymatic activities like dehydrogenase, alkaline phosphatase, and β-glucosidase were significant and exhibited specific trends. They concluded that overall FA has positive effect on the growth of *V. radiata* as well as physicochemical, microbial, and enzymatic activities. Thus the mixture of FA and press mud can be used for effective reclamation of sodic soil.

In the case of acidic FA application, Khan and Khan (1996) observed the impact of different levels of acidic FA on available plant nutrients, pH, and EC in the alkaline sandy loam soil. Their outcomes exhibited a reduction in soil pH and increase in soil EC with an increasing FA quantity. Kumar and Singh (2003) also investigated the efficiency of acidic FA and gypsum in the sodic soil reclamation as well as its paddy–wheat system. Water-logged sodic soils were subjected to acidic FA practices in combination with gypsum over a period of two years. The outcomes showed that for reclaiming sodic soils, gypsum could probably be substituted up to 40% of the gypsum need with 3.0% acidic FA. Thus FA is a feasible substitute to gypsum for the sodic soil amelioration, and its application depends on the FA nature and specific site conditions need to be considered when deciding FA doses required for ameliorating sodic soils. In another case, Tiwari et al. (1992) conducted a field experiment in sodic soil, Kanpur district, Uttar Pradesh, India, to explore the possibilities of using FA for reclamation purpose. FA efficiency is applied at the dose of 10 and 20 Mg ha$^{-1}$ on changes in soil physicochemical properties and its final effect on the yields of rice and wheat variety viz., IR-8 and Sonalika, respectively. They analyzed the changes in soil characteristics after each crop harvest. The outcomes showed that FA was an effective ameliorant for improving sodic soil physicochemical properties and increasing the rice and wheat yields. At the dose of 10 and 20 Mg ha$^{-1}$, FA application improved the yields of paddy and wheat. Pyrite is more effective than FA. The soil pH, soluble salt content, and Na saturation percentage of the exchange complex significantly reduced. Soil BD showed a well-marked decrease, while hydraulic conductivity and stability index improved significantly by the FA and pyrite application.

## 6.3.2 Acidic soil reclamation

The use of FA in reclamation of acidic soil is a cheapest disposal option, because it acts as a liming agent by increasing the pH of the soil. One thing should be in mind about soil pH before adding the FA in acidic soil that the soil pH must be within acceptable limits as the increase in soil pH makes certain elements like Zn, P, and Mn unavailable (Mulford and Martens, 1971; Page et al., 1979; Carlson and Adriano, 1993). Therefore it is confirmed that the amount of FA required to reclaim acidic soils will depend on the FA pH, the state of weathering FA, as well as the soil pH of the land to be reclaimed. FA when applied to a calcareous and acidic soil and planted with *Encelia farinosa* (native desert plant species) followed by barley resulted in increased pH, EC (electrical conductivity), Ca, Mg, Na, B, and SO$_4$ in soil (Elseewi et al., 1980). Warren (1992) stated a beneficial liming effect of FA up to 110 t ha$^{-1}$ in acidic soils. Likewise, Matsi and Keramidas (1999) reported that FA application on acidic soils (pH 4.5−5.8) increased the dry biomass yield of *Lolium perenne*, pH, electrical conductivity, and uptake of phosphorus at a recommended FA dose of 20 g kg$^{-1}$ soil. Most of the researchers also addressed to the concern of buildup of certain toxic elements in the soil-plant system. While it was noted that the accumulation of elements, that is, Fe, Mn, Mo, and others, occurred, and often these elements were within permissible limits. Still care is needed in use of FA, as boron toxicity is fretful due to excessive uptake of B in many cases. In conclusion, the outcomes suggest the use of FA of low B and salt content as a liming agent in acid soils, at amounts that depend on the acid-neutralizing capacity of the FA and the buffering capacity of the soils. Lagooning, stockpiling, and leaching of FA could significantly solve this problem, minimizing boron toxicity and other ill effects of unweathered FA (Mulford and Martens, 1971; Phung et al., 1978; Page et al., 1979). The cultivation of boron-tolerant crops is further projected to lessen these side effects. Finally, FA has been recognized well as a soil conditioner for acidic infertile soils due to its high pH and the presence of micro- and macroplant nutrients (Jala and Goyal, 2006; Basu et al., 2009).

## 6.3.3 Mine reclamation

Globally, several hectares of lands are a result of coal mining, and are physically, nutritionally, and biologically poor in nature. This type of degraded lands has infertility, poor WHC, high salinity, or acidity of the soil (Moffat and McNeill, 1994). The establishment of an ecologically and socioeconomically valuable vegetation cover on coal mine sites is a potential, effective, and permanent solution. But generally, natural succession on such type of degraded lands takes more time (Wali, 1987; Jha and Singh, 1992), because its low pH is a major negative determinant in natural revegetation of mine sites (Skousen et al., 1994). The potential of wide-ranging plant species has been identified to modify the characteristics of coal mine sites, and noted to differ significantly in their abilities (Alexander, 1989a, 1989b). An effective afforestation was established on the mine site of Amarkantak, India, and *Eucalyptus camaldulensis*, *Grevillea pteridifolia*, *Pongamia pinnata*, and *Pinus roxburghii* were identified as the most appropriate plant species on the basis of growth performance (Chaturvedi, 1983).

FA is a substitute of lime used for the neutralization and stabilization of the acidic coal mine spoils (Carlson and Adriano, 1993; Stehouwer et al., 1995). It increases the acidic soil pH and improves the soil structure due to addition of Ca and other cations in acidic soil, while incorporation of alkaline FA to mine site also reduces soil BD and compaction as well as increases the WHC (Taylor and Schumann, 1988). Numerous successful reclamations of coal mine spoils have been done by using FA in different countries of the world due to having favorable factors (Table 6.2). But always, there is a great concern for potential accumulation of toxic elements in the vegetation of coal mine spoils. Srivastava and Chhonkar (2000) demonstrated a pot-based experiment on *Sorghum* sp. and *Avena sativa*. At all levels of application, FA and lime were comparable significantly to the increase in pH, sulfur, available phosphorus, and exchangeable potassium in acidic mine spoils. No symptoms of elemental toxicity were observed at the same time. In contrast, Hammermeister et al. (1998) observed B toxicity symptoms and advised that B-tolerant plant species should be grown while using FA.

**Table 6.2** FA application in reclamation of mining sites across the nations.

| Mine spoil with country | Fly ash incorporation | Impact on soil | Impact on plant | Reference |
|---|---|---|---|---|
| Mine spoil, Edmonton, AB | FA at 0–400 t ha$^{-1}$ | Increased B, Na, K, Ca, Cr, Mg, Mo, P, Se, and Sr concentration | FA at up to 200 t ha$^{-1}$ increased yield of barley silage but at 400 t ha$^{-1}$ significantly reduced it | Hammermeister et al. (1998) |
| Mine sites, WV | FA–wood waste mixtures with 10% and 20% slope | FA alone increased erosion and reduced infiltration | FA–wood chips mixture is less erodable; increased vegetation (mosses and fungi) reduced erosion | Gorman et al. (2000) |
| Acidic coal mine spoils | Alkaline FA | FA decreased the bulk density and increased pH, P, K, and S | Increased yield and uptake of P, K, and S in *Sorghum sudanense* and *A. sativa* | Srivastave and Chhonkar (2000) |
| Mine spoil, PA | FA to fill the pits of about 10–20 acres | Mine spoil reclaimed with FA | Extensive wildlife habitat plantings established | Menghini et al. (2005) |
| Mine spoil in Mpumalanga province, South Africa | FA and sewage sludge | Increase in pH and available P and K | Revegetation with *Eragrostis Tef, Chloris gayana, Cynodon dactylon, Digitaria Eriantha,* and *Medicago sativa* | Truter and Rethman (2005) |
| Mine spoil of Jharia coalfield, India | FA incorporation | Improved the physicochemical characteristics of mine spoil as well as significantly conserved the soil erosion and water runoff | Improved photosynthetic activity and leaf area | Srivastava et al. (2014) |
| Mine I, Neyveli Lignite Corporation, Tamil Nadu, India | Lignite FA dosages (0, 5, 10, 20, 50, 100, and 200 t ha$^{-1}$) and biological amendments | Lignite FA acts as an excellent soil modifier and a source of essential plant nutrients for improving the texture and fertility of the mine spoil | With significant increase in the crop yield (around 42%) over the control, together with better residual effects, especially up to 20 t ha$^{-1}$ of Lignite FA, with and without press mud. The optimum dosage of Lignite FA was 20 t ha$^{-1}$ for both one time and repeat applications | Ram et al. (2006) |

The FA is suggested as a soil amendment for pyritic mine tailings deposits in the United States (Sonderegger and Donovan, 1984), where it supported the growth and establishment of vegetation on unproductive sites through the changes in the soil pH and offers as a source of plant nutrients (Capp, 1978). Pyrite ($FeS_2$) in mine spoils and wastes is oxidized to form sulfuric acid ($H_2SO_4$), and thereby results in extremely acidic conditions (Capp, 1978; Jastrow et al., 1981). FA has been noticed to suppress the pyrite oxidation (Keefer, 1993) by performing as a blanket to inhibit oxygen diffusion. Mine spoil areas with a pH of 4.4−5.0 were reclaimed by FA at 70 tons hectare$^{-1}$ (Fail and Wochok, 1977), while highly acidic mine spoils having a pH of 2.0−3.5 needed FA at 335−1790 tons hectare$^{-1}$ (Adams et al., 1972). In this regard, establishing of halo-tolerant plant species has been suggested to somewhat reduce this quantity (Capp et al., 1975; Capp, 1978). The best primary planting combination on mine spoil is a mixture of grasses and legumes, mainly *Festuca arundinacea*, *Agrostis alba*, *L. perenne*, *Dactylis glomerata*, and *Lotus corniculatus*. After many years, trees and shrubs such as *Malus* sp., *Pinus sylvestris* and *Picea abies* could be planted (Capp, 1978). Soil depth seems to be a favorable factor in planting *Pinus strobus* on mine spoil (Andrews et al., 1998). Therefore FA incorporation in increased soil depths or the rooting region could be advantageous. In addition to its plant growth promoting nature, FA also has many other applications. It was reported that FA can be used for filling up pits caused by surface mining at Pennsylvania, and the researchers observed that there was no groundwater contamination due to FA over a 15-year period of monitoring (Menghini et al., 2005). FA application for the reclamation of mining sites is presented in Table 6.2. It has also been noticed that when FA is mixed with wood chips and applied to abandon mine sites at West Virginia, it helped to reduce top soil erosion (Gorman et al., 2000). Therefore it is obviously evident from the above instances that FA can be used in several ways to restore mining sites.

## 6.3.4 Degraded land reclamation

Sometimes soil fertility and quality decline due to overexploitation of land by crop producers and environmental causes. It means that degraded land has lost some degree of its normal productivity due to anthropogenic activities. The agricultural lands are constantly degrading and the extent is increasing due to natural and anthropogenic activities. In this case, generally, soil physicochemical and biological characteristics are affected because of alteration in BD, hydraulic conductivity, poor aeration, osmoderegulation, and particular ion toxicities (Zia-ur-Rehman et al., 2016). Nutrient-enriched FA could be used as exploitable resource to increase soil productivity of the degraded lands, because FA has soil-like properties and contains essential macronutrients (Ca, Mg, K, P, and S) and micronutrients (Fe, Cu, Zn, Mn, Co, Mo, and B) (Pandey and Singh, 2010). A lot of field trials have suggested that the chemical components of FA can ameliorate the agronomic properties of soil in positive direction (Chang et al., 1977; Wong and Wong, 1989; Sikka and Kansal, 1995; Saxena and Asokan, 1998). Pandey et al. (2009a) also proposed the FA use in the reclamation of wastelands (i.e., acidic soil, sodic soil, and mine spoil) due to its lime and gypsum properties (Shainberg et al., 1989; Kumar and Singh, 2003). In addition, FA acts as soil conditioner due to the presence of macro- and micronutrients (Pandey et al., 2009a; Pandey and Singh, 2010).

Due to overexploitation of land in agriculture, thousands of hectares of land became degraded land worldwide. In these regions, massive FA could be used effectively for reclamation purposes (Pandey and Singh, 2010; Ram and Masto, 2014). The FA quantity, which is needed to reclaim such type of lands, depends mostly on the FA pH, soil pH, and state of weathered to be reclaimed. Huge amounts of FA can be cost-effectively used with the coapplication of an organic amendment for the reclamation of degraded land. For example, FA application with poultry biosolid restored the eroded land with no adverse effect on the soil nutrient status and environment (Punshon et al., 2002). Cost−benefit analysis is a main factor on which the use of FA depends on soil management. This is mostly governed by the costs of transporting and procuring the FA against those for fertilizer, lime, dolomite, and gypsum. Transportation is feasible to be a main cost even though the coal-based thermal power stations are situated within those areas where waste/degraded lands are dominant (Yunusa et al., 2006).

## 6.4 Fly ash application along with other ameliorants

Numerous studies suggest that the potential of FA application can be enhanced if pooled with organic materials such as press mud, farmyard manure, cow manure, paper factory sludge, crop residues, organic compost, and sewage sludge for amelioration of marginal/degraded soil

(Sajwan et al., 2003; Kumpiene et al., 2007; Shen et al., 2008; Pandey and Singh, 2014; Ram and Masto, 2014).

The coapplication of FA with organic amendments on degraded soil has many positive impacts such as killing pathogens and reduced heavy-metal availability in the sludge (Wong, 1995); has decreased the leaching of main nutrients (Sajwan et al., 2003); is favorable for plants (Rautaray et al., 2003); has increased soil biological activity (Kumpiene et al., 2007); has improved soils through better texture, higher porosity, higher moisture content, higher nutrient levels, lower BD, and higher fine-grained minerals content (Shen et al., 2008). Furthermore, the application of organic materials is also useful in an inhibitory effect to soil microbes caused by FA's toxic components.

The more availability of metals is associated with a lowering of FA pH due to the incorporation of organic amendment (Pandey and Singh, 2010). Boron toxicity is the most important limiting factor of FA use in agriculture. The coapplication of an eagerly oxidizable organic material is useful to prevent boron stimulated inhibition of microbial respiration (Page et al., 1979). A significant change was noticed in the soil physico-chemical properties and paddy yield by combined application of FA, farmyard manure, and paper factory sludge (Hill and Lamp, 1980). Coapplication of lime, sewage sludge, and FA at the rate of 10:30:60 had a positive ameliorating effect on soil system (Reynolds et al., 1999). Such addition in soil exhibited favorable effects on soil pH, certain macronutrient content, and lessening of the Ni and Cd translocation (Rethman and Truter, 2001) as well as better growth and yield for the crops of corn, potatoes, and beans in pot experiments. In addition, the application of organic amendment on the FA deposited site will offer a better substrate to anchorage and growth of the desirable plants to create designer FA ecosystem toward better livelihoods and environment. Therefore it is well proved that coapplication of FA with organic amendment is helpful to enhance multiple benefits with respect to soil system as well as agricultural crop production.

## 6.5 Opportunities and challenges of fly ash application

Agricultural market opportunities have been identified for the nutrient-enriched FAs more broadly instead of only soil amelioration via gypsum replacement. The FA potential in agriculture area covers off four main soil constraints for crop producers and their incomes: acidic soil, sodic soil, improvement in soil structural and hydrological properties, and nutrient delivery and minimization of nutrient losses. Pooled, these are most important soil problems to the growth and yield of plants, now costing the industry a profit of nearly \$3 billion per annum (Yunusa et al., 2007). Moreover, a wide range of other opportunities are noted within soil-based applications. The atmospheric carbon capture in soil system due to FA incorporation is an innovative area with significant potential in the reduction of carbon emissions (Yunusa et al., 2007). The main reason of behind this case is that FA tangibly protects every organic matter that is sorbed on mineral surfaces after microbial decay that would release $CO_2$ (Palumbo et al., 2004). Australian FA application in agriculture as a soil amendment also offers chances to increase food security along with carbon sequestration (Ukwattage et al., 2013). Such type of market opportunities should be identified to the massive use of FA in a sustainable manner globally. The coal-based industries should have a clear vision and opportunity to promote to the practitioners for the use of FA and their products. Market development and its expansions are well progressive with regulatory considerations, commercial implications, ash selections, and application options for FA use (Yunusa et al., 2012). Fig. 6.2 describes the highlights of about these dimensions (the strategies for the potential and bulk use of FA in the amelioration of wide-ranging soil systems).

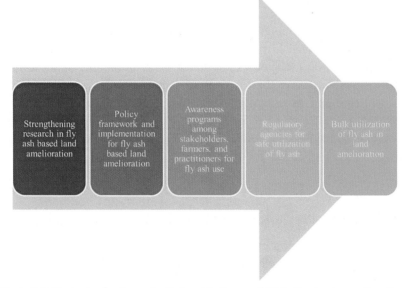

**Figure 6.2** Strategies for the potential and bulk use of FA in the lands amelioration.

The challenges of FA application in agriculture are mainly the presence of pollutants such as toxic heavy metals and its natural radioactivity. Numerous studies have proved that the effect of the pollutants is in the safe limit if FA is being used in lower doses. As the doses of FA application depend on soil to soil and crop to crop (Pandey and Singh, 2010), the study on these facts needs attention for better FA utilization in a sustainable manner. However, long-term research and development through field-based confirmatory demo are needed for assurance at the grass root level. The coapplication of FA and organic amendment has more potential in the soil amelioration and crop yields than the FA use only. Therefore a current research is needed to explore the coapplication of FA with locally available organic amendment for potential FA use in agriculture across the nations. Moreover, continuous monitoring should be given on long-term studies of FA impact on the soil health and the quality of crop productivity. Therefore the above-given apprehensions need more investigations toward the sustainability of FA use in agriculture.

## 6.6 Conclusions

The present chapter has explored the possibility of potential FA application in the amelioration of degraded soil. FA has some typical features that improve the soil physicochemical and biological properties. Coapplication of FA with organic or inorganic amendments is more beneficial to the improvement of soil quality and fertility as well as leads to plant growth and yield. Locally available organic amendments should be incorporated preferentially along with FA for soil amelioration due to their low cost and easy availability, and consequently, the benefits from their synergistic interaction could be exploited. Though numerous investigations recommend a significant potential for the FA use in amending the problematic soils, a combined approval from these studies is tough owing to heterogeneity in FA properties, soil types, crop types, and agroclimatic conditions. To develop and promote the FA blending technology, better understandings of all the above-mentioned research issues are needed.

## References

Adams, L.M., Capp, J.P., Gillmore, D.W., 1972. Coal mine spoil and refuse bank reclamation with power plant fly ash. Compost. Sci. 13, 20–26.
Adriano, D.C., Page, A.L., Elseewi, A.A., Chang, A., Straughan, I.A., 1980. Utilization and disposal of fly ash and other coal residues in terrestrial ecosystem: a review. J. Environ. Qual 9, 333–344.

Ajaz, S., Tiyagi, S., 2003. Effect of different concentrations of fly-ash on the growth of cucumber plant, *Cucumis sativus*. Arch. Agron. Soil Sci. 49, 457–461.

Alexander, M.J., 1989a. The long term effect of Eucalyptus plantations on tin mine spoil and its implication for reclamation. Landsc. Urban Plan. 17, 47–60.

Alexander, M.J., 1989b. The effect of *Acacia albida* on tin mine spoil and their possible use in reclamation. Landsc. Urban Plan. 17, 61–71.

Andrews, J.A., Johnson, J.E., Torbert, J.L., Burger, J.A., Kelting, D.L., 1998. Minesoil and site properties associated with early height growth of eastern white pine. J. Environ. Qual. 27, 192–199.

Anisworth, C.C., Rai, D., 1987. Chemical characterization of fossil fuel wastes. EPRI EA-5321. Electric Power Res. Inst., Palo Alto, CA.

Asokan, P., Saxena, M., Asolekar, S.R., 2005. Coal combustion residues environmental implications and recycling potentials. Resour. Conserv. Recycl. 43, 239–262.

Basu, M., Pande, M., Bhadoria, P.B.S., Mahapatra, S.S., 2009. Potential fly-ash utilization in agriculture: a global review. Prog. Nat. Sci. 19, 1173–1186.

Belyaeva, O.N., Haynes, R.J., 2012. Comparison of the effects of conventional organic amendments and biochar on the chemical, physical and microbial properties of coal fly ash as a plant growth medium. Environ. Earth Sci. 66, 1987–1997.

Bi, Y.L., Li, X.L., Christie, P., Hu, Z.Q., Wong, M.H., 2003. Growth and nutrient uptake of arbuscular mycorrhizal maize in different depths of soil overlying coal fly ash. Chemosphere 50, 863–869.

Bilski, J.J., Alva, A.K., Sajwan, K.S., 1995. Fly-ash. In: Reckcigl, J.E. (Ed.), Soil Amendments and Environmental Quality. CRC Press Inc, Boca Raton, FL, pp. 327–363.

Blissett, R.S., Rowson, N.A., 2012. A review of the multi-component utilisation of coal fly ash. Fuel 97, 1–23.

Capp, J.P., 1978. Power plant fly ash utilization for land reclamation in the eastern United States. In: Schaller, F.W., Sutton, P. (Eds.), Reclamation of Drastically Disturbed Lands, American Society of Agronomy. Crop Science Society of America and Soil Science Society of America, Madison, WI, pp. 339–353.

Capp, J.P., Gillmore, D.W., Simpson, D.G., 1975. Coal waste stabilization by enhanced vegetation. Min. Congr. J. 44–49.

Carlson, C.L., Adriano, D.C., 1993. Environmental impacts of coal combustion residues. J. Environ. Qual. 22, 227–247.

CEA (Central Electricity Authority), 2011. Report on fly ash generation at coal/lignite-based thermal power stations and its utilization in the country for the year 2010–11, New Delhi (18 December, 2011).

Chang, A.C., Lund, L.J., Page, A.L., 1977. Physical properties of fly ash amended soils. J. Environ. Qual. 6, 267–270.

Chaturvedi, J.K., 1983. Afforestation of bauxite mined area in central India. Indian For. 105, 465–484.

Chou, S.F.J., Chou, M.I.M., Stucki, J.W., et al., 2005. Plant growth in sandy soil/ compost mixture and commercial peat moss both amended with Illinois coal fly ash. In: Proc. World of Coal Ash Conf., Lexington, KE, pp 1–7.

Elseewi, A.A., Straughan, I.R., Page, A.L., 1980. Sequential cropping of fly ash-amended soils: effects on soil chemical properties and yield and elemental composition of plants. Sci. Total Environ. 15, 247–259.

FAO (Food and Agriculture Organization of the United Nations), 2013. FAO Statistical Yearbook 2013. World Food and Agriculture. Food and Agriculture Organization of the United Nations Rome, ISBN: 978-92-5-107396-4.

Fail Jr., J.L., 1987. Growth response of two grasses and a legume on coal fly ash amended strip mine spoils. Plant Soil 101, 149–150.

Fail Jr., J.L., Wochok, Z.S., 1977. Soyabean growth on fly ash amended strip mine spoils. Plant Soil 48, 473–484.

Fulekar, M.H., Dave, J.M., 1986. Disposal of fly ash—an enviornmental problem. Int. J. Environ.Stud. 26 (3), 191–215. <https://doi.org/10.1080/00207238608710257>.

Gill, J.S., Sale, P.W.G., Tang, C., 2008. Amelioration of dense sodic subsoil using organic amendments increases wheat yield more than using gypsum in a high rainfall zone of southern Australia.. Field Crops Res. 107, 265–275.

Gorman, J.M., Sencindiver, J.C., Horvath, D.J., Singh, R.N., Keefer, R.F., 2000. Erodibility of fly ash used as a topsoil substitute in mineland reclamation. J. Environ. Qual. 29, 805–811.

Gupta, D.K., Rai, U.N., Tripathi, R.D., Inouhe, M., 2002. Impacts of fly ash on soil and plant responses. J. Plant. Res. 115, 401–409.

Hammermeister, A.M., Naeth, M.A., Chanasyk, D.S., 1998. Implications of fly ash application to soil for plant growth and feed quality. Environ. Technol. 19, 143–152.

Haynes, R.J., 2009. Reclamation and revegetation of fly ash disposal sites—challenges and research needs. J. Environ. Manag 90, 43–53.

Hill, M.J., Lamp, C.A., 1980. Use of pulverized fuel ash from Victorian brown coal as a source of nutrients for pasture species. Aust. J. Exp. Agric. Anim. Husb. 20, 377–384.

Hooda, P.S., 2010. Trace Elements in Soils, first ed. John Wiley & Sons Ltd, Chichester, United Kingdom.

Izquierdo, M., Querol, X., 2012. Review article—leaching behavior of elements from coal combustion fly ash: an overview. Int. J. Coal Geol. 94, 54–66.

Jala, S., Goyal, D., 2006. Fly ash as a soil ameliorant for improving crop production − a review. Bioresour. Technol. 97, 1136–1147.

Jastrow, J.D., Zimmerman, C.A., Dvorak, A.J., Hinchman, R.R., 1981. Plant growth and trace element uptake on acidic coal refuse amended with lime or fly ash. J. Environ. Qual. 10, 154–160.

Jha, A.K., Singh, J.S., 1992. Influence of microsites on redevelopment of vegetation on coal mine spoils in a dry tropical environment. J. Environ. Manag. 36, 95–116.

Kalra, N., Harit, R.C., Sharma, S.K., 2000. Effect of fly ash incorporation on soil properties of textural variant soils. Bioresour. Technol. 75, 91–93.

Kassim, T.A., Williams, K.J., 2005. Environmental impact assessment of recycled wastes on surface and ground waters. Concepts; Methodology and Chemical analysis. Springer-Verlag:, Berlin, Heidelberg, p. 94.

Keefer, R.F., 1993. Coal ashes-industrial wastes or beneficial byproducts. Trace Elements in Coal and Coal Combustion Residues. Lewis Publishers, Boca Raton, FL, pp. 1–8.

Khan, M.R., Khan, M.W., 1996. The effect of fly-ash on plant growth and yield of tomato. Environ. Pollut. 92, 105–111.

Kumar, D., Singh, B., 2003. The use of coal fly ash in sodic soil reclamation. Land Degrad. Dev. 14, 285–299.

Kumari, A., Pandey, V.C., Rai, U.N., 2013. Feasibility of fern *Thelypteris dentata* for revegetation of coal fly ash landfills. J. Geochem. Explor. 128, 147–152.

Kumpiene, J., Lagerkvist, A., Maurice, C., 2007. Stabilization of Pb and Cu contaminated soil using coal fly ash and peat. Environ. Pollut. 145, 365–373.

Kutchko, B.G., Kim, A.G., 2006. Fly ash characterization by SEM-EDS. Fuel 85, 2537–2544.

Lal, J.K., Mishra, B., Sarkar, A.K., 1996. Effect of fly ash on soil microbial and enzymatic activity. J. Ind. Soc. Soil. Sci. 44, 77–80.

Lee, C.H., Lee, Y.B., Lee, H., Kim, P.J., 2007. Reducing phosphorus release from paddy soils by a fly ash—gypsum mixture. Bioresour. Technol. 98, 1980–1984.

Lee, S.B., Lee, Y.B., Lee, C.H., Hong, C.O., Kim, P.J., Yu, C., 2008. Characteristics of boron accumulation by fly ash application in paddy soil. Bioresour. Technol. 99, 5928–5932.

Lim, S.S., Choi, W.J., 2014. Changes in microbial biomass $CH_4$ and $CO_2$ emissions, and soil carbon content by fly ash co-applied with organic inputs with contrasting substrate quality under changing water regimes. Soil Biol. Biochem. 68, 494–502.

Lim, S.S., Choi, W.J., Lee, K.S., Ro, H.M., 2012. Reduction in $CO_2$ emission from normal and saline soils amended with coal fly ash. J. Soil. Sediment. 12, 1299–1308.

Lim, S.S., Choia, W.J., Chang, S.X., Arshad, M.A., Yoon, K.S., Kim, H.Y., 2017. Soil carbon changes in paddy fields amended with fly ash. Agr. Ecosyst. Environ. 245, 11–21.

Matsi, T., Keramidas, V.Z., 1999. Fly ash application on two acid soils and its effect on soil salinity, pH, B, P and on Ryegrass growth and composition. Environ. Pollut. 104, 107–112.

Matti, S.S., Mukhopadhyay, T.M., Gupta, S.K., Banerjee, S.K., 1990. Evaluation of fly ash as a useful material in agriculture. J. Ind. Soc. Soil Sci. 38, 342–344.

Menghini, M.J., Hornberger, R.J., Owen, T.D., Hill, S., Scheetz, B.E., 2005. Beneficial use of FBC coal ash for mineland reclamation in the anthracite region at the Wheelabrator Frackville and Mount Carmel Co-Gen Sites. In: Proc. World. Coal Ash Conf., Lexington, KE, USA, pp 1–14.

Mishra, M., Sahu, R.K., Padhy, R.N., 2007. Growth, yield and elemental status of rice (*Oryza sativa*) grown in fly-ash amended soil. Ecotoxicology 16, 271–278.

Mittra, B.N., Karmakar, S., Swain, D.K., Ghosh, B.C., 2005. Fly ash a potential source of soil amendment and a component of integrated plant nutrient supply system. Fuel 84, 1447–1451.

Moffat, A.J., McNeill, J.D., 1994. Restoring disturbed land for forestry, Forestry Commission Bulletin, 110. HMSO, London, UK, p. 103.

Moreno, N., Querol, X., Andres, J.M., Stanton, K., Towler, M., Nugtere, H., et al., 2005. Physico-chemical characteristics of European pulverized coal combustion flyashes. Fuel 84, 1351–1363.

Mulford, F.R., Martens, D.C., 1971. Response of alfalfa to boron in fly ash. Soil. Sci. Soc. Am. Proc. 35, 296–300.

Page, A.L., Elseewi, A.A., Straugham, I., 1979. Physical and chemical properties of fly ash from coal fired plants with reference to environmental impacts. Residue Rev. 71, 83–120.

Palumbo, A.V., McCarthy, J.F., Amonette, J.E., Fisher, L.S., Wullschleger, S.D., Lee Daniels, W., 2004. Prospects for enhancing carbon sequestration and reclamation of degraded lands with fossil-fuel combustion by-products. Adv. Environ. Res. 8, 425–438.

Pandey, V.C., 2013. Suitability of *Ricinus communis* L. cultivation for phytoremediation of fly ash disposal sites. Ecol. Eng. 57, 336–341.

Pandey, V.C., Singh, B., 2012. Rehabilitation of coal fly ash basins: current need to use ecological engineering. Ecol. Eng. 49, 190–192.

Pandey, V.C., Singh, N., 2010. Impact of fly ash incorporation in soil systems. Agric. Ecosyst. Environ. 136, 16–27.

Pandey, V.C., Abhilash, V.C., Singh, N., 2009a. The Indian perspective of utilizing fly ash in phytoremediation, photomanagement and biomass production. J. Environ. Manage. 90, 2943–2958.

Pandey, V.C., Abhilash, P.C., Upadhyay, R.N., Tewari, D.D., 2009b. Application of fly ash on the growth performance and translocation of toxic, heavy metals within *Cajanus cajan* L.: implication for safe utilization of fly ash for agricultural production. J. Hazard. Mater. 166, 255–259.

Pandey, V.C., Singh, J.S., Kumar, A., Tewary, D.D., 2010. Accumulation of heavy metals by chickpea grown in fly ash treated soil: effect on antioxidants. Clean. Soil. Air Water 38, 1116−1123.

Pandey, V.C., Singh, N., 2014. Fast green capping on coal fly ash basins through ecological engineering. Ecol. Eng. 73, 671−675.

Pandey, V.C., Prakash, P., Bajpai, O., Kumar, A., Singh, N., 2015. Phytodiversity on fly ash deposits: evaluation of naturally colonized species for sustainable phytorestoration. Environ. Sci. Pollut. Res. 22, 2776−2787.

Pandey, V.C., Bajpai, O., 2019. Phytoremediation: From theory toward practice. In: Pandey, V.C.V.C., Bauddh, K.K. (Eds.), Phytomanagement of Polluted Sites. Elsevier, Amsterdam, Netherlands, pp. 1−49.

Parab, N., Mishra, S., Bhonde, S.R., 2012. Prospects of bulk utilization of fly ash in agriculture for integrated nutrient management. Bull. Natl. Inst. Ecol. 23, 31−46.

Phung, H.T., Lund, L.J., Page, A.L., 1978. Potential use of fly ash as a liming material. In: Adriano, D.C., Brisbin, I.L. (Eds.), Environmental Chemistry and Cycling Processes, CONF-760429. US Department of Commerce, Springfield, VA, pp. 504−515.

Pichtel, J.R., Hayes, J.M., 1990. Influence of fly ash on soil microbial activity and populations. J. Environ. Qual. 19, 593−597.

Punshon, T., Adriano, D.C., Weber, J.T., 2002. Restoration of drastically eroded land using coal fly ash and poultry biosolid. Sci. Total Environ. 296, 209−225.

Ram, L.C., Masto, R.E., 2010. Review: an appraisal of the potential use of fly ash for reclaiming coal mine spoil. J. Environ. Manag. 91, 603−617.

Ram, L.C., Masto, R.E., 2014. Fly ash for soil amelioration: a review on the influence of ash blending with inorganic and organic amendments. Earth Sci. Rev. 128, 52−74.

Ram, L.C., Srivastava, N.K., Tripathi, R.C., Jha, S.K., Sinha, A.K., Singh, G., et al., 2006. Management of mine spoil for crop productivity with lignite fly ash and biological amendments. J. Environ. Manag. 79, 173−187.

Ram, L.C., Srivastava, N.K., Jha, S.K., Sinha, A.K., Masto, R.E., Selvi, V.A., 2007. Management of lignite fly ash through its bulk use via biological amendments for improving the fertility and crop productivity of soil. Environ. Manag. 40, 438−452.

Rautaray, S.K., Ghosh, B.C., Mittra, B.N., 2003. Effect of fly ash, organic wastes and chemical fertilizers on yield, nutrient uptake, heavy metal content and residual fertility in a rice−mustard cropping sequence under acid lateritic soils. Bioresour. Technol. 90, 275−283.

Rees, W.J., Sidrak, G.H., 1956. Plant nutrition on fly-ash. Plant Soil. 8, 141−159.

Rethman, N.F.G., Truter, W.F., 2001. Plant responses on soils ameliorated with waste products. In: 16th National Meeting of ASSMR, Albuquerque, NM, p. 425.

Reynolds, K.A., Kruger, R.A., Rethman, N.F.G., 1999. The manufacture and evaluation of an artificial soil prepared from fly ash and sewage sludge. Proc. Int. Ash Utiliz. Symp. Lexington, KY, pp. 378−385.

Sahin, U., Anapali, O., 2005. A laboratory study of the effects of water dissolved gypsum application on hydraulic conductivity of saline-sodic soil under intermittent ponding conditions. Ir. J. Agr. Food Res. 44 (2), 297−303.

Sajwan, K.S., Paramasivam, S., Alva, A.K., Adriano, D.C., Hooda, P.S., 2003. Assessing the feasibility of land application of fly ash, sewage sludge and their mixtures. Adv. Environ. Res. 8, 77−91.

Sarangi, P.K., Mahakur, D., Mishra, P.C., 2001. Soil biochemical activity and growth response of rice *Oryza sativa* in fly ash amended soil. Bioresour. Technol. 76, 199−205.

Saxena, M., Asokan, P., 1998. Fertility improvement by fly-ash application in wasteland. In: Fly-ash Disposal and Utilization, CBIP, International Conference I. Jan. 20-22, CBIP, New Delhi, pp. 90−104.

Scotti, A., Silva, S., Botteschi, G., 1999. Effect of fly ash on the availability of Zn, Cu, Ni and Cd to chicory. Agric. Eco. Env. 72, 159–163.

Shaheen, S.M., Hooda, P.S., Tsadilas, C.D., 2014. Opportunities and challenges in the use of coal fly ash for soil improvements---a review. J. Environ. Manag. 145, 249–267.

Shainberg, I., Sumner, M.E., Miller, W.P., Farina, M.P.W., Pavan, M.A., Fey, M.V., 1989. Use of gypsum on soils: a review. Adv. Soil. Sci. 9, 1–111.

Shen, J.F., Zhou, X.W., Sun, D.S., Fang, J.G., Liu, Z.J., Li, Z., 2008. Soil improvement with coal ash and sewage sludge: a field experiment. Environ. Geol. 53, 1777–1785.

Shende, A., Juwarkar, A.S., Dara, S.S., 1994. Use of fly ash in reducing heavy metal toxicity to plants. Resour. Conserv. Recycl. 12, 221–228.

Sikka, R., Kansal, B.D., 1995. Effect of fly-ash application on yield and nutrient composition of rice, wheat and on pH and available nutrient status of soil. Bioresour. Technol. 51, 199–203.

Singh, J.S., Pandey, V.C., 2013. Fly ash application in nutrient poor agriculture soils: impact on methanotrophs population dynamics and paddy yields. Ecotoxicol. Environ. Saf. 89, 43–51.

Singh, S., Ram, L.C., Masto, R.E., Verma, S.K., 2011a. A comparative evaluation of minerals and trace elements in the ashes from lignite, coal refuse, and biomass fired power plants. Int. J. Coal Geol. 87, 112–120.

Singh, J.S., Pandey, V.C., Singh, D.P., 2011b. Coal fly ash and farmyard manure amendments in dry-land paddy agriculture field: effect on N-dynamics and paddy productivity. Appl. Soil. Ecol. 47, 133–140.

Singh, K., Pandey, V.C., Singh, B., Patra, D.D., Singh, R.P., 2016. Effect of fly ash on crop yield and physico-chemical, microbial and enzyme activities of sodic soils. Environ. Eng. Manag. J. 15 (11), 2433–2440.

Skousen, J.G., Johnson, C.D., Garbutt, K., 1994. Natural revegetation of 15 abandoned mine land sites in West-Virginia. J. Environ. Qual. 23, 1224–1230.

Skousen, J., Ziemkiewicz, P., Yang, J.E., 2012. Use of coal combustion by-products in mine reclamation: review of case studies in the USA. Geosyst. Eng 15, 71–83.

Sonderegger, J.L., Donovan, J.J., 1984. Laboratory simulation of fly ash as an amendment to pyrite-rich tailings. Ground Water Monit. Rev. 4, 75–80.

Srivastava, A., Chhonkar, P.K., 2000. Amelioration of coal mine spoils through fly ash application as liming material. J. Sci. Ind. Res. 59 (4), 309–313.

Srivastava, N.K., Ram, L.C., 2010. Reclamation of coal mine spoil dump through fly ash and biological amendments. Int. J. Ecol. Dev. 17 (F10), 17–33.

Srivastava, N.K., Ram, L.C., Masto, R.E., 2014. Reclamation of overburden and lowland in coal mining area with fly ash and selective plantation: a sustainable ecological approach. Ecol. Eng. 71, 479–489.

Srivastave, A., Chhonkar, P.K., 2000. Amelioration of coal mine spoils through fly ash application as liming material. J. Sci. Ind. Res. 59, 309–313.

Stehouwer, R.C., Sutton, P., Fowler, R.K., Dick, W.A., 1995. Minespoil amendment with dry flue gas desulfurization by-products: element solubility and mobility. J. Environ. Qual. 24, 165–174.

Taylor Jr., E.M., Schumann, G.E., 1988. Fly ash and lime amendment of acidic coal spoil to aid revegetation. J. Environ. Qual. 17, 120–124.

Tejada, M., Garcia, C., Gonzalez, J.L., Hernandez, M.T., 2006. Use of organic amendment as a strategy for saline soil remediation: influence on the physical, chemical and biological properties of soil. Soil. Biol. Biochem. 38, 1413–1421.

TIFAC, 2001. Technology linked business opportunity publications. Non Conventional Sources of Plant Nutrient & Soil Conditioners to Enhance Agricultural Productivity, Code no. TMS1551.

Tiwari, K.N., Sharma, D.N., Sharma, V.K., Dingar, S.M., 1992. Evaluation of fly ash and pyrite for sodic soil rehabilitation in Uttar Pradesh, India. Arid. Soil. Res. Rehabil. 6 (2), 117–126. Available from: https://doi.org/10.1080/15324989209381304.

Truter, W.F., Rethman, N.F.G., 2005. Revegetating mine land that has been ameliorated with alternative soil ameliorants. In: Proc. World Coal Ash Conf., Lexington, KE, pp. 1–12.

Ukwattage, N.L., Ranjith, P.G., Bouazza, M., 2013. The use of coal combustion fly ash as a soil amendment in agricultural lands (with comments on its potential to improve food security and sequester carbon). Fuel 109, 400–408.

UNCCD (United Nations Convention to Combat Dessertification), 2012. Message of UNCCD Executive Secretary on World Day to Combat Desertification, 17 June 2012. <http://www.unccd.int/en/media-center/MediaNews/Pages/highlightdetail.aspx?>.

UNWPP (United Nations World Population Prospects), 2013. World Population Prospects: The 2012 Revision. <www.unpopulation.org>.

Wali, M.K., 1987. The structure, dynamics and rehabilitation of drastically disturbed ecosystems. In: Khoshoo, T.N. (Ed.), Perspectives in Environmental Management. Oxford Publications, New Delhi, pp. 163–183.

Warren, C.J., 1992. Some limitations of sluiced fly ash as a liming material for acidic soils. Waste Manag. Res. 10, 317–327.

Wong, J.W.C., 1995. The production of artificial soil mix from coal fly ash and sewage sludge. Environ. Technol. 16, 741–751.

Wong, M.H., Wong, J.W.C., 1986. Effects of fly ash on soil microbial activity. Environ. Pollut. Ser. A 40, 127–144.

Wong, J.W.C., Wong, M.H., 1987. Co-recycling of fly ash and poultry manure in nutrient-deficient sandy soil. Resour. Conserv. 13, 291–304.

Wong, M.H., Wong, J.W.C., 1989. Germination and seedling growth of vegetable crops in fly-ash amended soils. Agric., Ecosyst. Environ. 26, 23–35.

Wong, J.W.C., Wong, M.H., 1990. Effects of fly-ash on yields and elemental composition of two vegetables, *Brassica parachinensis* and *B. chinensis*. Agric. Ecosys. Env. 30, 251–264.

Yunusa, I.A.M., Eamus, D., DeSilva, D.L., Murray, B.R., Burchett, M.D., Skilbeck, G. C., et al., 2006. Fly-ash: an exploitable resource for management of Australian agricultural soils. Fuel 85, 2337–2344.

Yunusa, I.A.M., Manoharan, V., Burchett, M.D., Eamus, D., Skilbeck, C.G., 2007. Utilization of coal combustion products in agriculture. In: Coal Combustion Products Handbook, Chapter 12. In: Gurba, L., Heidrich, C., Ward, C. (Eds.), Cooperative Research Centre for Coal in Sustainable Development. Australian Black coal utilization Research Limited, Australia, pp. 373–414. , 2007.

Yunusa, I.A.M., Loganathan, P., Nissanka, S.P., Manoharan, V., Burchett, M.D., Skilbeck, C.G., et al., 2012. Application of coal fly ash in agriculture: A strategic perspective. Crit. Rev. Environ. Sci. Technol. 42, 559–600.

Zia-ur-Rehman, M., Murtaza, G., Qayyum, M.F., Saifulla, Rizwan, M., Ali, S., et al., 2016. Degraded soils: origin, types and management. In: Hakeem, K., Akhtar, J., Sabir, M. (Eds.), Soil Science: Agricultural and Environmental Prospectives. Springer, Cham, Switzerland, <https://doi.org/10.1007/978-3-319-34451-5-2>.

## Further reading

Sahin, U., Eroglu, S., Sahin, F., 2011. Microbial application with gypsum increases the saturated hydraulic conductivity of saline-sodic soils. Appl. Soil Ecol. 48, 247–250.

# CHAPTER 7

# Afforestation on fly ash catena: an adaptive fly ash management

## Contents

## 7.1 Introduction

Fly ash (FA) generation is unabatedly increasing worldwide due to our dependence on coal-fired thermal power plants for energy purposes. FA produced in huge quantities is not being disposed safely. FA is generally utilized as building materials and nutrients for poor agricultural soils, while the rest is transported to disposal sites (Pandey et al., 2009a, b; Singh and Pandey, 2012; Singh et al., 2011a,b). There are several terms used for FA disposal sites such as landfill, basin, lagoon, dump, deposit, and pond. However, the term "FA catena" has been used in this chapter for disposal sites, as several thousand hectares of land near thermal power plants are overloaded with FA across the world. The global FA generation is

*Phytomanagement of Fly Ash*
ISBN: 978-0-12-818544-5
DOI: https://doi.org/10.1016/B978-0-12-818544-5.00007-9

estimated at around 600 million tons annually (Ram et al., 2008), which would require about 3235 km$^2$ of land (equal to threefold area of Hong Kong) by 2015 (Pandey and Singh, 2012). In addition, a number of pollutants like metal(loid)s (Cr, Cd, Pb, Hg, As, B, Ni, etc.), polycyclic aromatic hydrocarbons, radioactive elements (Celik et al., 2007; Ruhl et al., 2009, Pandey et al., 2011a), and a few greenhouse gases like carbon dioxide, methane, nitrous oxide, and sulfur dioxide are added in the atmosphere as a result of coal combustion. Coal-fired power stations release pollutants in the environment by two pathways: (1) atmospheric emissions through stacks; and (2) leaching of heavy metals from FA catena to the soil.

FA catena needs to be vegetated properly in order to mitigate such pollution effects. Ash catena can be rehabilitated with an intensive vegetation cover by using microbial inoculants and organic manures in the planting pits. Preliminary efforts have been made to establish various tree species of diverse nature. However, the establishment and growth of plants on the substrate of FA catena is inhibited on account of some adverse physicochemical properties of the substrate (Mulhern et al., 1989). The possible limitations include high pH, negligible nitrogen and available phosphorus, lack of organic carbon (OC), absence of microbial activities, and presence of some toxic heavy metals (Pandey and Singh, 2010). Physicochemical properties of FA dumping site can be improved by the addition of suitable amendments for the initial establishment of seedlings. Furthermore, biofertilizer cultures may be applied for the purpose of accelerating the growth of plants. Application of organic amendments provides nutrients such as carbon, nitrogen, phosphorus, and potassium to support plant growth and to help in decreasing the toxicity of heavy metals owing to the enhanced organic matter content that complexes the heavy metals by precipitation (Haynes, 2009; Pandey and Singh, 2010). FA dumping sites near thermal power plants are usually considered as an environmental nuisance. Revitalization processes, the formation of vegetation cover on such bare lands through herbaceous and woody plant species, are usually adopted to prevent the dispersion of suspended particulate matter from FA catena (Bogdanović, 1990; Dželetović and Filipović, 1995). Eijsackers (2010) reviewed earthworms as one of the first colonizers in FA catena. Establishment of plantations on such sites is still a challenging task to the ecologists, environmentalists, and policy makers. Nevertheless, it can be done through repeated afforestation attempts (Singh et al., 2012a), involving latest biotechnological tools (Pandey et al.,

2011b; Singh et al., 2012b). Following this approach, some plantations have been done on FA catena, but the growth potential of species is yet to be determined at temporal scale. Preliminary growth performances of these species were found satisfactory, depending upon the quality of the site (Juwarkar and Jambhulkar, 2008; Ram et al., 2008; Mitrović et al., 2012).

New forests are being created worldwide on different types of degraded lands with various silvicultural approaches. Transformation of an abandoned FA catena in biodiversity rich forest ecosystems helps to alleviate different types of pollution. Furthermore, the rehabilitation of FA catena under productive land use systems is the need of the hour in view of the dwindling forest resources all over the world. Therefore afforestation on FA catena, apart from providing various goods and services to the society, plays a great role in the abatement of pollution (Fig. 7.1). Carbon sequestration in FA substrate controls runoff and maintains congenial watershed hydrology, facilitating to ground flora and accommodating a rich biodiversity at spatial scale. Despite the availability of a plethora of literature, as shown by several review articles, there is still a wide scope for addressing the afforestation issues on FA catena for its safe management. Thus this chapter deals to understand the development of tree plantations of suitable species on an abandoned FA catena to derive a number of tangible and intangible benefits.

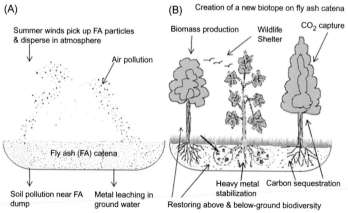

**Figure 7.1** Conceptual design represents the role of afforestation for the mitigation of environmental threats of FA catena along with secondary environmental benefits. (A) FA catena causes soil, air, and water pollution through different modes. (B) Creation of a new biotope on FA catena.

## 7.2 Limiting factors of plantation on fly ash catena

Plantation plays a key role to enhance the substrate quality of FA catena in order to develop FA ecosystems as well as environmental quality of nearby areas of FA catena. It stabilizes the ash against water and wind erosion, and provides a habitat for wildlife (Hodgson and Townsend, 1973; Hodgson and Buckley, 1975). Landscape quality can also be upgraded visually by vegetation. However, natural colonization of vegetation on FA catena is a slow process. For example, it took 40−50 years for the transformation of passive FA catena to resemble the soil in England (Shaw, 1992). As weathering progressed on FA catena, vegetation richness and its diversity increased. Phytotoxic symptoms have been reported on some plants growing on FA catena. Almost certainly, physicochemical and biological conditions of FA catena are harmful to plant survival, establishment, and growth. Thus the establishing flora on FA catena is a tough job owing to FA physicochemical limitations (Mulhern et al., 1989). Such adverse physicochemical and biological conditions of FA catena could be resolved by applying suitable waste amendments in the substrate of FA catena for plant establishment and to achieve a good and healthy green cover. Some amendments [i.e., press mud, farmyard manure (FYM), cowdung manure, sewage sludge, sugar mill waste, etc.] are suitable for the amelioration of FA physicochemical and biological properties (for more details, see Pandey and Singh, 2010; Ram and Masto, 2014).

*Physical problems*—FA physical characteristics can inhibit effective plant establishment on FA catena, owing to the FA pozzolanic properties that cement FA particles when wetted in the presence of $CaSO_4$ (liming material) in FA. Thus the compacted layers are formed in the FA catena that lessens root penetration, aeration, and water infiltration (Hodgson and Townsend, 1973; Bradshaw and Chadwick, 1980). Lack of humus colloids in FA catena makes the ash structure poor. FA physical properties can be improved by the addition of suitable amendments for the initial establishment of plants. Some suitable amendments are press mud, organic wastes, FYM, composts, and sewage sludge for the amelioration of FA physical properties (Ram and Masto, 2014).

*Nutrient problems*—Some essential plant nutrients are deficient in FA. N is absent in FA, as most of the N is volatilized during the process of coal combustion (Adriano et al., 1980). P in FA is high $(400−800 \ \mu g \ g^{-1})$ (Page et al., 1979), but it is not present in FA in a readily available form to plants possibly owing to complexation with ash

Fe and Al (Hodgson and Townsend, 1973; Bradshaw and Chadwick, 1980). Ca, Mg, K, Na, and other micronutrients are sufficient. N and P can be added in the substrate of FA catena through the blending of various suitable amendments (Ram and Masto, 2014). The problem of N can be avoided by the selecting N-fixing plant species during the plantation programs on FA catena (Pandey and Singh, 2011).

*Toxicity problems*—The most important factors limiting plant establishment on FA catena are too much concentrations of B and soluble salts (Carlson and Adriano, 1993). B content of FA can exceed 250 $\mu$g g$^{-1}$, which is beyond the concentration of 30 $\mu$g g$^{-1}$ considered highly toxic to vegetation (Hodgson and Townsend, 1973; Hodgson and Buckley, 1975). FA soluble salt content can result in electrical conductivity values of $>$4 mS cm$^{-1}$, at which level plant growth is adversely affected (Townsend and Gillham, 1975). The high pH of FA can also inhibit plant growth on FA catena, causing paucities of essential nutrients, that is, P and trace elements, for instance, Cu, Fe, Zn, and Mn. In contrast, FA alkalinity can increase the accumulation of some nonessential trace elements, that is, Se, V, and As (Page et al., 1979; Adriano et al., 1980). Most of the toxicity problems (i.e., high pH, metal(loid)s, etc.) of FA can be reduced by the amelioration of the substrate of FA catena through the blending of suitable amendments (for detail, see Pandey and Singh, 2010; Ram and Masto, 2014).

*Biological problems*—Freshly deposited FA is a biologically inactive site (microbiologically sterile). Invasion of microbes to FA catena starts when it exposed to air and water. Though the microbial population and diversity usually increase as FA weathering proceeds and nutrients accumulate, the low content of organic matter, possibly, is a factor limiting microbial growth (Klubek et al., 1992). The progress of a vital microbial community can enhance the suitability of the substrate of FA catena for plant growth (Rippon and Wood, 1975). In this regard, several organic materials have been suggested to increase the growth of microbial population through the blending with FA (for detail, see Pandey and Singh, 2010; Ram and Masto, 2014).

## 7.3 Flora on fly ash catena

Several plant species have been suggested for effective revegetation of passive FA catena worldwide. The keystone species for fast revegetation of FA catena may be divided in two categories. The first category is based on naturally grown vegetation, and the second one is assisted vegetation on FA catena. In the first category, several naturally grown potential plant species belonging to grasses, legumes, shrubs, and trees have been identified and suggested for the application of green cover development on abandoned FA catena through revegetation and phytoremediation programs with ecologically and socioeconomically benefits. In the second category, several assisted vegetation/induced plant species on FA catena have been advised for the use of green cover development with multipurpose benefits through potential rehabilitation programs on the passive FA catenas worldwide. The majority of vegetation studies revealed that a wide range of plant species that are planted or naturally colonized abandoned FA catenas for revegetation and regeneration belong to the families *Poaceae, Rosaceae, Asteraceae, Brassicaceae, Chenopodiacea,* and *Fabaceae* (Pandey et al., 2015b; Gajić et al., 2019). It is also noted that inactive FA catenas were effectively revegetated with plant species that come from the mostly xerophytic and halophytic families.

## 7.3.1 Natural vegetation on fly ash catena

Naturally growing plant species on FA dumps seem to be a vital tool for the stabilization and restoration of newly FA catenas, because naturally growing plant species on FA catena response better and can easily survive than the introduced species from other areas (Pandey and Singh, 2011). Naturally growing native grass species act as a primary colonizer due to their adapted potential against stress conditions of passive FA catena. They multiply quickly and rapidly establish a thick green cover in a nutrient poor (i.e., unavailable phosphorus and nitrogen deficiency) substrate of FA catena (Pandey and Singh, 2011). Native grass species offer many benefits, such as enhanced soil fertility, and deliver sustainable microclimates for the establishment of secondary colonizers. The massive fibrous root system of grass species catches FA particles, prevents FA erosion, preserves FA moisture, and controls heavy metal leaching. Even after the death of the grass, drying of their roots and shoots forms biomass called mulches that provide OC and nutrients to the substrate of FA catena (Pandey and Singh, 2014; Pandey et al., 2015b, 2016; Gajić et al., 2016). The naturally growing plant species on FA catena have a great capability to tolerate the FA toxicity and deficiency of some nutrients in FA (Pandey et al., 2012a,b; Gajić et al., 2016). A number of naturally growing plant species on FA dumps have been reported for the use of revegetation/rehabilitation programs worldwide (Table 7.1).

**Table 7.1** Earthworms as early colonizers of FA catena.

| Earthworm species | References |
| --- | --- |
| *L. rubellus* | Ma and Eijsackers (1989); Satchell and Stone (1977) |
| *L. castaneus* | Satchell and Stone (1977) |
| *L. terrestris* | Satchell and Stone (1977) |
| *D. rubida* | Satchell and Stone (1977) |
| *D. octaedra* | Satchell and Stone (1977) |
| *A. chlorotica* | Ma and Eijsackers (1989); Satchell and Stone (1977); Tamis and Udo de Haes (1995) |
| *A. rosea* | Satchell and Stone (1977) |
| *A. caliginosa* | Ma and Eijsackers (1989); Satchell and Stone (1977) |

In India, there is a number of naturally colonizing potential plant species that have been identified for the revegetation of FA catenas (Figs. 7.2 and 7.3). These are *Acacia nelotica*, *Amaranthus deflexus*, *Azolla caroliniana*, *Cassia tora*, *Cannabis sativa*, *Calotropis procera*, *Chenopodium album*, *Cynodon dactylon*, *Croton bonplandium*, *Eclipta alba*, *Ipomea carnea*, *Limnanthe*, *Prosopis juliflora*, *Parthenium hysterophorus*, *Saccharum munja*, *Saccharum bengalense*, *Saccharum revennae*, *Saccharum spontaneum*, *Solanum nigrum*, *Sida cordifolia*, *Typha latifolia*, and *Thelypteris dentata* (Gupta and Sinha, 2008; Maiti and Jaiswal, 2008; Rau et al., 2009; Pandey, 2012a,b; Pandey and Singh, 2014; Pandey, 2015; Pandey et al., 2015b; Pandey et al., 2016; Kumari et al., 2013). Figs. 7.2 and 7.3 show the naturally colonized potential plant species on the Indian FA deposits. In Serbia, the naturally growing plant species that are tolerant to hostile conditions on FA catenas are *Amorpha fruticosa*, *Ambrosia artemisifolia*, *Calamagrostis epigeios*, *Crepis setosa*, *C. album*, *Cirsium arvense*, *Cichorium intybus*, *Carduus acanthoides*, *Daucus carota*, *Echium vulgare*, *Eupatorium cannabinum*, *Euphorbia cyparissias*, *Epilobium collinum*, *Epilobium hirsutum*, *Erigeron canadensis*, *Festuca rubra*, *Hypericum perforatum*, *Linaria vulgaris*, *Papaver rhoes*, *Populus alba*, *Rosa canina*, *Salsola kali*, *Sinapis arvense*, *Silene vulgaris*, *Tamarix gallica*, and *Verbascum phlomoides* (Pavlović et al., 2004; Djurdjević et al., 2006; Mitrović et al., 2008; Gajić and Pavlović, 2018; Gajić et al., 2019). Fig. 7.4 depicts the spontaneous colonized potential plant species on the Serbian FA deposits. In Hong Kong, naturally growing plant species that are able to grow on FA catenas are *Chenopodium acuminatum*, *Eleusine indica*, *Fimbristylis polytrichoides*, *Neyraudia reynaudiana*, *Panicum repens*, *Pteridium aquiilinum*, and *Tamarix chinensis* (Chu, 2008). Furthermore, in Poland, naturally growing plant species

**Figure 7.2** Spontaneous colonized potential plant species on the FA deposits of Feroze Gandhi Unchahar Thermal Power Station (FGUTPP), Raebareli, Uttar Pradesh, India. (A) *S. spontaneum* colonized on the FA deposits as a spontaneous native tall grass invader. (B) *P. juliflora* and *S. munja* grown spontaneously on the FA deposits. (C) *S. munja* and *C. dactylon* grown naturally on the FA deposits. (D) *C. dactylon* grown on the FA deposits as a spontaneous native grass colonizer. Photo courtesy: *V.C. Pandey.*

with a high ecological potential to colonize on FA deposits are *Festuca arundinacea*, *F. rubra*, *Lolium perenne*, *Poa annua*, *Rumex acetosa*, *Taraxacum officinale*, *Poa trivialis*, and *Viccia cracca* (Jasionkowski et al., 2016). Worldwide naturally colonizing plant species with a high potential to grow on FA catena are given in Table 7.2.

## 7.3.2 Assisted vegetation on fly ash catena

Ash dumping in open places causes many disturbances in the ecosystem owing to the dispersion of ash particles in the atmosphere by wind and water erosion. The leaching of heavy metals from FA catena is also a serious problem for the surrounding terrain and underground water. Planting trees on FA catena inhibits erosion, stabilizes the leaching of heavy metals from plant uptake, binds the ash particles by their fine roots, and develops

**Figure 7.3** Naturally growing potential plant species on the FA deposits in India. (A) *Rumex acetosella* colonized as a perennial herb on the FA deposits of Patratu Thermal Power Station, Ranchi, Jharkhand, India. (B) *I. carnea* is a shrub plant species grown on the FA dumps of Patratu Thermal Power Station. (C) *C. dactylon* with *I. carnea* and *C. procera* grown on the FA dumps of Patratu Thermal Power Station. Photo courtesy: *Deblina Maiti*.

an aesthetically pleasant landscape. Phytoremediation has been used as an effective tool in rehabilitation and restoration of FA catena (Ram et al., 2008; Pandey, 2012b). From the ecological point of view, the development of a new biotope on FA catena is the most effective and desirable approach in alleviating the hazardous effects of ash along with other secondary benefits (Pandey and Singh, 2012). FA catena can be used for

**Figure 7.4** Spontaneous colonized plant species on the largest thermal power plant Nikola Tesla (TENT—A). (A) *Calamagrostis epigejos* grown as a spontaneous native grass colonizer on the FA deposits of the passive cassette 11 years old. (B) *A. fruticosa* is a nonnative shrub and spontaneous colonizer on the FA deposits of the passive cassette 11 years old. (C) *P. alba*—spontaneous colonized native tree on the FA deposits on the passive cassette 3 years old at TENT—A, Obrenovac, [D] Landscape and spontaneous species *P. alba, C. epigejos, Oenothera biennis* colonized on the fly ash deposits on the passive cassette 11 years old at the TENT—A in Obrenovac, Belgrade. Photo courtesy: *Gordana Gajić.*

biomass production and energy conservation. FA contains soil like physicochemical characteristics and is rich in various essential and nonessential elements, yet poor in both nitrogen and available phosphorus (Pandey and Singh, 2010). Numerous experiments based on fields and greenhouses have demonstrated that many chemical constituents of FA are beneficial to plant growth and even improve the physicochemical and nutritional qualities of degraded soil (Pandey and Singh, 2010). Afforestation is difficult on FA catena unless organic amendments, microbial inoculants, and biofertilizers are applied with a standardized silvicultural technology. A number of field trials were executed to select plant species appropriate for the revegetation of passive FA catena worldwide (Table 7.3).

In India, a study was conducted on a 10-ha area of FA catena at Khaperkheda Thermal Power Station, Khaperkheda, Maharashtra, in order to vegetate with suitable plants, namely *Annona squamosa*,

**Table 7.2** Plant species naturally growing on FA catena all over the world.

| S. No. | Naturally growing plant species | Country | References |
|---|---|---|---|
| 1 | *A. nelotica, A. deflexus, A. caroliniana, C. procera, C. sativa, C. tora, C. album, C. bonplandium, C. dactylon, E. alba, Limnanthe, I. carnea, P. hysterophorus, P. juliflora, S. bengalense, S. munja, S. spontaneum, S. revennae, S. cordifolia, S. nigrum, T. dentate,* and *T. latifolia* | India | Gupta and Sinha (2008); Maiti and Jaiswal (2008); Pandey and Singh (2014); Pandey (2012a,b, 2015); Pandey et al. (2015b, 2016a); Ray and Adholeya (2009); Kumari et al. (2016) |
| 2 | *A. artemisifolia, A. fruticosa, E. cannabinum, C. setosa, E. collinum, V. phlomoides, C. arvense, F. rubra, E. canadensis, C. epigeios, T. gallica, C. album, S. kali, S. arvense, D. carota, P. rhoes, H. perforatum, E. cyparissias, C. intybus, C. acanthoides, L. vulgaris, E. hirsutum, E. vulgare, S. vulgaris, R. canina,* and *P. alba* | Serbia | Pavlović et al. (2004); Djurdjević et al. (2006); Mitrović et al. (2012); Gajić and Pavlović (2018); Gajić et al. (2019); Gajić (2014) |
| 3 | *E. indica, N. reynaudiana, T. chinensis, C. acuminatum, F. polytrichoides, P. aquiilinum,* and *P. repens* | Hong Kong | Chu (2008) |
| 4 | *F. arundinacea, F. rubra, L. perenne, P. annua, R. acetosa, T. officinale, P. trivialis,* and *V. cracca* | Poland | Jasionkowski et al. (2016) |

*Azadirachta indica, Cassia siamea, Dalbergia sissoo, Dendrocalamus strictus, Emblica officinalis, Eucalyptus hybrid, Pongamia pinnata,* and *Tectona grandis* (Juwarkar and Jambhulkar 2008). Another plantation was carried out on the temporary FA catena at Ramagundam Super Thermal Power Station, Karimnagar, Telangana, India. This plantation was done on about a 1-ha area with *Eucalyptus globulus, Acacia auriculiformis, Casuarina equisetifolia, Ipomoea carnosa,* and *Leucaena glauca.* These species have shown a good growth performance (Zevenbergem et al., 2000). A similar activity was

**Table 7.3** Afforestation programs on passive FA deposits through an assisted vegetation approach over the world.

| S. no. | Assisted vegetation | Amendments | Observation | Country | References |
|---|---|---|---|---|---|
| 1 | A. indica, C. siamea, E. hybrid, P. pinnata, E. officinalis, T. grandis, A. squamosa, D. strictus, and D. sissoo | Soil, FYM (50 t ha$^{-1}$), lime slurry (2.2 t ha$^{-1}$), VAM, Rhizobium (leguminous), and Azatobacter (nonleguminous). | FA amendment with FYM and biofertilizer assisted in profuse root expansion showing 15 times higher growth in D. strictus than the control. Thus the amendment and biofertilizer are better supportive materials for establishing plants on FA catena. | India | Juwarkar and Jambhulkar (2008) |
| 2 | E. globulus, A. auriculiformis, and C. equisetifolia | | Plantation can be developed directly on the FA catena without amendment with fertilizers. | India | Zevenbergem et al. (2000) |
| 3 | A. indica, C. siamea, E. hybrida, E. officinalis, T. grandis, D. strictus, D. sissoo, and P. pinnata | Soil, FYM (50 t ha$^{-1}$), lime slurry (2.2 t ha$^{-1}$), VAM, Rhizobium (leguminous), and Azatobacter (nonleguminous). | It was noticed that C. siamea accumulated a;l metals at higher concentrations compared with other plants. C. siamea could be used as a hyperaccumulator species for phytoremediation of FA catena. | India | Jambhulkar and Juwarkar (2009) |
| 4 | D. sissoo, E. tereticornis, M. azadirachta, Populus deltoides, S. robusta, T. grandis, A. procera, C. equisetifolia, D. strictus, P. euphratica, B. ceiba, G. arborea, and P. aculeata | FYM and mycorrhizal biofertilizer. | Marked alterations in nutritional status, microbial population, and microbial activities were observed in reclaimed FA catenas. Thus mycorrhiza technology has a great potential to develop a strong, dense, and resilient green cover. | India | Das et al. (2013) |
| 5 | T. tentandra, P. alba, R. pseudoacacia, and A. fruticosa | | Naturally colonized woody species can be used in the first step to stabilize the FA catena. | Serbia | Kostić et al. (2012) |

| # | Species | Amendment | Description | Location | Reference |
|---|---------|-----------|-------------|----------|-----------|
| 6 | *Pinus taeda*, *Pinus palustris*, *L. styraciflua*, and *P. occidentalis* | | Hardwood species (*P. occidentalis* and *L. styraciflua*) survived and grew acceptably on FA catena. | United States | McMinn et al. (1982) |
| 7 | *Populus* sp., *L. styraciflua*, *P. occidentalis*, *Acer rubrum*, *P. taeda*, and *Myrica cerifera* | | This case study showed that FA catena may provide conditions more favorable for tree growth than local soils. Trees can grow on FA catena without using topsoil. FA physicochemical characteristics are important factors to determine which tree species can become established on a specific site. *P. occidentalis* and *L. styraciflua* grew the best on the wet FA catena. | United States | Carlson and Adriano (1991) |
| 8 | *Betula* spp. and *Salix* spp. | | Results revealed that soil development on FA catena correlated more closely with site age than it did with the plant community composition. | England | Shaw (1992) |
| 9 | *A. glutinosa* and *A. incana* | CFA (control), CFA + MS, CFA + L. Sewage sludge (4 Mg ha$^{-1}$) mixed with grass seedling (200 kg ha$^{-1}$), and NPK mineral fertilization (N−60, P−36, and K−36 kg ha$^{-1}$) were applied by hydroseeding. | The best amendment was lignite culm for FA substrate improvement. So, the goal of biological stabilization of FA catena would be the greatest increase of tree biomass of energy plantations. Yet, the introduction of alders directly on FA catena using NPK fertilizing and hydroseeding with seed sludge may be suggested due to economic reasons, especially when the introduced alders are to have mainly protective and phytomelioration functions and thus prepare the substrate for the afforestation and the next generation of target species. | Poland | Krzaklewski et al. (2012) |

(*Continued*)

**Table 7.3** (Continued)

| S. no. | Assisted vegetation | Amendments | Observation | Country | References |
|---|---|---|---|---|---|
| 10 | *A. auriculiformis* and *L. leucocephala*. | LA–VC–N fertilizer, LA–SSC–N fertilizer, LA–VC–RI–N fertilizer, and LA–SSC–RI–N fertilizer. | Both ameliorants improved the physical properties of lagoon ash by increasing the porosity and aeration for root development, but only sewage sludge compost (SSC) increased nutrient levels (N, P, and K) of the infertile lagoon ash. Even both species showed the potential to establish on mended lagoon ash, but *A. auriculiformis* being the best adapted. | China | Cheung et al. (2000) |
| 11 | *R. pseudoacacia* and *Fraxinus* spp. | | 30 years of pedogenesis changed the original properties of the substrate of fresh FA catena, as the decrease of pH and salinity, the increase of total organic carbon contents and total nitrogen due to the accumulation of organic matter, as well as the formation of pedogenic carbonates. | Poland | Uzarowicz et al. (2018) |
| 12 | *Secale cereale, Arrhenatherum elatius, Lolium multiflorum, F. rubra, Dactylis glomerata, Vicia villosa, Lotus corniculatus, Medicago sativa, Tamarix tetrandra, R. pseudoacacia, Populus euramericana,* and *Salix alba* | No use topsoil, application of agronomic measures (800 kg ha$^{-1}$ of 15 N:15 P:15 K mineral fertilizer and the wetting of the seeded area until establishing plant cover), and grass-legume mixture (sowing density: 270–300 kg ha$^{-1}$). | Study offers knowledge on the effects of weathering and the development of vegetation directly on ash itself and on changes in the FA physicochemical properties as significant indicators of pedogenesis on this substrate, with these processes being of special importance for the effective ecological reclamation of FA catena globally. | Serbia | Kostić et al. (2018) |

| # | Species | Treatment | Country | Notes | Reference |
|---|---------|-----------|---------|-------|-----------|
| 13 | Alnus viridis | FA (control), FA + L. Sewage sludge (4 Mg ha⁻¹), grass seeds (Dactylis glomerata and L. multiflorum) (200 kg ha⁻¹), NPK mineral fertilizer (N−60, P−36, and K−36 kg ha⁻¹). | Poland | Application of lignite culm in planting pits and preliminary surface preparation by hydroseeding and mineral fertilization had the most positive effect on green alder seedling parameters. Results suggest that green alder can be used as a potential biological stabilizer on FA catena. | Pietrzykowski et al. (2015) |
| 14 | A. glutinosa, A. incana, and A. viridis | CCW, CCW + L. start-up mineral fertilization (N: 60, P: 36, and K: 36 kg ha⁻¹) and initial stabilization by hydroseeding with sewage sludge (4 Mg ha⁻¹ dry mass), and a mixture of grasses (D. glomerata and L. multiflorum) (200 kg ha⁻¹). | Poland | The obtained results supported findings of a previous study (Pietrzykowski et al., 2015) on alder growth on combustion waste disposal sites and showed effectiveness of alders, mainly A. glutinosa, for biological reclamation of these kinds of barrens. | Pietrzykowski et al. (2018) |
| 15 | A. glutinosa and A. incana | CCW, CCW + L, CCW + MS hydroseeding with sewage sludge (4 Mg dry mass ha⁻¹) mixed with the seeds (200 kg ha⁻¹) of D. glomerata and L. multiflorum, and NPK mineral fertilization (N: 60, P:36, and K 36 kg ha⁻¹). | Poland | Despite the good growth attentions, the N:P and N:K ratios in alder leaves largely differed from the optimal values, showing poor P and K supply at FA catena. This may pose a threat to further development of the introduced tree plantings. The introduction of alders along with the lignite addition into the planting holes seems to be an effective technique of FA catena revegetation. | Pietrzykowski et al. (2018) |
| 16 | Casuarina equstifolia, D. sisso, Bauhnia bauhni, Delonix regia, Michelia champaca, Hibiscus rosa-sinensis, Bougainvillea glabra, Chamaedorea elegans, Thuja occidentalis, and Lawsonia inermis | Soil + cow manure (2:1 ratio), chemical fertilizers (diammonium phosphate; 25 g pit⁻¹), Rhizobium and phosphobacterium biofertilizers (10 g pit⁻¹), and humic acid (5 g pit⁻¹). | India | The microbial activity of the FA catena could be successfully upgraded through the plantation of D. sisso, B. glabra, C. equstifolia, D. regia, and T. occidentalis to a large extent and well reclaimed for social forestry purposes. | Ram et al. (2008) |

(Continued)

**Table 7.3** (Continued)

| S. no. | Assisted vegetation | Amendments | Observation | Country | References |
|---|---|---|---|---|---|
| 17 | *Alder ssp.*, *A. glutinosa*, *A. viridis*, and *A. incana* | Hydroseeding with sewage sludge (4 Mg ha$^{-1}$ dry mass) mixed with the seeds (200 kg ha$^{-1}$) of *D. glomerata* and *L. multiflorum* and mineral fertilizer (NPK 60:36:36 kg ha$^{-1}$). | Study showed that *A. glutinosa*, *A. incana*, and *A. viridis* growing on infertile FA catena have a similar fine root biomass increment. The results also indicate a significant role of fine roots in the development of the carbon and nutrient pools in FA catena. | Poland | Świątek et al. (2019) |
| 18 | Herbaceous legumes (*Rhynchosia minima*, *Macroptilium atropurpureum*, and *Crotalaria juncea*), Stoloniferous perennial grasses (*C. dactylon* and *Cynodon nlemfuensis*), and indigenous nitrogen-fixing leguminous trees (*Acacia gerrardii* and *Sesbania sesban*) | Soil, nitrogen fertilizers, and herbicide. | The proposed low input approach was not successfully proved on FA catena due to the greater toxicity problems and constraints imposed by the active status of the FA catena. In the longer term, however, after the ash catena is no longer active and weathering can reduce the toxicities, it is predicted that this approach may yet achieve the ultimate goal of a self-sustaining ecosystem. | Zimbabwe | Piha et al. (1995) |
| 19 | *Q. robur*, *Q. rubra*, *P. sylvestris*, *Betula verrucosa*, *Populus tremula*, *Populus nigra*, *Salix fragilis*, *S. alba*, *A. glutinosa*, *Q. robur*, and *P. sylvestris* | *Stabilizate*—a mixture of FA treated with 1%–2% CaO, 25% water, slag, and FDG gypsum. Physicochemical properties contribute to the origination of an adverse soil environment (high alkalinity and cementing effects). | Mainly *R. pseudacacia*, *B. verrucosa*, *S. alba*, *S. fragilis*, *P. tremula*, *P. nigra*, *P. alba*, and *P. sylvestris* can be considered as the tree species taxon of very high reclamation significance characterized by a good performance already in the first vegetation series of primary succession on FA catena only. | Czech Republic | Čermák (2008) |

| | | | | | |
|---|---|---|---|---|---|
| 20 | *F. arundinacea* and *Onobrychis viciifolia* | Incorporating fertilizers and amendments as composts and modified indigenous volcanic tuff. | The strategy chosen is very significant toward obtaining green armor that rapidly and capably covers the FA catena, as well as to permit the wildlife habitat development. The fertilizer and modified indigenous volcanic tuff provided establishing conditions for the plant supplying the nutrients. Limiting the transfer of heavy metals in plant tissues is in accordance with a healthy habitat for the wildlife. | Romania | Anca et al. (2011) |
| 21 | Leguminous species (*O. viciifolia, L. corniculatus,* and *Melilotus officinalis*) | Untreated FA, FA amended with modified/unmodified volcanic tuff, FA treated with modified/ unmodified volcanic tuff mixed with organic fertilizer, and anaerobically stabilized municipal sludge. | The suggested approach for vegetating FA deposit was by seeding leguminous crop species *O. viciifolia* followed by the installation of some invasive more resistant species, which allow the ecological restoration of the FA deposit with an effective stabilization against the wind, erosion, and leaching. | Romania | Morariu et al. (2013) |
| 22 | *A. excelsa, A. lebbeck, A. procera, D. sissoo, E. tereticornis, H. integrifolia, L. leucocephala, P. dulce, P. pinnata, P. juliflora,* and *Z. mauritiana* | Soil layer. | On the basis of importance value index, *P. juliflora, P. dulce,* and *P. pinnata* have been noticed as dominant species in FA ecosystem. In this regeneration status study, the above three tree species (68.91%) have been found in a good regeneration category that can be used for revegetation of a new FA catena. | India | Pandey et al. (2016) |

(Continued)

**Table 7.3** (Continued)

| S. no. | Assisted vegetation | Amendments | Observation | Country | References |
|---|---|---|---|---|---|
| 23 | *Atriplex holocarpa, Atriplex lindleyi, Atriplex vesicaria, Enchylaena tomentosa, Halosarcia halocnemoides, Halosarcia pergranulata, Saevola colloris, Nitraria billardieri,* and *Mesembryanthemum aitonis* | A range of surface amelioration treatments for FA lagoons were designed and used. | The soil or compost layer offered a seed bed for initial germination against FA toxicity, while the stabilization treatment barred wind-blown abolition of both seed and seed bed, though also conferring certain safety to seedlings. Compost layer stabilized with mulch was the ideal amelioration toward biomass production from seed or transplants. | Australia | Jusaitis and Pillman (1997) |
| 24 | *Chloris gayana, C. dactylon, Conyza bonariensis, Eragrostis* sp., *Schkuhria pinnata, Pogonarthria squarrosa, Pseudognaphalium undulatum, Pennisetum clandestinum, Hyparrhenia hirta, Chamaecrista biensis, Plantago lanceolata, Walafrida densiflora,* and *M. sativa* | Topsoil cover. | Flora composition recognized the species that are adapted to the limiting factors posed by the FA deposits. Thus the composition outcomes can be used to lead the next rehabilitation programs. | South Africa | Van Rensburg et al. (2003) |

*CCW*, Coal combustion waste; *CM*, cow dung manure; *FA* fly ash; *FDG*, flue gas desulphurization; *FYM*, farmyard manure; *L*, lignite culm; *LA*, lagoon ash; *MS*, miocene acidic sand; *RI*, *Rhizobium* inoculation; *SSC*, sewage sludge compost; *VAM*, vesicular-arbuscular mycorrhizae; *VC*, vermiculite.

successfully carried out on an area of 3.4 ha of FA catena at Korba Super Thermal Power Station, Korba, Chhattisgarh, India, with *Albizia procera*, *Bombex ceiba*, *C. equisetifolia*, *D. sissoo*, *D. strictus*, *Eucalyptus tereticornis*, *Gmelina arborea*, *Melia azadirachta*, *Populus euphratica*, *Shorea robusta*, and *T. grandis* (Das et al., 2013). Another plantation was done on an area of 0.4 ha of FA catena at Badarpur Thermal Power Station, Delhi, India, where *D. sissoo*, *E. tereticornis*, *M. azadirachta*, and *Populus deltoids* were used. Likewise, *Jatropha curcas* and *Parkinsonia aculeata* were planted on a 4-ha area of FA catena at Vijayawada Thermal Power Station, Vijayawada, Andhra Pradesh, India (Das et al., 2013). A successful tree planting action was carried on the FA deposits of Panki Thermal Power Station, Kanpur, Uttar Pradesh, India, through some plant species such as *Ailanthus excelsa*, *Albizia lebbeck*, *A. procera*, *D. sissoo*, *E. tereticornis*, *Holoptelea integrifolia*, *Leucaena leucocephala*, *Pithecellobium dulce*, *P. pinnata*, *P. juliflora*, and *Ziziphus mauritiana* (Pandey et al., 2016). Fig. 7.5 represents the potential and dominant plant species in the rehabilitated FA site of Panki Thermal

**Figure 7.5** Luxurious growth of dominant and potential planted woody tree species with a good regeneration capacity on the FA dumps of Panki Thermal Power Station. (A) *P. dulce* (Roxb.) Benth. (B) *Prossopis juliflora* (Sw.) DC. (C) *P. pinnata* (L.) Pierre. (D) *L. leucocephala* (Lam.) de Wit. Photo courtesy: *V.C. Pandey.*

Power Station, which have a good regeneration capacity for the revegetation of fresh FA dumpsites (Pandey et al., 2016).

In Serbia, two planted woody species (*Tamarix tentandra* and *Robinia pseudoacacia*) and two naturally colonized woody species (*A. fruticosa* and *P. alba*) were assessed for 3 and 11 years on the weathered FA catenas of "Nikola Tesla-A" for their photosynthetic efficiency, photosynthetic pigments, and leaf damage symptoms. The survival of plants growing on ash was low except in the case of *A. fruticosa*, which was naturally colonized. Therefore naturally colonized woody species can be used in the first step to stabilize the FA catena (Kostić et al., 2012). Fig. 7.6 depicts the planted grass, shrub, and tree on the largest thermal power plant Nikola Tesla (TENT−A), Obrenovac, Belgrade, Serbia. Likewise, the grass *S. munja* and its other species as native colonizers have also been reported to be very effective in binding loose ash particles and in stabilizing FA substrate along with multiple uses with a high societal value (Pandey and Singh, 2011;

**Figure 7.6** Planted vegetation on the largest thermal power plant Nikola Tesla (TENT−A). (A) *F. rubra*—sown native grass on the FA deposits on the passive cassette 11 years old. (B) *Robinia pseudoacacia*—planted nonnative tree on the FA deposits on the passive cassette 11 years old. (C) *Tamarix tetandra*—Planted nonnative shrub on the FA deposits on the passive cassette 11 years old. (D) Planted nonnative shrub *T. tetandra* with the spontaneous colonized nonnative shrub *A. fruticosa* on the FA deposits on the passive cassette 3 years old. Photo courtesy: *Gordana Gajić.*

Pandey et al., 2012a,b). A suitable combination of naturally colonized woody and grass species can rehabilitate abandoned FA catena to a productive forest ecosystem. The potential of *Ipomoea carnea*, which is an invasive species in the phytoremediation of FA catena, has recently been demonstrated (Pandey, 2012a). Doren et al. (2009) proposed a comprehensive ecological model to control the spread of invasive species. In North America, few studies on FA were carried out as a substrate for growing woody tree species, including nitrogen-fixing species like alders and maples (Scanlon and Duggan, 1979). In South Carolina, *Platanus occidentalis* and *Liquidambar styraciflua* were successfully grown on the abandoned alkaline and acidic FA catena, respectively (McMinn et al., 1982; Carlson and Adriano, 1991). In England, successional changes in vegetation cover and soil development were studied on abandoned FA catena, covering a time range of 7−24 years. The plant succession was found with a mixed ruderal community, leading to woodland dominated by *Betula* spp. and *Salix* spp. Multivariate analyses revealed that soil development on FA catena correlated more closely with the site age than it did with the plant community composition (Shaw, 1992). In Europe, several species of trees such as *Pinus sylvestris*, *Betula pendula*, *R. pseudoacacia*, *Quercus rubra*, *Quercus robur*, *Alnus glutinosa*, and *Salix* spp. were introduced on FA catena (Čermák, 2008; Pietrzykowski et al., 2010). Nitrogen-fixing tree species like *Elaeagnus angustifolia*, *Colutea arborescens*, *Hippophae rhamnoides*, and *Gleditsia triacanthos* were also grown on FA catena, which performed well due to the high tolerance capacity (Hodgson and Townsend, 1973). In Poland, a study was conducted for the survival and growth of *A. glutinosa* and *Alnus incana* on FA technosols at different substrate improvements. After 5 years of investigation and having compared the growth of alder species and the applied experimental variants, it was noticed that the best results were found in the case of *A. glutinosa* in the variant using lignite culm. As shown by experiments, the introduction of both alder species directly on FA catena using hydroseeding and NPK fertilizer may be recommended for economic motives. The introduced alder species have mainly ameliorating and protective roles, and thus prepare the substrate for the afforestation and for the next generation of desired species (Krzaklewski et al., 2012).

As a result of plantation, changes in the physicochemical, nutritional status, heavy metal content, microbial populations, and activities were observed in reclaimed FA catena. Significant differences in pH, available N, P, K, and OC were observed in reclaimed FA catena against the control (Das et al., 2013). Continuous litter fall from the growing stands on

FA catena and natural decomposition of the same may be responsible for the amelioration of substrate quality. Furthermore, vegetation cover also diminishes the risk of FA erosion and can restore the organic matter (Sinoga et al., 2012). Phenolic compounds were increased in the reclaimed FA catena due to an increase in humus content of leaf litter and also due to the organic debris emanating from the vegetation (Das et al., 2013). Phenolics have a marked effect on plant—soil systems. For example, they regulate not only the residue decomposition rate but also the nutrient release from plant residues. They regulate the plant growth as well, which increases the stability of long-term soil structure (Martens, 2002).The development of flora on FA catena not only leads to rejuvenation of barren landscape in an aesthetically pleasing environment but also improves the substrate structure of FA catena. This is necessary for further establishment of native flora and for the improvement of existing ecosystem. Vegetation may be established on FA catena in a short time by fertilization with municipal sludge alone or by mixing the indigenous volcanic tuff with clinoptilolite. The synergistic effect of municipal sludge (biosolids) with a volcanic tuff determines the vegetation cover of up to 60%—65% and reduces the bioavailability of toxic metals (Cr, Pb, Ni, and Cu) in the substrate (Mâsu et al., 2010). Van Rensburg et al. (2003) showed a comparative analysis of the vegetation and nutrient status of the top substrate between two similarly rehabilitated ash disposal sites in the Mpumalanga coalfield, South Africa. Application of organic amendment and biofertilizer can provide better support for the anchorage and growth of the plant on the bare FA catena. In this context, Juwarkar and Jambhulkar (2008) conducted a field experiment on a 10-ha area of FA catena to restore and revegetate it by using biological interventions. They indicated that amendment with FYM at 50 ton $ha^{-1}$ improved the physical properties of FA substrate. The content of nitrogen improved by four and a half times due to the addition of nitrogen-fixing strains of *Azotobacter* and *Bradyrhizobium* species, whereas the phosphate content augmented by ten times owing to the addition of vesicular-arbuscular mycorrhizal (VAM) that supports in phosphate immobilization. Various microbial groups such as *Azotobacter*, *Rhizobium*, and VAM spores developed their colonies due to biofertilizer inoculation. These colonies were absent in FA. Application of biofertilizer and FYM helped reduce the toxicity of metals by 25%, 46%, 48%, and 47% for Cd, Cu, Ni, and Pb, respectively. Amendment of FA with biofertilizer and FYM assisted profusely in the development of root system, showing 15 times higher growth in the *D. strictus* plant than the control.

Therefore a number of tree species have been successfully grown on FA catena, which, apart from the stabilizing substrate, produced good biomass. Biomass production usually depends on the dimensional growth, that is, height and diameter, of a tree that determine the amount of harvestable wood (Chaturvedi et al., 1991). Therefore the screening of species for a high biomass yield on FA catena is desired for economic returns. For example, planting *Eucalyptus* on FA catena for biomass production is a good choice, because its growth is not adversely affected by the pH of the substrate. In this regard, a conceptual framework represents the role of afforestation for the mitigation of environmental threats of FA catena along with the secondary benefits of resource generation (Fig. 7.2A and B).

## 7.4 Microbial and macrofaunal tools

### 7.4.1 Fly ash tolerant fungi

Mycorrhizal fungi (symbiotic fungi) show a synergistic effect for the survival and growth of majority of the plant species in the barren areas. Mycorrhizal fungi facilitate the transformation of soil properties for increasing the plant growth through enhanced nutrient uptake and through the cycling of N, P, K, and other trace elements (Selvam and Mahadevan, 2002; Rillig and Mummey, 2006). Mycorrhizal hyphae can influence the key ecosystem process of soil aggregation through the binding of soil particles that possibly produce polysaccharides (Rillig and Mummey, 2006). Selvam and Mahadevan (2002) showed the effectiveness of a native fungus *Glomus mosseae* in increasing the growth of 20 plant species in an abandoned FA catena. Recently, Babu and Reddy (2011) showed that the inoculation of FA-adapted arbuscular mycorrhizal (AM) fungi in combination with *Aspergillus tubingensis* increased the growth and nutrient uptake [P (150%), K (67%), Ca (106%), and Mg (180%)] in bamboo (*D. strictus*) plants grown on FA, and it also reduced the metal translocation. Dual inoculation of AM fungi and *A. tubingensis* might be a promising strategy to promote the vegetation on FA catenas. Furthermore, Ray and Adholeya (2009) reported that the ability of metal uptake by ectomycorrhizal fungi is substrate specific, and fungal organic acid exudation is strongly influenced by the FA-associated metal uptake. These research findings can be effectively used for the rehabilitation of FA dumpsites. Hrynkiewicz et al. (2009) showed the use of inoculation with a mycorrhiza-associated *Sphingomonas* sp. 23L (bacterial strain) to boost the mycorrhizal formation and willow growth on FA substrate. The high pH value (8.7) and low nitrogen content ($N_t = 0.01\ \mu g\ g^{-1}$) of FA are the reasons for the low growth of the plant. Inoculated willow clones with the bacterial strain enhanced the nitrogen uptake, improved the plant growth, and promoted the formation of ectomycorrhizae with the *Geopora* sp. strain on all willows. They concluded that inoculation with mycorrhiza helper bacterial strains might be a promising strategy to promote mycorrhizal formation with autochthonous site-adopted ectomycorrhizal fungi in hostile conditions of FA catena, thereby improving revegetation on FA catena with willow plants (Hrynkiewicz et al., 2009). Selvam and Mahadevan (2002) surveyed and reported the presence of 15, 4, and 13 AM fungal species in an abandoned FA catena, overburden dumps, and reclaimed overburden dumps, respectively. In all these cases, the AM fungal species G. *mosseae* was the dominant species that seems to be an obvious choice for inhospitable conditions (for detail, see Selvam and Mahadevan, 2002). Jambhulkar and Juwarkar (2009) also used VAM spores of *Gigaspora* and *Glomus* species in mixture with site-specific specialized nitrogen-fixing strains of biofertilizers and organic amendments. Therefore mycorrhizal fungi protect the plants from soil-borne diseases and detoxify soils contaminated by FA toxic metals. The mycorrhizal biofertilizer in plant—root association turns FA catena into rich soil like substratum. Thus mycorrhizal technology is expected to turn waste into wealth by developing afforestation on FA catena. Grasses of the abandoned FA catena represent vital agents of AM distribution, which can facilitate the establishment of mycorrhizal association of tree seedlings that are planted during the rehabilitation processes (Enkhtuya et al., 2005).

### 7.4.2 Fly ash tolerant bacteria

Bioremediation of degraded lands are carried out with different approaches, in which bacteria play a prominent role to reduce the concentration/toxicity of various chemical pollutants such as heavy metals (Singh et al., 2011a,b). Jambhulkar and Juwarkar (2009) conducted an experiment on a 10-ha area of FA catena that involved isolation and inoculation of site-specific particular nitrogen-fixing strains of biofertilizers (*Azotobacter* and *Bradyrhizobium* species). Currently, Kumari and Singh (2011) isolated and reported 11 bacterial consortia from FA, in which a mixture of two strains, that is, *Bacillus pumilus* NBRFT9 (C7) + *Paenibacillus macerans* NBRFT5, when inoculated in the rhizospheric soil of *Brassica juncea*, increased the accumulation of metals, such as Pb (278%), Mn (75%), Zn (163%), Cr (226%), and Ni (414%), compared with the control. Bacterial activities such as nitrogen fixation, P solubilization, and micronutrient uptake promote the growth of plant, thereby resulting in the accumulation of high metals. In addition to it, a mixture of four bacteria, *Bacillus endophyticus* NBRFT4 + *B. pumilus* NBRFT9 (C4) + *P. macerans* NBRFT5 + *Micrococcus roseus* NBRFT2, increased the Cd uptake by 237%. Generally, bacterial secretion systems involve enzymes, protons, organic acids, and siderophores, which enhance the mobilization of metals and boost the metal accumulation from FA. Therefore this technique may be used for the phytoextraction of metals from FA through suitable plant—rhizosphere interactions. Tiwari et al. (2008) investigated an eco-friendly and cost-effective technology to bioremediate toxic metals related to FA deposits that pollute surface and ground water in and around FA dumpsites. They noticed that most of the FA tolerant bacterial strains either immobilized Cd, Cr, Pb, and Cu or induced the bioavailability of Zn, Fe, and Ni in the FA. However, they also observed exceptions in Ni case, where eight bacterial strains increased metal mobility, whereas others triggered metal immobilization. They suggested that metal solubilization and immobilization are specific to bacterial strains. Metal solubilization by bacteria may be implemented to speed up the metal phytoextraction from FA basins through hyper accumulator plants. Immobilization of metals can be used to hinder their migration to water bodies (Tiwari et al., 2008).

### 7.4.3 Fly ash tolerant earthworm

Earthworm and FA catena reclamation are interesting topics from FA revegetation point of view. Eijsackers (2010) reviewed in detail the earthworm colonization of various waste deposits including FA deposit. But, here, we briefly described the earthworm colonization of FA deposits. Earthworms live in the burrows of land and have the ability to correct soil environment significantly by their burrowing, litter decomposing, and soil structuring activities. Various species of earthworms have been revealed as colonizers of FA deposits, and are given in Table 7.1. Eijsackers et al. (1983) reported *Lumbricus rubellus* as the first colonizer from FA basins over a total period of 8 years, followed by *Aporrectodea* caliginosa, *Allolobophora chlorotica*, and *Eiseniella tetraeda*. But, *Lumbricus casta-neus* is not seen as an invader species from the surrounding clayey area after 8 years. Experiments showed that inoculated *L. rubellus* survived bet-ter in comparison with *A. caliginosa* in alkaline conditions. Satchell and Stone (1977) studied FA deposits of different ages that contain harsh envi-ronmental conditions (high pH and elevated heavy metal levels). They noticed *L. rubellus*, *L. castaneus*, and some *Dendrobaena* species in dry areas, whereas *E. tetraeda* and *A. chlorotica* were observed in wet areas of the FA deposits. In 4.5−7 years, old FA basins, *L. rubellus*, *L. castaneus*, and *Lumbricus festivus* had dispersed from 6 to over 20 m, while *A. caliginosa*, *Aporrectodea rosea*, *Allolobophora longa*, and *A. chlorotica* had dispersed just over 2 m. Satchell and Stone (1977) studied in detail the mean burrowing rates of various earthworm species under field conditions of FA deposits. They noted the increasing order of mean burrowing rates (m year$^{-1}$) of various earthworm species on FA deposit as *Lumbricus terrestris* (2.2 m year$^{-1}$), *L. rubellus* (3 m year$^{-1}$), *A. chlorotica* (5 m year$^{-1}$), *A. rosea* (5.5 m year$^{-1}$), *A. longa* (6.5 m year$^{-1}$), and *A. caliginosa* (7 m year$^{-1}$). Ma and Eijsackers (1989) showed the clear avoidance of FA amended soil by *L. rubellus* at high percentages ($> 10\%$), while earthworms preferred FA added substrate at lower percentages ($< 4\%$). This may be the reason for the colonization of FA deposits. Satchell and Stone (1977) also studied the earthworm survival on the different ages of FA deposits and noticed that young FA deposits (1−2 years) were lethal. *A. longa*, *A. caliginosa*, and *L. terrestris* could survive but less well in old FA deposits ($> 15$ years) in comparison with the control soil. They recorded low mortalities ($< 10\%$) of *L. terrestris*, *A. caliginosa*, and *A. longa* in 10%, 25%, and 50% FA treated soil instead of the age or concentration of FA. Ma and Eijsackers (1989) noticed that the rate of survival of earthworms slowly decreased with increasing FA percentages (2%−32%), while cocoon production rapidly reduced.

## 7.4.4 Fly ash tolerant macrofauna

Leaf litter is the food and shelter of detritivores. Usually, detritivore macrofauna favor plants with a high N and a low C: N ratio that tends to be more palatable (Zimmer, 2002). Recently, Podgaiski and Rodrigues (2010) conducted a litter bag experiment, spanning for a period of 140 days. They studied the role of macrofauna (invertebrates) in leaf-litter decomposition of three pioneer plant species, such as grass (*C. dactylon*), shrub (*Ricinus communis*), and tree (*Schinus terebinthifolius*), in a system affected by two coal ash deposits (FA and boiler slag). They reported that the degree of decomposition was higher (80% faster) in *R. communis* than the other plant species due to its high N (%), low C:N ratio, and enhanced richness of detritivores during the first days of decomposition. In addition, this leaf litter involved the lowest richness of invertebrate species. *C. dactylon* and *S. terebinthifolius* leaf litters were almost the same in the diversity of macrofauna and decomposition rates (Podgaiski and Rodrigues, 2010). They reported a total of 2573 individuals and 126 morphospecies of soil macroinvertebrates, colonizing the litter bags. In the studied macrofauna, the order of macroarthropod groups was Isopoda (38%) > Oligochaeta (19%) > Hymenoptera (16%). Coleoptera had the maximum morphospecies richness (54), followed by Hymenoptera (24). They concluded that the leaf-litter decomposition and detritivore densities were not affected by the type of ash disposal system. Shaw (2003) extracted Collembola from the surface layers of four FA lagoons (ranged from 4 to 40 years) in East London (United Kingdom). Colonization by Collembola on FA catena was first studied by him. A small community of early successional (usually epiedaphic) species develops the first, which is replaced by a denser and richer community of generalists, including euedaphic species. It is reflected in the improving level of soil development, as the sites begin to mature, thereby creating new microhabitats within the soil.

## 7.5 Bioindicators of fly ash catena revegetation

Bioindicators denote the health status of any ecosystems. They have been recognized as a potential tool in restoration ecology. Bioindicators can be used to monitor the patterns and changes in the improvement of substrate quality. Bioindicators may be micro- and macroorganisms, and include their activities or functions. Microbial biomass, fungi, actinomycetes, lichens, as well as the population of earthworms, nematodes, termites, and ants can be used as bioindicators on account of their important role in nutrient cycling, specific soil fertility, soil development, and soil conservation (Anderson, 2003). Microbial community structure and the size of population are sensitive to the changes in soil chemical properties (Tokuda and Hayatsu, 2002). Microbial biomass carbon is one of the most important microbiological properties. It has been identified as an early and sensitive indicator of soil quality change. Numerous studies have indicated that the microbial biomass carbon reacts more quickly to the changes, resulting from forest management activities than from the soil organic matter (Sparling, 1992; Bosatta and Agren, 1994). Furthermore, the ratio of microbial biomass carbon relative to soil OC (microbial quotient) is used as a bioindicator for carbon availability (Insam and Domsch, 1988; Yan et al., 2003). Thus microbial biomass carbon and microbial quotient may be used as an effective early indicator to assess the improvement in substrate quality during the rehabilitation of FA catena through afforestation.

Bioindicators also include biological (metabolic) processes such as soil respiration (evolution of $CO_2$). It mainly comes from microbial respiration, and is used for measuring microbial activity related to organic matter decomposition in soil system. A normally used index, that is, metabolic quotient ($qCO_2$) defined as the soil respiration to microbial biomass ratio, is related to the mineralization of organic substrate per unit of microbial biomass (Bastida et al., 2008). The metabolic quotient ($qCO_2$) has been used in numerous studies as an eco-physiological measurement for ecosystem succession (Anderson and Domsch 1993; Singh et al., 2012a). Many studies reported higher values of $qCO_2$ under harsh conditions rather than the favorable conditions (Anderson and Domsch 1993; Wardle and Ghani 1995). As on date, no specific studies have been carried out on these aspects in relation to the rehabilitation of FA catena. Soil respiration and metabolic quotient might be used as a bioindicator of total soil microbial activity and as an eco-physiological measure for the rehabilitation of FA catena.

Other bioindicators that have been widely used are: metabolic products of organisms, particular enzymes such as β-glucosidase, dehydrogenase, protease, catalase, urease, arylsulfatase, and phosphatase, associated to specific functions of substrate degradation, or the mineralization of organic C, N, S, or P (Singh et al., 2012c). Soil enzyme assays act as potential indicators of ecosystem stability and are described as "biological fingerprints" of past soil management (Dick, 2000).

The presence of phenolic compounds (secondary plant metabolites such as ferulic, vanillic, p-coumaric, p-hydroxybenzoic, and syringic acid), formed as a result of biological process, can be used as reliable bioindicators for the influence of plants in the restoration of FA catena. Phenolic compounds are found in all plant components either in free forms or in association with other compounds, for example, lignin and polysaccharides of cell walls (Djurdjević et al., 2006). Phenolic compounds play a significant role in the process of humus formation and soil fertility, and therefore have a marked effect on plant—soil systems. They affect cell and plant growth, act as constitutive defense against invading organisms, and are the functional and structural components of soil organic matter. Furthermore, they control the decomposition rate of residue, release the nutrients in soil system from the plant residues, and enhance a long-term soil structure (Martens 2002). After 7 years of revegetation, Djurdjević et al. (2006) studied the population abundance, floristic composition, and the cover of pioneer species of naturally growing plant communities and total phenolics and phenolic acid contents as humus constituents of FA catena. The results showed that the amount of total phenolics $(38.07-185.16\ \mu g\ g^{-1})$ and phenolic acids $(4.12-27.28\ \mu g\ g^{-1})$ in FA increased from the center to the edges of the FA catena in correlation with the increase in the plant abundance and cover. Phenolic compounds may be used as a credible sign (bioindicators) for assessing the success of the revegetation on FA catena (Djurdjević et al., 2006). Recently, Das et al. (2013) reported that total phenolic contents were significantly higher in the reclaimed FA catena than the control. This is owing to an increase in humus content from organic debris and leaf litter from the established vegetation cover.

## 7.6 Ecological engineering for effective afforestation on fly ash catena

It is suggested that the site should be covered by a fine layer of soil or organic amendments and may be merged into the dumping surface for the successful plantation on the FA catena. The latter option is more common in which FYM/compost manure mixed with VAM culture and plant growth promoting rhizobacteria culture are applied for effective plantations. Afforestation on abandoned FA catena can be achieved by using ecological engineering as a tool (for detail, see Pandey and Singh, 2012). An important recipe for the production of tree species on FA catena is extracted by some important articles (Haynes, 2009; Pandey et al., 2009a,b; Pandey and Singh, 2010; Pandey and Singh, 2012) that include the following major considerations:

- Selection of tree species is a key factor in determining the success of plantation on the FA catena depending on the main purpose of plantations.

- Preference should be given to tree species with their commercial applicability, as the primary colonizers may be of little or no economic importance.
- Tolerance to metal toxicity with enriched levels of trace elements.
- Incorporation of organic amendments (FYM, sewage sludge, press mud, compost manure, etc.) into FA catena to improve the physico-chemical conditions of the site for the colonization of soil biota and plants.
- Effective measures for the formation of an organic mulch to reduce evaporative losses of $H_2O$ and settlement of weathered FA particles.
- Plantation of inoculated leguminous trees with FA tolerant *Rhizobium* strains.
- Inoculation with FA tolerant ecto and endomycorrhizal fungi to increase the nutrient use efficiency of plants that grow on poor sites.
- Use of FA tolerant plant growth promoting rhizobacteria for high bio-mass production on FA catena.

## 7.7 Why to remediate fly ash catena through afforestation? We suggest it due to the following benefits

- It stabilizes heavy metal pollutants.
- Mitigates air, soil, and water pollution.
- It is a cost effective and eco-friendly technology.
- Afforestation on FA catena will increase forest cover.
- Enables biomass production and resource generation.
- Potential sink for carbon sequestration.
- Enhancement of above- and below-ground biodiversity.
- Formation of soil-like substrate from FA.
- Utilization of abandoned FA disposal sites.
- Maintains the watershed hydrology.
- Source of revenue.
- Conservation of flora and fauna.
- Provides shelter for wildlife and improves bioaesthetic environment.

## 7.8 Conclusions

FA catena may be used as a good substrate for the rehabilitation programs with assorted tree species. Edible crops may not be advocated, as there is a risk of heavy metal contamination. Tree plantation on abandoned FA catena is desirable to stabilize the surface against water and wind erosion and to provide a pleasant bioaesthetic landscape. There are indications that show that FA catena can be used for biomass production with a wide range of tree species for multipurpose uses. It can be used along with secondary environmental benefits such as carbon sequestration, biodiversity conservation, pollution abatement, and balanced hydrology. Different types of plants can be grown on FA catena. However, the members of the Leguminosae family should be preferred for achieving good success due to nitrogen deficiency in FA substrate. The selection of appropriate plant species with suitable organic amendments using biofertilizer cultures would improve plant establishment and growth rate. It would be better to fix the crop rotation for intermittent harvesting and the removal of biomass as selection felling. Clearing felling on such a site will expose the FA catena for repeated threats of pollution and environmental degradation.

## 7.9 Future research recommendations

Rehabilitation of abandoned FA catena has been identified as a potential field for the current research priority in environmental management (Pandey and Singh, 2012). Both flora and fauna play a vital role in the creation of new ecosystems on FA catena. Periodical monitoring of the FA catena for the species succession and for the improvement in the substrate quality should be undertaken for a better understanding of rehabilitation process toward sustainability and stability. Bioindicators of the revegetation on FA catena are less studied. The integration of energy plants for biofuel production in rehabilitation/phytoremediation of FA catena should be considered for further trials. Likewise, carbon sequestration and the mitigation of environmental pollution should also be monitored periodically for the strategic changes in the rehabilitation program of FA catena. Isolation of efficient microorganisms from the rehabilitated FA ecosystem with the use of genomics-based markers for their identification and functional activities is the new dimension of future research. This will enhance our understanding of physiological, biological, and molecular levels for taking appropriate scientific measures for efficient rehabilitation of FA catenas.

## References

Adriano, D.C., Page, A.L., Elseewi, A.A., Chang, A.C., Straughan, I., 1980. Utilization and disposal of fly ash and other coal residues in terrestrial ecosystem: A review. J. Environ. Qual 9, 333–344.

Anca, P., Masu, S., Lixandru, B., Morariu, F., Dragomir, N., Laffont-Schwob, I., et al., 2011. Strategies for covering fly ash dumps with plant species suitable for phytostabilization. Anim. Sci. Biotechnol. 44 (2), 229–234.

Anderson, T., 2003. Microbial eco-physiological indicators to asses soil quality. Agric. Ecosyst. Environ. 98, 285–293.

Anderson, T.H., Domsch, K.H., 1993. The metabolic quotient for $CO_2$ ($qCO_2$) as a specific activity parameter to assess the effects of environmental conditions, such as pH, on the microbial biomass of forest soils. Soil Biol. Biochem. 25, 393–395.

Babu, A.G., Reddy, M.S., 2011. Dual inoculation of arbuscular mycorrhizal and phosphate solubilizing fungi contributes in sustainable maintenance of plant health in fly ash ponds. Water Air Soil Pollut. 219, 3–10.

Bastida, F.Z.A., Hernandez, H., Garcia, C., 2008. Past, present and future of soil quality indices: a biological perspective. Geoderma 147, 159–171.

Bogdanović, V., 1990. The number of some microorganisms in ash deposit under *Robinia pseudoacacia* of the "Lazarevac" thermoelectric power plant. Soil. Plant. 39, 139–145 (in Serbian).

Bosatta, E., Agren, G.I., 1994. Theoretical analysis of microbial biomass dynamics in soils. Soil Biol. Biochem. 26, 143–148.

Bradshaw, A.D., Chadwick, M.J., 1980. The Restoration of Land: The Ecology and Reclamation of Derelict and Degraded Land. Blackwell Scientific Publications, Oxford, p. 317.

Carlson, C.L., Adriano, D.C., 1991. Growth and elemental content of two tree species growing on abandoned coal fly ash basins. J. Environ. Qual. 20, 581–587.

Carlson, C.L., Adriano, D.C., 1993. Environmental impacts of coal combustion residues. J. Environ. Qual 22, 227–247.

Celik, M., Donbak, L., Unal, F., Yuzbasioglu, D., Aksoy, H., Yilmaz, S., 2007. Cytogenic damage in workers from a coal-fired power plant. Mutat. Res. 627, 158–163.

Čermák, P., 2008. Forest reclamation of dumpsites of coal combustion by-products (CCB). J. For. Sci. 54 (6), 273–280.

Chaturvedi, A.N., Bhatia, S., Behl, H.L., 1991. Biomass assessment for shrubs. Indian For. 117, 1032–1035.

Cheung, K.C., Wong, J.P.K., Zhang, Z.Q., Wong, J.W.C., Wong, M.H., 2000. Revegetation of lagoon ash using the legume species *Acacia auriculiformis* and *Leucaena leucocephala*. Environ. Pollut. 109, 75–82.

Chu, M., 2008. Natural revegetation of coal fly ash in a highly saline disposal lagoon in Hong Kong. Appl. Veg. Sci. 11, 297–306. Available from: https://doi.org/10.3170/2008-7-18427.

Das, M., Agarwal, P., Singh, R., Adholeya, A., 2013. A study of abandoned ash ponds reclaimed through green cover development. Int. J. Phytoremediat. 15, 320–329.

Dick R., 2000. Soil enzyme stability as an ecosystem indicator. Oregon, United States <http://cfpub.epa.gov/ncer_abstracts>.

Djurdjević, L., Mitrović, M., Pavlović, P., Gajić, G., Kostić, O., 2006. Phenolic acids as bioindicators of fly ash deposit revegetation. Arch. Environ. Contam. Toxicol. 50, 488–495.

Doren, R.E., Volin, J.C., Richards, J.H., 2009. Invasive exotic plant indicators for ecosystem restoration: an example from the Everglades restoration program. Ecol. Indic. 9, 29–36.

Dželetović, S., Filipović, R., 1995. Grain characteristics of crops grown on power plant ash and bottom slag deposit. Resour. Conserv. Recy. 13, 105–113.

Eijsackers, H., Lourijsen, N., Mentink, J., 1983. Effects of fly ash on soil fauna. In: Lebrun, Ph., André, H.M., deMedts, A., Gregoire Wibo, C., Wauthy, G. (Eds.), New Trends in Soil Biology. Dieu-Brichart, Louvain-la-Neuve, pp. 680–681.

Eijsackers, H., 2010. Earthworms as colonisers: primary colonisation of contaminated land, and sediment and soil waste deposits. Sci. Total Environ. 408, 1759—1769.

Enkhtuya, B., Poschl, M., Vosatka, M., 2005. Native grass facilitates mycorrhizal colonization and p uptake of tree seedlings in two anthropogenic substrates. Water Air Soil Pollut. 166 217—236.

Gajić, G., 2014. Ecophysiological Adaptations of Selected Species of Herbaceous Plants at the Fly Ash Landfill of the Thermal Power Plant 'Nikola Tesla-A' in Obrenovac (Ph.D. thesis). Faculty of Biology, University of Belgrade, Belgrade, 406 p (in Serbian).

Gajić, G., Pavlović, P., 2018. The role of vascular plants in the phytoremediation of fly ash deposits. In. In: Matichenkov, V. (Ed.), Phytoremediation: Methods, Management and Assessment. Nova Science Publishers, Inc, New York, NY, pp. 151—236.

Gajić, G., Djurdjević, L., Kostić, O., Jarić, S., Mitrović, M., Stevanović, B., et al., 2016. Assessment of the phytoremediation potential and an adaptive response of *Festuca rubra* L. sown on fly ash deposits: native grass has a pivotal role in ecorestoration management. Ecol. Eng. 93, 250—261.

Gajić, G., Mitrović, M., Pavlović, P., 2019. Ecorestoration of fly ash deposits by native plant species at thermal power stations in Serbia. In: Pandey, V.C., Bauddh, K. (Eds.), Phytomanagement of Polluted Sites. Elsevier, Amsterdam, The Netherlands.

Gupta, A.K., Sinha, S., 2008. Decontamination and/or revegetation of fly ash dykes through naturally growing plants. J Hazard. Mater. 153, 1078—1087.

Haynes, R.J., 2009. Reclamation and revegetation of fly ash disposal sites — challenges and research needs. J. Environ. Manag. 90, 43—53.

Hodgson, D.R., Buckley, G.P., 1975. A practical approach towards the establishment of trees and shrubs on pulverized fuel ash. In: Chadwick, M.J., Goodman, G.T. (Eds.), The Ecology of Resource Degradation and Renewal. Blackwell Scientific, Oxford, pp. 305—329.

Hodgson, D.R., Townsend, W.N., 1973. The amelioration and revegetation of pulverized fuel ash. In: Hutnik, R.J., Davis, G. (Eds.), Ecology and Reclamation of Devastated Land, vol. 2. Gordon and Breach, London, pp. 247—270.

Insam, H., Domsch, K.H., 1988. Relationship between soil organic carbon and microbial biomass on chronosequences of reclamation sites. Microb. Ecol. 15, 177—188. <http://www.kuenvbiotech.org/casestudy/.htm>.

Jambhulkar, H.P., Juwarkar, A.A., 2009. Assessment of bioaccumulation of heavymetals by different plant species grown on flyash dump. Ecotoxicol. Environ. Saf. 72, 1122—1128.

Jasionkowski, R., Wojciechowska, A., Kamiński, D., Piernik, A., 2016. Meadow species in the early stages of succession on the ash settler of power plant EDF Toruń SA in Toruń, Poland. Ecol. Quest. 23, 79—86. Available from: https://doi.org/10.12775/EQ.2016.008.

Jusaitis, M., Pillman, A., 1997. Revegetation of waste fly ash lagoons. I. Plant selection and surface amelioration. Waste Manage. Res. 15 (3), 307—321.

Juwarkar, A.A., Jambhulkar, H.P., 2008. Restoration of fly ash dump through biolog-ical interventions. Environ. Monit. Assess. 139, 355—365.

Kostić, O., Mitrović, M., Knežević, M., Jarić, S., Gajić, G., Djurdjević, L., et al., 2012. The potential of four woody species for the revegetation of fly ash deposits from the 'Nikola Tesla-A' thermoelectric plant (Obrenovac, Serbia). Arch. Biol. Sci. 64, 145—158.

Kostić, O., Jarić, S., Gajić, G., Pavlov́, D., Pavlović, M., Mitrović, M., et al., 2018. Pedological properties and ecological implications of substrates derived 3 and 11 years after the revegetation of lignite fly ash disposal sites in Serbia. Catena 163, 78—88. Available from: https://doi.org/10.1016/j.catena.2017.12.010.

Krzaklewski, W., Pietrzykowski, M., Wos, B., 2012. Survival and growth of alders (*Alnus glutinosa* (L.) Gaertn. and *Alnus incana* (L.) Moench) on fly ash technosols at different substrate improvement. Ecol. Eng. 49, 35–40.

Klubek, B., Carlson, C.L., Oliver, J., Adriano, D.C., 1992. Characterization of microbial abundance and activity from three coal ash basins. Soil Biol. Biochem. 24, 1119–1125.

Kumari, B., Singh, S.N., 2011. Phytoremediation of metals from fly ash through bacterial augmentation. Ecotoxicology 20, 166–176.

Kumari, A., Pandey, V.C., Rai, U.N., 2013. Feasibility of fern *Thelypteris dentata* for revegetation of coal fly ash landfills. J. Geochem. Explor. 128, 147–152. Available from: https://doi.org/10.1016/j.gexplo.2013.02.005.

Ma, W.-C., Eijsackers, H., 1989. The influence of substrate toxicity on soil fauna return in reclaimed land. In: Majer, J.D. (Ed.), Animals in Primary Succession: The Role of Fauna in Reclaimed Land. Cambridge University Press, Cambridge, pp. 223–243.

Maiti, S.K., Jaiswal, S., 2008. Bioaccumulation and translocation of metals in the natural vegetation growing on fly ash lagoons: a field study from Santaldih thermal power plant, West Bengal, India. Environ. Monit. Assess. 136, 355–370.

Martens, D.A., 2002. Identification of phenolic composition of alkali extracted plants and soil. Soil Sci. Soc. Am. J. 66 (4), 1240–1248.

Mâsu, S., Morariu, F., Pricop, A.D., 2010. Revegetation of fly ash sites by using municipal sludge. Bull. Univ. Agric. Sci. Vet. Med. 67, 76–79.

McMinn, J.W., Berry, C.R., Horton, J.H., 1982. Ash basin reclamation with commercial forest species. Reclam. Reveg. Res. 1, 359–365.

Mitrović, M., Pavlović, P., Lakušić, D., Stevanović, B., Djurdjevic, L., Kostić, O., et al., 2008. The potential of *Festuca rubra* and *Calamagrostis epigejos* for the revegetation on fly ash deposits. Sci. Total Environ. 72, 1090–1101.

Mitrović, M., Jarić, S., Kostić, O., Gajić, G., Karadžić, B., Djurdjević, L., et al., 2012. Photosynthetic efficiency of four woody species growing on fly ash deposits of a Serbian 'Nikola Tesla - A' thermoelectric plant. Pol. J. Environ. Stud. 21, 1339–1347.

Morariu, F., Mâsu, S., Lixandru, B., Popescu, D., 2013. Restoration of ecosystems destroyed by the fly ash dump using different plant species. Sci. Pap. Anim. Sci. Biotechnol. 46 (2), 180–184.

Mulhern, D.W., Robel, R.J., Furness, J.C., Hensley, D.L., 1989. Vegetation of waste disposal areas of a coal fired power plant in Kansas. J. Environ. Qual. 18, 285–292.

Page, A.L., Elseewi, A.A., Straughan, I., 1979. Physical and chemical properties of fly-ash from coal fired power plants with reference to environmental impacts. Residue Rev 71, 83–120.

Pandey, V.C., 2012a. Invasive species based efficient green technology for phytoremediation of fly ash deposits. J. Geochem. Explor. 123, 13–18.

Pandey, V.C., 2012b. Phytoremediation of heavy metals from fly ash pond by *Azolla caroliniana*. Ecotoxicol. Environ. Saf. 82, 8–12.

Pandey, V.C., 2015. Assisted phytoremediation of fly ash dumps through naturally colonized plants. Ecol. Eng. 82, 1–5.

Pandey, V.C., Singh, N., 2010. Impact of fly ash incorporation in soil systems. Agric. Ecosyst. Environ. 136, 16–27.

Pandey, V.C., Singh, K., 2011. Is *Vigna radiata* suitable for the revegetation of fly ash basins? Ecol. Eng. 37, 2105–2106.

Pandey, V.C., Singh, B., 2012. Rehabilitation of coal fly ash basins: current need to use ecological engineering. Ecol. Eng. 49, 190–192.

Pandey, V.C., Abhilash, P.C., Singh, N., 2009a. The Indian perspective of utilizing fly ash in phytoremediation, phytomanagement and biomass production. J. Environ. Manag. 90, 2943−2958.

Pandey, V.C., Abhilash, P.C., Upadhyay, R.N., Tewari, D.D., 2009b. Application of fly ash on the growth performance, translocation of toxic heavy metals within *Cajanus cajan* L.: implication for safe utilization of fly ash for agricultural production. J. Hazard. Mater. 166, 255−259.

Pandey, V.C., Singh, J.S., Singh, R.P., Singh, N., Yunus, M., 2011a. Arsenic hazards in coal fly ash and its fate in Indian scenario. Resour. Conserv. Recy. 55, 819−835.

Pandey, V.C., Singh, K., Singh, B., Singh, R.P., 2011b. New approaches to enhance eco-restoration efficiency of degraded sodic lands: critical research needs and future pros-pects. Ecol. Restor. 29 (4), 322−325.

Pandey, V.C., Singh, K., Singh, R.P., Singh, B., 2012a. Naturally growing *Saccharum munja* on the fly ash lagoons: a potential ecological engineer for the revegetation and stabi-lization. Ecol. Eng. 40, 95−99.

Pandey, V.C., Singh, K., Singh, J.S., Kumar, A., Singh, B., Singh, R.P., 2012b. *Jatropha curcas*: a potential biofuel plant for sustainable environmental development. Renew. Sustain. Energ. Rev. 16, 2870−2883.

Pandey, V.C., Singh, N., 2014. Fast green capping on coal fly ash basins through ecologi-cal engineering. Ecol. Eng. 73, 671−675.

Pandey, V.C., Prakash, P., Bajpai, O., Kumar, A., Singh, N., 2015b. Phytodiversity on fly ash deposits: evaluation of naturally colonized species for sustainable phytorestoration. Environ. Sci. Pollut. Res. 22, 2776−2787. Available from: https://doi.org/10.1007/s11356-014-3517-0.

Pandey, V.C., Bajpai, O., Nandita Singh, N., 2016. Plant regeneration potential in fly ash ecosystem. Urban For. Urban Gre. 15, 40−44.

Pavlović, P., Mitrović, M., Djurdjevic, L., 2004. An ecophysiological study of plants growing on the fly ash deposits from the "Nikola Tesla-A" thermal power station in Serbia. Environ. Manage. 33, 654−663.

Pietrzykowski, M., Krzaklewski, W., Gaik, G., 2010. Assessment of forest growth with plantings dominated by Scots pine (*Pinus sylvestris* L.) on experimental plots on a fly ash disposal site at the Bełchatów power plant. Environ. Eng. 137 (17), 65−74 (In Polish, English summary).

Pietrzykowski, M., Krzaklewski, W., Woś, B., 2015. Preliminary assessment of growth and survival of green alder (*Alnus viridis*), a potential biological stabilizer on fly ash dis-posal sites. J. For. Res. 26, 131−136. Available from: https://doi.org/10.1007/s11676-015-0016-1.

Pietrzykowski, M., Wos, B., Pajak, M., Wanic, T., Krzaklewski, W., Chodak, M., 2018. Reclamation of a lignite combustion waste disposal site with alders (*Alnus* sp.): assess-ment of tree growth and nutrient status within 10 years of the experiment. Environ. Sci. Pollut. Res. 1−9. Available from: https://doi.org/10.1007/s11356-018-1892-7.

Piha, M.I., Vallack, H.W., Michael, N., Reeler, B.M., 1995. A low input approach to vegetation establishment on mine and coal ash wastes in semi-arid regions. II. Lagooned pulverized fuel ash in Zimbabwe. J. Appl. Ecol. 32, 382−390.

Podgaiski, L.R., Rodrigues, G.G., 2010. Leaf-litter decomposition of pioneer plants and detritivore macrofaunal assemblages on coal ash disposals in southern Brazil. Eur. J. Soil Biol. 46, 394−400.

Ram, L.C., Masto, R.E., 2014. Fly ash for soil amelioration: a review on the influence of ash blending with inorganic and organic amendments. Earth Sci. Rev. 128, 52−74.

Ram, L.C., Jha, S.K., Tripathi, R.C., Masto, R.E., Selvi, V.A., 2008. Remediation of fly ash basins through plantation. Remediation 18, 71−90.

Rau, N., Mishra, V., Sharma, M., Das, M.K., Ahaluwalia, K., Sharma, R.S., 2009. Evaluation of functional diversity in rhizobacterial taxa of a wild grass (*Saccharum ravennae*) colonizing abandoned fly ash dumps in Delhi urban ecosystem. Soil Biol. Biochem. 41, 813−821.

Ray, P., Adholeya, A., 2009. Correlation between organic acid exudation and metal uptake by ectomycorrhizal fungi grown on pond ash in vitro. Biometals 22, 275−281.

Rillig, C.M., Mummey, D.L., 2006. Mycorrhizas and soil structure. N. Phytol. 171, 41−53.

Rippon, J.E., Wood, M.J., 1975. Microbiological aspects of pulverized fuel ash. In: Chadwick, M.J., Goodman, G.T. (Eds.), The Ecology of Resource Degradation and Renewal. John Wiley and Sons, New York, pp. 331−349.

Ruhl, L., Vengosh, A., Dwyer, G.S., Hsu-Kim, H., Deonarine, A., Bergin, M., et al., 2009. Survey of the potential environmental and health impacts in the immediate aftermath of the coal ash spill in Kingston, Tennessee. Environ. Sci. Technol. 43, 6326−6333.

Satchell, J.E., Stone, D.E., 1977. Colonization of pulverished fuel ash sites by earthworms. Publ. Del. Cent. Pirenacio de. Biol. Exper. 9, 59−74.

Scanlon, D.H., Duggan, J.C., 1979. Growth and element uptake of woody plant on fly ash. Environ. Sci. Technol. 13 (3), 311−315.

Selvam, A., Mahadevan, A., 2002. Distribution of mycorrhizas in an abandoned fly ash pond and mined sites of Neyveli Lignite Corporation, Tamil Nadu, India. Basic. Appl. Ecol. 3, 277−284.

Shaw, P.J.A., 1992. A preliminary study of successional changes in vegetation and soil development on unamended fly ash (PFA) in southern England. J. Appl. Ecol. 29, 728−736.

Shaw, P., 2003. Collembola of pulverised fuel ash sites in east London. Eur. J. Soil Biol. 39, 1−8.

Singh, J.S., Pandey, V.C., 2012. Fly ash application in nutrient poor agriculture soils: impact on methanotrophs population dynamics and paddy yields. Ecotoxicol. Environ. Saf. 89, 43−51.

Singh, J.S., Pandey, V.C., Singh, D.P., Singh, R.P., 2011a. Coal fly ash and farmyard manure amendments in dry-land paddy agriculture field: effect on N−dynamics and paddy productivity. Appl. Soil Ecol. 47, 133−140.

Singh, J.S., Pandey, V.C., Singh, D.P., 2011b. Efficient soil microorganisms: a new dimension for sustainable agriculture and environmental development. Agric. Ecosyst. Environ. 140, 339−353.

Singh, K., Pandey, V.C., Singh, B., Singh, R.R., 2012a. Ecological restoration of degraded sodic lands through afforestation and cropping. Ecol. Eng. 43, 70−80.

Singh, K., Pandey, V.C., Singh, B., (2012b). Soil biology: soil enzymology. In: Shukla, G., Varma A. (Eds.). Applied Soil Ecology, Springer, London, NY, 384, p. 9783642142246. 62, 50−51.

Sinoga, J.D.R., Pariente, S., Diaz, A.R., Murillo, J.F.M., 2012. Variability of relationships between soil organic carbon and some soil properties in Mediterranean rangelands under different climatic conditions (South of Spain). Catena 94, 17−25.

Sparling, G.P., 1992. Ratio of microbial biomass carbon to soil organic carbon as a sensitive indicator of changes in soil organic matter. Aust. J. Soil Res. 30, 195−207.

Tamis, W.L.M., Udo de Haes, H.A., 1995. Recovery of earthworm communities (Lumbricidae) in some thermally and biologically cleaned soils. Pedobiologia 39, 351−369.

Tiwari, S., Kumari, B., Singh, S.N., 2008. Microbe-induced changes in metal extractability from fly ash. Chemosphere 71, 1284−1294.

Tokuda, S., Hayatsu, M., 2002. Nitrous oxide emission potential of 21 acidic tea field soils in Japan. Soil Sci. Plant. Nutr. 47, 637−642.

Townsend, W.N., Gillham, E.W.F., 1975. Pulverized fuel ash as a medium for plant growth. In: Chadwick, M.J., Goodman, G.T. (Eds.), In The Ecology of Resource Degradation and Renewal. Blackwell Scientific, Oxford, pp. 287−304.

Uzarowicz, Ł, Skiba, M., Leue, M., Zagórski, Z., Gąsiński, A., Trzciński, J., 2018. Technogenic soils (Technosols) developed from fly ash and bottom ash from thermal power stations combusting bituminous coal and lignite. Part II. Mineral transformations and soil evolution. Catena 162, 255−269.

Van Rensburg, L., Morgenthal, T.L., Van Hamburg, H., Michael, M.D., 2003. A comparative analysis of the vegetation and topsoil cover nutrient status between two similarly rehabilitated ash disposal sites. Environmentalist 23, 285−295. Available from: https://doi.org/10.1023/B:ENVR.0000031359.70523.4a.

Wardle, D.A., Ghani, A., 1995. A critique of the microbial metabolic quotient (qCO₂) as a bioindicator of disturbance and ecosystem development. Soil Biol. Biochem. 27, 1601−1610.

Świątek, B., Woś, B., Chodak, M., Maiti, S.K., Józefowska, A., Pietrzykowski, M., 2019. Fine root biomass and the associated C and nutrient pool under the alder (*Alnus* spp.) plantings on reclaimed technosols. Geoderma 337, 1021−1027.

Yan, T., Yang, L., Campbell, C.D., 2003. Microbial biomass and metabolic quotient of soils under different land use in the Three Gorges Reservoir area. Geoderma 115, 129−138.

Zevenbergem, C., Bradley, J.P., Shyam, A.K., Jenner, H.A., Platenburg, R.J.P.M., 2000. Sustainable ash pond development in India−a resource for forestry and agriculture. Waste Mater. Ser. 1, 533−540.

Zimmer, M., 2002. Nutrition in terrestrial isopods (Isopoda: Oniscidea): an evolutionary-ecological approach. Biol. Rev. 77, 455−493.

## Further reading

Adholeya, A., Bhatia, N.P., Kanwar, S., Kumar, S., 1998. Fly ash source and substrate for growth of sustainable agro-forestry system. In: Proceedings of Regional Workshop cum Symposium on Fly Ash Disposal and Utilization. Organized by Kota Thermal Power Station, RSEB, Kota, India.

Anonymous, 1997. Study on sustainable bamboo management. Final report. Luso Consult, Hamburg, Germany.

Becker, S., Soukup, J.M., Gallagher, J.E., 2002. Differential particulate air pollution induced oxidant stress in human granulocytes, monocytes, and alveolar macrophages. Toxicol. Vitro 16, 209−218.

Borm, P.J., 1997. Toxicity and occupational health hazards of coal fly-ash (CFA). A review of data and comparison to coal mine dust. Ann. Occup. Hyg. 41, 659−676.

Bryan Jr., A.L., Hopkins, W.A., Baionno, J.E., Jackson, B.P., 2003. Maternal transfer of contaminants to eggs of Common Grackles (Quiscalus quiscala) nesting on coal fly ash basins. Arch. Environ. Contam. Toxicol. 45, 273−277.

Bryan, A.L., Hopkins, W.A., Parikh, J.H., Jackson, B.P., Unrine, J.M., 2012. Coal fly ash basins as an attractive nuisance to birds: parental provisioning exposes nestlings to harmful trace elements. Environ. Pollut. 161, 170−177.

Chakraborty, R., Mukherjee, A., 2009. Mutagenicity and genotoxicity of coal fly ash water leachate. Ecotoxicol. Environ. Saf. 72, 838−842.

Chaudhary, M.K., 1992. Kendbona ecodevelopment project. A novel approach to waste-land reclamation. Indian For. 118, 879−888.

Chizhikov, V., Chikina, S., Gasparian, A., Zborovskaya, I., Steshina, E., Ungiadze, G., et al., 2002. Molecular follow-up of preneoplastic lesions in bronchial epithelium of former chernobyl clean-up workers. Oncogene 21, 2398—2405.

Costa, D.L., Dreher, K.L., 1997. Bioavailable transition metals in particulate matter mediate cardiopulmonary injury in healthy and compromised animal models. Environ. Health Perspect. 105, 1053—1060.

Dwivedi, S., Saquib, Q., Al-Khedhairy, A.A., Ali, A.S., Musarrat, J., 2012. Characterization of coal fly ash nanoparticles and induced oxidative DNA damage in human peripheral blood mononuclear cells. Sci. Total Environ. 437, 331—338.

Fairbrother, A., Brix, K.V., Toll, J.E., McKay, S., Adams, W.J., 1999. Egg selenium concentrations as predictors of avian toxicity. Hum. Ecol. Risk Assess. 5, 1229—1253.

Garg, V.K., Jain, R.K., 1992. Influence of fuel wood trees on sodic soils. Can. J. For. Res. 22, 729—735.

Goel, V.L., Behl, H.M., 1995. Fuelwood production potential of six *Prosopis* species on an alkaline soil site. Biomass Bioenerg. 8, 17—20.

Greenwood, E.F., Gemmill, R.P., 1978. Derelict industrial land as a habitat for rare plants in S. Lancs (v.c. 59) and W. Lancs (v.c. 60). Watsonia 12, 33—40.

Heinz, G.H., 1996. Selenium in birds. In: Beyer, W.N., Heinz, G.H., Redmon-Norwood, A.W. (Eds.), Environmental Contaminants in Wildlife: Interpreting Tissue Concentrations. CRC Press, Boca Raton, FL, pp. 447—458.

Hopkins, W.A., Mendonca, M.T., Rowe, C.L., Congdon, J.D., 1998. Elevated trace element concentrations in southern toads, *Bufo terrestris*, exposed to coal combustion wastes. Arch. Environ. Contam. Toxicol. 35, 325—329.

Hopkins, W.A., Rowe, C.L., Congdon, J.D., 1999. Elevated trace element concentrations and standard metabolic rate in banded water snakes (*Nerodia fasciata*) exposed to coal combustion wastes. Environ. Toxicol. Chem. 18, 1258—1263.

Hopkins, W.A., Congdon, J., Ray, J.K., 2000. Incidence and impact of axial malformations in larval bullfrogs (*Rana catesbeiana*) developing in sites polluted by a coal-burning power plant. Environ. Toxicol. Chem. 19, 862—868.

Hopkins, W.A., Snodgrass, J.W., Staub, B.P., Jackson, B.P., Congdon, J.D., 2003. Altered swimming performance of a benthic fish (*Erimyzon sucetta*) exposed to contaminated sediments. Arch. Environ. Contam. Toxicol. 44, 383—389.

Hrynkiewicz, K., Baum, C., Niedojadło, J., Dahm, H., 2009. Promotion of mycorrhiza formation and growth of willows by the bacterial strain *Sphingomonas* sp. 23L on fly ash. Biol. Fertil. Soil. 45, 385—394.

Jamil, S., Abhilash, P.C., Singh, N., Sharma, P.N., 2009. *Jatropha curcas*: a potential crop for phytoremediation of coal fly ash. J. Hazard. Mater. 172, 269—275.

Janz, D.M., DeForest, D.K., Brooks, M.L., Chapman, P.M., Gilron, G., Hoff, D., et al., (Eds.), 2010. Ecological Assessment of Selenium in the Aquatic Environment. CRC Press, Boca Raton, FL.

Kapur, S.L., Shyam, A.K., Soni, R., 2002. Greener management practices ash mound reclamation. TERI Inf. Dig. Energ. Environ 1, 559—567.

King, K.A., 1988. Elevated selenium concentrations are detected in wildlife near a power plant. U.S. Dep. Interior, Fish. Wildl. Serv. Res. Inf. Bull. 88-31.

King, K.A., Custer, T.W., Weaver, D.A., 1994. Reproductive success of barn swallows nesting near a selenium-contaminated lake in east Texas, USA. Environ. Pollut. 84, 53—58.

Kumar, B.M., Kumar, S.S., Fisher, R.F., 1998. Intercropping teak with leucaena increases tree growth and modifies soil characteristics. Agrofor. Syst. 42, 81—89.

Lemly, D.A., 1996. Selenium in aquatic organisms. In: Beyer, W.N., Heinz, G.H., Redmon-Norwood, A.W. (Eds.), Environmental Contaminants in Wildlife: Interpreting Tissue Concentrations. CRC Press, Boca Raton, FL, pp. 427—455.

Lemly, D.A., 2002. Symptoms and implications of selenium toxicity in fish: the Belews Lake case example. Aquat. Toxic. 57, 39−49.

Linak, W.P., Yoo, J.I., Wasson, S.J., Zhu, W., Wendt, J.O.L., Huggins, F.E., et al., 2007. Ultrafine ash aerosols from coal combustion: characterization and health effects. Proc. Combust. Inst. 31, 1929−1937.

Lopez, F., Garcia, M.M., Yanez, R., Tapias, R., Fernandez, M., Diaz, M.J., 2008. *Leucaena* species valoration for biomass and paper production in 1 and 2 year harvest. Bioresour. Technol. 99, 4846−4853.

Love, A., Tandon, R., Banerjee, B.D., Babu, C.R., 2009. Comparative study on elemental composition and DNA damage in leaves of a weedy plant species, *Cassia occidentalis*, growing wild on weathered fly ash and soil. Ecotoxicology 18, 791−801.

MA, 2005. Ecosystems and human well-being: current state and trends, Millennium Ecosystem Assessment, vol. 1. Island Press, Washington, DC.

Malewar, G.U., Adsul, P.B., Ismail, S., 1998. Effect of different combinations of fly-ash and soil on growth attributes of forest and dryland fruit crops. Indian J. For. 21, 124−127.

Markad, V.L., Kodam, K.M., Ghole, V.S., 2012. Effect of fly ash on biochemical responses and DNA damage in earthworm, *Dichogaster curgensis*. J. Hazard. Mater. 215−216, 191−198.

Nagle, R.D., Rowe, C.L., Congdon, J.D., 2001. Accumulation and selective maternal transfer of contaminants in the turtle *Trachemys scripta* associated with coal ash deposition. Arch. Environ. Contamin. Toxicol. 40, 531−536.

Neenan, M., Steinbeck, K., 1979. Caloric values for young sprouts of hardwood species. For. Sci. 25, 455−461.

Olivares, A.J., Carrillo-González, R., González-Chávez, Ma. del, C.A., Hernández, R.M. S., 2012. Potential of castor bean (*Ricinus communis* L.) for phytoremediation of mine tailings and oil production. J. Environ. Management. . Available from: https://doi.org/10.1016/j.jenvman.2012.10.023.

Pandey, V.C., 2013. Suitability of *Ricinus communis* L. cultivation for phytoremediation of fly ash disposal sites. Ecol. Eng. 57, 336−341.

Pandey, V.C., Kumar, A., 2012. *Leucaena leucocephala*: an underutilized plant for pulp and paper production. Genet. Resour. Crop Evol. 60, 1165−1171.

Pandey, V.C., Pandey, D.N., Singh, N., 2015a. Sustainable phytoremediation based on naturally colonizing and economically valuable plants. J. Clean. Prod. 86, 37−39. Available from: https://doi.org/10.1016/j.jclepro.2014.08.030.

Pietrzykowski, M., Krzaklewski, W., 2017. Reclamation of mine lands in Poland. Bio-Geotechnologies for Mine Site Rehabilitation Chapter 27. Elsevier, Amsterdam-Oxford-Cambridge, ISBN: 978-0-12-812986-9, pp. 493−513.

Pietrzykowski, M., Krzaklewski, W., Woś, B., 2013. Concentration of trace elements (Mn, Zn, Cu, Cd, Pb, Cr) in alder (*Alnus* sp.) leaves used as phytomelioration species on fly ash disposal (in Polish). University of Zielona Góra, scientific reports. Environ. Eng. 151, 26−34. 31.

Pietrzykowski, M., Woś, B., Pająk, M., Wanic, T., Krzaklewski, W., Chodak, M., 2017. Alders (*Alnus* sp.) as a potential biological stabilizer on fly ash disposal sites. In: Hu, Z. (Ed.), Land Reclamation in Ecological Fragile Areas. CRC Press, Boca Raton,FL, pp. 465−471.

Pillman, A., Jusaitis, M., 1997. Revegetation of waste fly ash lagoons. IL Seedling transplants and plant nutrition. Waste Manag. Res. 15 (4), 359−370.

Poliakova, V.A., Suchko, V.A., Tereshchenko, V.P., Bazyka, D.A., Golovinia, O.M., Rudavskaia, G.A., 2001. Invasion of microorganisms in the bronchial mucosa of liquidators of the Chernobyl accident consequences. Mikrobiology 63, 41−50.

Rai, U.N., Pandey, K., Sinha, S., Singh, A., Saxena, R., Gupta, D.K., 2004. Revegetating fly-ash landfills with *Prosopis juliflora* L.: impact of different amendments and *Rhizobium* inoculation. Environ. Int. 30, 293–300.

Reash, R.J., Lohner, T.W., Wood, K.V., 2006. Selenium and other trace metals in fish inhabiting a fly ash stream: implications for regulatory tissue thresholds. Environ. Pollut. 142, 397–408.

Riekerk, H., 1984. Coal–ash effects on fuelwood production and runoff water quality. South. J. Appl. For. 8, 99–102.

Rowe, C.L., 1998. Elevated standard metabolic rate in a freshwater shrimp (*Palaemonetes paludosus*) exposed to trace element-rich coal combustion waste. Comp. Biochem. Physiol. Part A 121 (4), 299–304.

Rowe, C.L., Kinney, O.M., Fiori, A.P., Congdon, J.D., 1996. Oral deformities in tadpoles (*Rana catsbeiana*) associated with coal ash deposition: effects on grazing ability and growth. Freshw. Biol. 36, 723–730.

Ruhl, L., Vengosh, A., Dwyer, G.S., Hsu-Kim, H., Deonarine, A., 2010. Environmental impacts of the coal ash spill in Kingston, Tennessee: an 18-month survey. Environ. Sci. Technol. 44, 9272–9278.

Sahu, K.C., 1994. Power plant pollution: cost of coal combution. Survey of the Environment. The Hindu, Madras, pp. 47–51.

Schilling, C.J., Tams, I.P., Schilling, R.S.F., Nevitt, A., Rossiter, C.E., Wilkinson, B., 1988. A survey into the respiratory effects of prolonged exposure to pulverized fuel ash. Br. J. Ind. Med. 45, 810–817.

Shaw, P.J.A., 1996. Role of seedbank substrates in the revegetation of fly ash and gypsum in the United Kingdom. Restor. Ecol. 4, 61–70.

Shrivastava, S., 2007. Tribal dependence on fly ash in Korba. J. Ecol. Anthropol. 11, 69–73.

Shukla, A.K., Misra, P.N., 1993. Improvement of sodic soil under tree cover. Indian. For. 119, 43–52.

Singh, J.S., Singh, D.P., 2012. Reforestation: a potential approach to mitigate excess atmospheric $CH_4$ build-up. Ecol. Manag. Restor . Available from: https://doi.org/10.1111/emr.12004.

Singh, A., Singh, N., Behl, H.M., 2005. Utilization of fly ash for biomass production: a case study on *Jatropha curcas* L. proceeding abstract. In: Third International Conference on Plants and Environmental Pollution, 28 Nov–2 Dec.

Singh, V.K., Behal, K.K., Rai, U.N., 2000. Comparative study on the growth of mulberry (*Morus alba*) plant at different levels of fly ash amended soil. Biol. Mem. 26, 1–5.

Staub, B.P., Hopkins, W.A., Novak, J., Congdon, J.D., 2004. Respiratory and reproductive characteristics of eastern mosquitofish (*Gambusia holbrooki*) inhabiting a coal ash settling basin. Arch. Environ. Contam. Toxicol. 46, 96–101.

Tripathi, R.D., Yunus, M., Singh, S.N., Singh, N., Kumar, A., 2000. Final technical report – reclamation of fly-ash landfills through successive plantation, soil amendments and or through integrated biotechnological approach. Directorate of Environment, U.P, India.

US EPA 2007. Human and Ecological Risk Assessment of Coal Combustion Wastes (draft).

US EPA 2008. Particulate Matter. <http://www.epa.gov/air/particlepollution/health.html>.

# CHAPTER 8

# Fly ash deposits—a potential sink for carbon sequestration

## Contents

## 8.1 Introduction

### 8.1.1 Fly ash deposits

Coal-based thermal power stations are contributing around 40% of world's electricity with the generation of 600 million tons of fly ash (FA) and 12,000 million tons of $CO_2$ annually (Montes-Hernandez et al., 2009). Only 16% of the FA is used in manufacturing of cement and concrete, while the rest is disposed of in basins and landfills, which accounts for thousands of hectares of land nearby thermal power plants. FA generation is expected to increase in the event of global energy demand, and it is difficult to find a safe disposal. Globally, some of the countries that generate huge amounts of FA are United States, India, China, South Africa,

*Phytomanagement of Fly Ash*
ISBN: 978-0-12-818544-5
DOI: https://doi.org/10.1016/B978-0-12-818544-5.00008-0
**235**

Europe, Australia, Italy, Greece, and Japan. The Ministry of Science and Technology, Government of India established a "FLY ASH UNIT" (FAU) in the Department of Science and Technology for the promotion of FA research. A number of projects have been governed by this FAU. The protocol on FA was signed by the Secretary of the Department of Science and Technology, India, and the Chairman of the Inter-Regional Association of the Economic Cooperation of the Constituent Entities of the Russian Federation in the presence of the Prime Minister of India and the President of Russia at the Kremlin on December 16, 2011. The goals of this protocol were the implementation of the mechanism similar to "FA Mission—India" in the Siberian Federal District, to facilitate the development and application of technologies for the utilization and safe management of FA, including import of technologies from India. This type of protocol against other countries should be singed for a better promotion of FA. The World Bank has apprised that FA disposal in India would require 1000 km$^2$ of land, while in global aspect, FA generation would require 3235 km$^2$ of land by the year 2015 (Pandey and Singh, 2012). Therefore the abandoned FA basins may be used as a "potential missing sink" for carbon sequestration through vegetation development, where plant species succession leads to sequester carbon into the substrate-vegetation system. Thus FA basins appear to be a new sink or area for atmospheric carbon sequestration at global level. Native vegetation or afforestation on FA basins may alleviate the level of atmospheric $CO_2$ and generate the additional bioresources. It is well known that FA contains a relatively high concentration of calcium, including plant nutrients, and has a proven effect in terms of carbon capture (Sun et al., 2012) as well as environmental conservation through different ways: (1) providing a bioaesthetic vegetation cover on FA disposal sites; (2) stabilization of heavy metals to avoid the leaching; and (3) carbon sequestration in the substrate-vegetation system of FA basins and generation of bioresources useful to local people (Pandey and Singh, 2012).

## 8.1.2 Carbon sequestration

Coal-based thermal power stations have a great impact on the environment due to the emissions of carbon dioxide. The carbon dioxide emissions through coal-based thermal power stations have been taken into consideration. Carbon dioxide is one of the greenhouse gases, and the reduction of carbon dioxide emissions is an important target according to the Kyoto Protocol and may lead to lessening of global warming (Grigore et al., 2016). A 500 MW coal-based thermal power station emits roughly 8000 tons of $CO_2$ per day. The anticipated global change due to the increasing carbon dioxide in the atmosphere has prompted the scientific community to develop mechanisms to reduce as well as stabilize the atmospheric carbon dioxide. In this regard, terrestrial carbon sequestration processes could make a significant contribution to abate carbon dioxide increases (Pausstian et al., 1998; Reichle et al., 1999). Several studies reported carbon sequestration in different types of places such as sodic soil (Kaur et al., 2002), saline land (Lim et al., 2012), mine soil (Shrestha and Lal, 2006), red mud spill (Renforth et al., 2012), volcanic soil (Zehetner, 2010), phytoliths (Parr and Sullivan, 2005), deep-sea site (Thistle et al., 2006), ocean (Israelsson et al., 2009), Indo-Gangetic Plain (Grace et al., 2012), tropical dry land (Farage et al., 2007), pine stands (Parajuli and Chang, 2012), and forest (Wang et al., 2012). However, there has been no reported research on carbon sequestration on FA basins, which is the current need of the hour. The rehabilitation of abandoned FA disposal sites that are devoid of vegetation covers presents an opportunity for carbon sequestration through natural and planted vegetation. To the best of my knowledge, no information is available on the carbon flux throughput across the substrate-vegetation system developed during the restoration of FA basins. It is assumed that the carbon input to the substrate-vegetation system will significantly contribute to the restoration of soil carbon stock in abandoned FA basins. The substrate-plant systems of FA deposits have the ability to capture carbon dioxide that comes during the coal combustion process and the present cost-effective carbon capture technologies. So, carbon capture sequestration should be implemented at abandoned FA disposal sites of coal-based thermal power stations through afforestation. The soil—plant system of FA deposits will be able to consume all of the carbon dioxide captured on a daily basis from fossil fuel burning power stations.

One feature of the terrestrial carbon sequestration approach used the soil and vegetation factors as long-term storage pools for the atmosphere-derived carbon. To accomplish this, the carbon sequestration can be enhanced by improving the natural biological processes that assimilate carbon dioxide, that is, increasing the plant productivity and pools of soil organic matter (SOM) that is resistant to microbial decomposition. This strategy would require a plant—soil-based program that can be successfully implemented across different land-use categories. The approach of this study is to evaluate that rehabilitation of FA basins presents a unique opportunity for carbon sequestration through native vegetation and its enhancement through man-made plantation with a new commodity resource generation. Thus FA carbon economics is a new area of research, in which carbon sequestration will be observed as carbon stocks in above- and below-ground biomass and in the organic detritus mass of the substrate.

## 8.2 Why carbon sequestration in fly ash deposits is so important?

The proliferation of coal-fired power stations over the world is the main culprit in global warming and contributed about one-quarter of total emissions of carbon dioxide. Where coal-fired power station is source, FA deposits may play as a potential sink. Therefore coal-fired power stations are significant drivers of global warming. First time, the carbon dioxide emissions of 50,000 power stations worldwide, the globe's most concentrated source of greenhouse gases, have been assembled into a massive new data base, called Carbon Monitoring for Action (*Science Daily*, 2007). The issue with the Indian coal-fired stations is the higher emissions of carbon dioxide, which is around 980 g of $CO_2$ $kWh^{-1}$. These stations add 60% of the total carbon dioxide emissions in India (Aswathanarayana, 2008). Hence, coal-based electricity generation has been an extremely aggressive issue regarding climate change.

The huge areas of FA disposal sites offer an opportunity to use them as "potential sinks" of atmospheric carbon dioxide (mostly that was emitted by coal-based thermal power stations), but it needs a proper rehabilitation plan. Carbon sequestration is a process of changing atmospheric carbon dioxide to biomass carbon through photosynthesis, and the incorporation of the same biomass into the soil as an organic matter. Soils possess around 75% of the terrestrial carbon pool (750 Pg) and three times the amount fixed by plants (Saleska et al., 2003; Farage et al., 2007), which shows that soils play an essential role in the carbon cycle. Soils deserve a close examination due to their large carbon mitigation potential. It is important to determine that sinks of atmospheric carbon dioxide are available near the sources. Then, such a system might be termed "carbon neutral." It would be better, if one proverb can be made for the popularity of carbon dioxide sequestration worldwide. In my opinion, it should be "capture carbon dioxide where it emits." This proverb is suitable for coal-based thermal power stations (major source of carbon dioxide emissions) and FA deposits (a "potential sink" through effective vegetation development). Hence, it is important to define the sources and sinks of atmospheric carbon dioxide to decide which steps have practical values to mitigate the global climate change due to carbon dioxide emissions.

## 8.3 Limitations of carbon sequestration in fly ash deposits

The three most important factors, namely physical, chemical, and microbial factors, hinder plant growth on freshly deposited FA disposal sites. Initial high pH, high soluble salt, boron toxicity, nutrient deficiencies (e.g., N and P), and some heavy metals are the chemical limitations of FA deposits (Haynes, 2009; Pandey and Singh, 2010). A high pH (generally from 8 to 12) reduces the bioavailability of some nutrients (P, Fe, Mn, Zn, and Cu) to the plants growing on FA deposits. Physical barriers include natural compaction of fine ash particles due to the pozzolanic nature that restricts plant root growth in FA deposits. The heterotrophic microorganisms use substrate carbon as an energy source. Usually, the lack of microbial activity in FA is due to the absence of substrate carbon and adequate nitrogen supply that limit plant growth (Haynes, 2009; Pandey and Singh, 2010). The readers interested in the details of the limitations of plant growth on FA deposits are suggested to access the review articles of Haynes (2009) and Pandey and Singh (2010).

## 8.4 Revival of fly ash ecosystems to enhance carbon sequestration pool by a low-input tool

An extensive Scopus (http://www.scopus.com) and library-based survey revealed that a number of workers reported a wide range of naturally growing plant species for their phytoremediation potential of FA deposits worldwide. Some of these, *Calotropis gigantea* (Selvam and Mahadevan, 2002), *Calotropis procera* (Maiti et al., 2005), *Cassia occidentalis* (Love et al., 2009), *Cassia tora* (Gupta and Sinha, 2008), *Cynodon dactylon* (Maiti and Nandhini, 2006), *Lantana camera* (Maiti et al., 2005), *Pteris vittata* (Kumari et al., 2011), *Saccharum ravennae* (Rau et al., 2009), *Saccharum spontaneum* (Maiti and Jaiswal, 2008; Pandey et al., 2015), *Saccharum munja* (Pandey et al., 2012), *Typha latifolia* (Tiwari et al., 2008; Babcock et al., 1983; Pandey et al., 2014), *Zizyphys jujuba* (Maiti et al., 2005), *Ziziphus mauritiana* (Pandey and Mishra, 2018), and *Ipomoea carnea* (Pandey et al., 2012), have more potential for the accumulation of heavy metals from FA deposits. These screened species having high FA tolerant nature, fast growth rate, high biomass productivity, and dominant nature are the key factors for selecting effective species for carbon sequestration on FA deposits with ecologically and economically significance.

But, nobody has estimated the carbon sequestration potential of native vegetation and planted species on FA deposited sites. If we vegetate bare FA deposits through potential species, then carbon sequestration may increase as the percentage of terrestrial biosphere carbon sequestration. As the discussed species above, *S. munja* and *S. spontaneum* appear as a potential species of Gramineae family to sequester carbon dioxide in the substrate-vegetation system of FA deposits due to the dense cover of ash disposal sites and must be assessed for carbon sequestration efficiency in future (Fig. 8.1). Recently, Pandey (2012a,b) reported that *I. carnea* is an efficient plant for the phytoremediation of FA deposits and recommended it as a promising species for carbon sequestration on barren FA basins. Presently, the terrestrial biosphere is reported to sequester 20%−30% of the global anthropogenic carbon dioxide emissions (Saleska et al., 2003). Rehabilitation of degraded lands through agroforestry has been suggested in the current policies of India. FA disposal sites need their safe environmental management and should be considered as a missing sink for carbon sequestration through an active revegetation approach, because FA deposit is a soil-like substrate where plant succession can be accelerated through anthropogenic interventions. The residual soil organic carbon (SOC) is a dynamic quantity, as it depends on the carbon fluxes between the

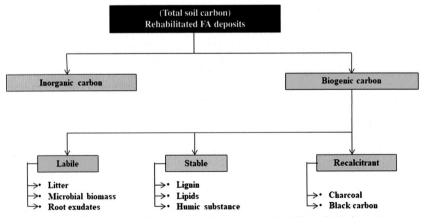

**Figure 8.1** Distribution of various carbon fractions in rehabilitated FA deposits.

atmosphere and lithosphere. The mechanisms that influence SOC pool is the net primary productivity and the allocation of photosynthates across the plant body. Despite of the above controlling factors, the activity of soil microbial community also regulates the carbon storage through the processes like respiration and immobilization of SOM (Zhu and Miller, 2003). Hence, it is asserted that the development microbial community along with the plants can effectively mitigate the global climate change (Lal, 2003). Carbon sequestration in soils is also affected by root productivity, root turnover rates, exudation, and mycorrhizal colonization (Matamala et al., 2003). We also know that root production and its turnover can directly impact biogeochemical cycling through catering carbon and energy to the microbes. The current research needs to screen out the potential of naturally growing and planted species for carbon sequestration on different FA landfills along with their heavy metal phytoremediation potential. Finally, this study will provide a phytotechnology for carbon sequestration along with phytoremediation of heavy metals of FA basins. If we further add economic species in the said agenda, then additional resource generation can achieve three goals by one action plan during the rehabilitation program of FA basins. Plant—microbe interactions have been suggested as a low-input biotechnology for ecosystem revitalization and a novel application for exploitation in multipurpose remediation technologies (Abhilash et al., 2012).

## 8.4.1 Role of enzymes in carbon sequestration of ash—soil system

Soil enzyme analyses provide an opportunity to measure the optimum rate of organic matter decomposition, since most of the processes occurring in soil are microbial-mediated. The enzyme activity in soil that oxidizes the recalcitrant fractions of SOM and hence controls the soil carbon sequestration is indomitably related to the pH of the soil. Arid and semiarid site, for example, sodic soils having a low SOM and high soil pH showed variations in relation to enzymes like dehydrogenase, β-glucosidase, leucyl aminopeptidase, phenol oxidase, and peroxidases (Sinsabaugh et al., 2008). Phenol oxidase has been considered as an "enzymatic latch" to hold SOM to provide good protection from decomposition (Freeman et al., 2001). Upon release of oxygen stress by drought in summer, phenol oxidase activity will increase, thereby accelerating the decomposition of soluble polyphenols, the compound toxic to hydrolytic enzymes (Shi et al., 2009). Freeman et al. (2001) reported that phenol oxidase activity regulated the SOM mineralization as well as the carbon dioxide efflux. Moreover, the SOM is the lowest in arid ecosystems, which also have lesser degree of primary production, due to alkaline pH, and increased solubility of polyphenols (Collins et al., 2008). These studies are from other than ash–soil systems. What is happening on the carbon cycle during the rainy, winter, and summer seasons in relation to phenol oxidase and hydrolytic enzymes has to be documented. Thus analyses of enzyme activities shall suggest a broader context for dehydrogenase, β-glucosidase, phosphatase, leucyl aminopeptidase, phenol oxidase, and peroxidase in OM storage/carbon sequestration in FA deposits.

## 8.4.2 Plant–microbe interactions in carbon sequestration of ash-vegetation system

Plant–microbe interactions are an important factor, influencing carbon sequestration potential in ash–soil system. The estimation of plant–microbe interactions in the rhizosphere is important for understanding the processes like nutrient cycling, ecosystem functioning, as well as carbon sequestration (Singh et al., 2004). On the other hand, carbon sequestration in soil depends on microbial activity, root productivity and turnover rates, root exudation, soil properties, plant community structure, as well as mycorrhizal colonization (Matamala et al., 2003); but, the complexity of interactions between root turnover and microbial activity is not adequately reported (Singh et al., 2004). Therefore the potential exists for studies to determine how plant–microbe interactions maximize the carbon stocks in FA basins.

## 8.5 Using fly ash as an amendment for carbon sequestration

Despite the limitations, FA substrate has a significant agronomic potential to plant growth. Numerous studies have suggested that FA improves soil fertility and plant growth when used as a soil amendment (Pandey and Singh, 2010), due to the properties like high water-holding capacity, presence of macro- and micronutrients, and liming capacity (Haynes, 2009; Pandey et al., 2009). Agricultural lime plays an important role in carbon dioxide flux; thus several studies have been done at this end to decrease global warming through agronomic practices (Robertson et al., 2000). As per the Intergovernmental Panel on

Climate Change (IPCC), 954 million metric tons of $CO_2$ was emitted due to the application of 2120 million metric tons of lime in agricultural fields in 2001 (West and McBride, 2005). At this end, some researchers are working on the utilization of FA as a replacement of lime in agricultural fields and thus reduce global warming (Bernoux et al., 2003; McBride and West, 2005).

Alternatively, FA can be used for mineral carbon sequestration nearby coal-based thermal power stations, as it contains free lime and because of its ready availability in such places. The carbon sequestration potential of FA has been comparatively similar to municipal solid waste ash (Bobicki et al., 2012). Mineral carbon sequestration is the only technique of permanent carbon storage by capturing carbon dioxide in a single step (Bobicki et al., 2012). It is accomplished by capturing carbon either in geologic formations (in situ) or in a chemical processing plant (ex situ) manually (Gerdemann et al., 2004). Naturally, the carbonic acid, generated through carbon dioxide dissolution in rain water, reacts with alkaline minerals to form carbonates (Lackner, 2002; Huijgen and Comans, 2003). There is no possibility of carbon dioxide release after mineral carbonation, because significant energy is needed to trigger carbon dioxide production from carbonates (Lackner et al., 1995). Advantageously, it is a permanent method of carbon dioxide disposal over other carbon storage practices (Goff and Lackner, 1998), while also generating economically valuable products (Maroto-Valer et al., 2005). In addition, mineral carbon sequestration can be used in areas where other storage techniques (such as geologic carbon sequestration) are not feasible (Zevenhoven and Fagerlund, 2010).

Carbon sequestration assisted by FA amendments has been widely shown to be an effective way to reduce both FA disposal and carbon dioxide from the environment. Some researchers have reported their works on enhancing carbon sequestration through reclamation of degraded lands with coal combustion products (Palumbo et al., 2004), like indirect mineralization using brown coal FA (Sun et al., 2012). Reduction in carbon dioxide emissions from normal and saline soils amended with coal FA was reported by Lim et al. (2012). FA offers sites for the formation of organomineral complexes due to its high specific area because of its silt-sized particles (Jala and Goyal, 2006). Moreover, FA can fix carbon dioxide generated from soil respiration through carbonation of calcium and magnesium that are already present in FA, which also increases the soil carbon content (Lim et al., 2012; Lim and

Choi, 2014). FA application has also been found to be effective in increasing soil carbon sequestration without decreasing rice yields (Lim et al., 2017).

Montes-Hernandez et al. (2009) reported 82% of carbonate conversion and demonstrated the potential of FA to sequester 26 kg of carbon dioxide per ton of FA. Uliasz-Boche_nczyk et al. (2009) reported 7.85 g of carbon sequestration per 100 g FA, and reported that sequestration depends on FA characteristics. Both studies confirmed the possibility to use alkaline FA for carbon dioxide mitigation. Soong et al. (2006) analyzed FA carbonation in brine obtained from oil and gas wastes that increased the FA carbonate content to 53% at 20°C and sequestered 1.36 MPa of $CO_2$ in 2 h, which was also contributed by the Ca present in both waste streams (Soong et al., 2006). On the basis of general estimates, FA with the carbon sequestration capacity of 5% could sequester 0.25% of carbon dioxide emitted from coal-based thermal power stations (Montes-Hernandez et al., 2009; Pan et al., 2012; Sanna et al., 2012). Therefore FA utilization in replacement of lime as a soil ameliorant can decrease net carbon dioxide release and thus decline global warming.

FA application in concrete mixture in place of cement helps in decreasing carbon dioxide production, which would have generated due to generation of cement. For example, 1 ton of FA application in concrete can decrease the release of 2 tons of carbon dioxide (Naik and Tyson, 2000; Krishnamoorthy, 2000). This carbon dioxide is assumed to be a most important contributor to the global warming of the planet (Tietenberg, 2003; Ferreira et al., 2003). Thus the application of FA in the above industries can also have the following benefits apart from carbon sequestration or reduction, which are: (1) use of a zero-cost raw material; (2) elimination of waste; (3) minimization of global warming; and (4) conservation of topsoil and other natural resources like water, coal, and lime (Pandey and Singh, 2010).

This is a win—win solution and shows that FA is at the top regarding all the green technologies being adopted globally. Subbituminous and lignite FAs can stabilize the soil enzyme required for carbon sequestration (tyrosinase) and enhance humification (Amonette et al., 2003a). Exclusively, FA catalyzes soil humification reaction through three steps that are physical stabilization of tyrosinase, oxidation of monomers, and enhancing the oxidation and condensation steps through alkaline pH (Amonette et al., 2003b).

## 8.6 Carbon sequestration and its calculation in ash-vegetation system

Carbon sequestration has received much attention in view of the climate change effects on natural flora and fauna. Vegetation and soils are considered as the major sink for the atmosphere carbon dioxide level. The plants sequester carbon into the soil, biomass, and plant residues. This section deals with the potential for the sequestration of atmospheric carbon on FA disposal sites by natural vegetation and man–made plantation. Initially, the FA deposits do not contain organic carbon, and hence the developed organic carbon in the substrate is the result of plant growth. The annual global FA generation is nearly 600 million tons (Ram et al., 2008) and is recognized as an environmental hazard. In addition, the huge ranges of FA disposal tracts are regarded as a "missing sink" to carry out carbon sequestration. In this direction, the rehabilitation of FA deposits by growing tolerant tree species can effectively conserve the site in due course of time, which will also contribute to carbon balance. Assessment of carbon inputs in FA deposits through natural vegetation or man–made plantations and their standing root—shoot biomass will determine the carbon sequestration potential of the vegetation.

Soil carbon is a key component of the terrestrial biosphere pool in the carbon cycle. The amount of carbon in the soil is a function of the historical vegetative cover and productivity, which in turn is dependent in part upon climatic variables. The potential of sequestering carbon by different terrestrial ecosystems depends mainly on land-use, land type, and forestry practices. The rate of sequestration can be increased by increasing the reforestation activities (dealt in article 3.3 of the Kyoto Protocol; IPCC 2000). Some other techniques to increase the carbon sequestration rate are conversion of cropland to grassland, and shift to conservation agriculture with minimum tillage, which essentially increases the SOM (FAO 2001). Likely, harvested wood containing high carbon (e.g., 1 ton of dry wood contains 1.8 tons of fixed carbon dioxide) will return carbon to the atmosphere if used as a fuel, but if incorporated into construction material, it can help in sequestering the carbon over many years. Long-term storage of atmospheric $CO_2$ in the substrate-vegetation system of FA deposits may help to slow down the accumulation of $CO_2$ in the atmosphere and avoid further pollution of the natural environment nearby coal-based thermal power stations. Usually, mature trees grow slowly and thus uptake less carbon dioxide later in the life cycle (*Science Daily, Nov.ember*

*15, 2007*, www.sciencedaily.com/releases/2007/11/071114163448.htm). Hence, perennial grasses may play a major role toward carbon sequestration in the substrate-vegetation system of FA deposits. Terrestrial ecosystems are considered the next highest global carbon pool after oceans and geological layers. Terrestrial ecosystem locks the atmospheric $CO_2$ as carbohydrates in plants and SOM along with other carbonate minerals in the soil until decomposition. This carbon pool contains both organic and inorganic carbon. The soil inorganic carbon (SIC) includes elemental carbon and carbonate salts of Ca and Mg, while SOC contains plant-derived carbon and humus. Recycling of this carbon in the ecosystem is accomplished mainly by photoautotrophs.

A study undertaken to measure the in-situ $CO_2$ flux via an automated soil $CO_2$ flux system on naturally colonized and barren FA deposits showed higher $CO_2$ flux in the vegetated site than the nonvegetated site due to root density and respiration of the plants in the former. This study helped in recognizing potential plant species for carbon sequestration. The presence of organic carbon, root biomass, and microbial activity helped in $CO_2$ sequestration in the naturally vegetated areas. The vegetated sites also showed that the $CO_2$ efflux rates were the lowest in the *S. spontaneum* (by 84%) and *Prosopis juliflora* (by 92%) association, compared with the *T. latifolia* association, which proved that *S. spontaneum* and *P. juliflora* can be used for sequestering atmospheric $CO_2$ on FA dumps (Pandey et al., 2016).

The distribution of various carbon fractions in rehabilitated FA deposits is presented in Fig. 8.1, which describes the biogenic carbon (originated from recent biological inputs, easily mineralized, and contributes maximum percentage of total soil carbon). The generation of $CO_2$ from coal-based thermal power stations is a critical concern in the context of climate change and well known as a potential source of atmospheric $CO_2$ emission, while the FA deposits have the capability to play a potential sink for carbon sequestration in ash-vegetation system through rehabilitation program (Fig. 8.2). Likewise, the development of park on FA deposits near coal-based thermal power stations is another excellent example of source—sink at one place (Fig. 8.3). The FA ecosystem carbon pool can be calculated by adding the carbon stocks from different ecosystem components:

Total FA ecosystem C pool (Mg C ha$^{-1}$) = Vegetation C $_{(AGB+BGB)}$ + SOC $_{(0-30\ cm)}$ + MBC $_{(0-30\ cm)}$ + C$_{(Litter)}$.

Abbreviations: *C*, carbon; *FA*, fly ash; *MBC*, microbial biomass carbon; *SIC*, soil inorganic carbon; *SOC*, soil organic carbon.

**Figure 8.2** Source of carbon dioxide emissions (Feroze Gandhi Unchahar Thermal Power Station, Unchahar, Raebareli, Uttar Pradesh, India) and sink for carbon sequestration (naturally vegetated FA deposits at the Arkha site). *Courtesy V. C. Pandey.*

**Figure 8.3** Public park development on the old FA deposits adjacent to the Renusagar Thermal Power Plant, Renukoot, Sonbhadra, Uttar Pradesh, is a brilliant instance of source−sink at one place. The power plant is operated by Hindalco Industries. *Courtesy V. C. Pandey.*

## 8.7 Carbon indices

Several carbon indices are helpful to evaluate the carbon accumulation in soil of different land uses (Rovira and Vallejo, 2002; Vieira et al., 2007; Silveira et al., 2010). Here, some carbon indices are given, such as recalcitrant index, carbon pool index (CPI), liability index (LI), and carbon management index, to assess carbon accumulation in rehabilitated FA deposits:

$$\text{Recalcitrant index} = \frac{Unhydrolyzable\ carbon}{Total\ organic\ carbon} \times 100 \qquad (8.1)$$

$$\text{Carbon pool index (CPI)} = \frac{TOC\ in\ rehabilitated\ FA\ deposit}{TOC\ in\ forest\ soil} \qquad (8.2)$$

$$\text{Lability of carbon} = \frac{KMnO_4\ oxidizable\ carbon}{Nonlabile\ carbon} \qquad (8.3)$$

$$\text{Liability index (LI)} = \frac{Lability\ of\ carbon\ in\ rehabilitated\ FA\ deposit}{Lability\ of\ carbon\ in\ forest\ soil} \qquad (8.4)$$

$$\text{Carbon management index} = \text{CPI} \times \text{LI} \times 100 \qquad (8.5)$$

## 8.8 Research recommendations

Newly rehabilitated area has not been considered generally in the carbon cycle, and the satellite images are not frequently updated in determining the global carbon budget. As such newly rehabilitated sites are often ignored, which, however, contributes insignificantly to total carbon budget, their little contribution is meaningful. In this chapter, my aim is to explore carbon sequestration potential under the two types of vegetation, natural succession and man-made plantations, on coal FA deposits. This study focused on the determination of above- and below-ground carbon sequestration on the coal FA deposits. The FA deposits are presumed to be free of organic carbon at first, and thus the organic carbon level can be initiated by plant growth. These deposit sites are usually unmanaged because of their inception, but these overlooked sites provide a tremendous opportunity for carbon sequestration. More widely, the management strategies implemented on such sites in regard to carbon sequestration can be effectively expected on any such site that is unmanaged like FA deposits. There are some important research gaps, which are given below:

- Assessment of the carbon sequestration potential of native vegetation and planted species at FA deposits globally.
- Assessment of carbon sequestration in rehabilitated FA deposits by measuring biogenic carbon such as vegetation carbon, SOC, labile carbon, microbial biomass carbon, and litter carbon.
- Investigating the role of soil enzymes in carbon sequestration of FA deposits.
- Carbon influx and efflux mechanisms in the substrate-vegetation system of FA deposits.
- Establishing an efficient rehabilitation approach for carbon sequestration in FA deposits.
- Role of organic amendments in carbon sequestration of ash–soil system.
- Role of nanotechnology in carbon sequestration of FA deposits.
- The project on carbon sequestration should be carried out at bench scale from the experiences gained earlier at lab scale. Furthermore, it may be tested at pilot scale in different biogeographical regions of India. Country-level index may be generated through the mathematical modeling and computer simulations.
- Plant–microbe interactions play an important role in carbon sequestration, yet present status of research activities is not well defined. Therefore the potential exists for studies to determine how plant–microbe interactions maximize the carbon stocks in FA deposits. Recently, plant–microbe interactions have suggested as a low-input biotechnology for ecosystem revitalization and a novel application for exploitation in multipurpose remediation technologies (Abhilash et al., 2012).
- Fungi are organisms that can induce long-term carbon sequestration in soil due to their priming effect (Fontaine et al., 2011). For example, Iqbal et al. (2012) reported increased carbon sequestration in tall fescue stands infected with fungal endophytes. Arbuscular mycorrhizal fungi inoculation can stimulate microbial activity and improve soil fertility and soil enzyme activity, and has an important role in enhancing soil carbon sequestration of reclaimed mine soil (Kuimei et al., 2012). Thus it is imperative to elucidate the role of fungi in this aspect with an emphasis on naturally colonized FA deposits.

## 8.9 Conclusions

Stabilization of heavy metals and carbon sequestration in FA disposal sites is a critical concern for the sustainability of ecosystems. Capping carbon dioxide emissions by the rehabilitation of FA disposal sites is a most effective and eco-friendly approach for absorbing carbon dioxide near coal-based thermal power stations and increasing potential carbon sinks in terrestrial ecosystems of FA deposits, mitigating global warming and the beginning of ecological restoration. Finally, it can be concluded that FA deposits have the potential as a sink for carbon sequestration in the substrate-vegetation system of ash dumping sites and may be a resource rather than a waste. In this regard, designer ecosystem should be implemented globally on FA deposits to enhance carbon sequestration in ash-vegetation system with multiple benefits to achieve sustainable livelihoods and ecosystem services.

Moreover, it is also reported that the FA application in the saline soil inhibits microbial respiration because of its high pH, EC, and B concentration. However, it was also noticed that $CO_2$ emissions reduce by FA application in normal soil mostly via the carbonation (physicochemical mechanisms via the formation of $CaCO_3$ and$MgCO_3$) of $CO_2$ with soluble Ca and Mg in the FA, and the increased pH will facilitate the carbonation process. Thus FA must be considered as a potential soil amendment that decreases $CO_2$ emission from the soil systems.

## References

Abhilash, P.C., Powell, J.R., Singh, H.B., Singh, B.K., 2012. Plant—microbe interactions: novel applications for exploitation in multipurpose remediation technologies. Trends Biotechnol. 30 (8), 416—420.

Amonette, J.E., Kim, J., Russell, C.K., Palumbo, A.V., Daniels, W.L., 2003a. Enhancement of soil carbon sequestration by amendment with fly ash. International ash utilization symposium. Centre for applied energy research, University of Kentucky, Paper#47 http://www.flyash.info.

Amonette, J.E., Kim, J., Russell, C.K., Palumbo, A.V., Daniels, W.L., 2003b. Fly ash catalyzes carbon sequestration. In Proceedings of the second annual conference on carbon sequestration, May 5-8, 2003. Alexandria VA.

Aswathanarayana, U., 2008. A low-carbon, technology-driven strategy for India's energy security. Current Science 94 (4), 440—441.

Babcock, M.F., Evans, D.W., Alberts, J.J., 1983. Comparative uptake and translocation of trace elements from coal ash by Typha latifolia. The Science of Total Env. 28, 203—214.

Bernoux, M., Volkoff, B., Carvalho, M.S., 2003. $CO_2$ emissions from liming of agricultural soils in Brazil. Global Biogeochem. Cycles 17 (2), 18—21.

Bobicki, E.R., Liu, Q., Xu, Z., Zeng, H., 2012. Carbon capture and storage using alkaline industrial wastes. Prog. Energ. Combust. Sci. 38, 302−320.

Collins, S.L., Sinsabaugh, R.L., Crenshaw, C., Green, L.E., Porras-Alfaro, A., Stursova, M., et al., 2008. Pulse dynamics and microbial processes in aridland ecosytems. J. Ecol 96, 413−420.

Farage, P.K., Ardo, J., Olsson, L., Rienzi, E.A., Ball, A.S., Pretty, J.N., 2007. The potential for soil carbon sequestration in three tropical dryland farming systems of Africa and Latin America: a modelling approach. Soil. Till. Res. 94, 457−472.

Ferreira, C., Ribeiro, A., Ottosen, L., 2003. Possible applications for municipal solid waste fly ash. J. Hazard. Mater. 96, 201−216.

Fontaine, S., Henault, C., Aamor, A., Bdioui, N., Bloor, J.M.G., Maire, V., et al., 2011. Fungi mediate long term sequestration of carbon and nitrogen in soil through their priming effect. Soil Biol. Biochem. 43, 86−96.

Freeman, C., Ostle, N., Kang, H., 2001. An enzymic 'latch' on a global carbon store. Nature 409, 149.

Gerdemann S.J., Dahlin D.C., O'Connor W.K., Penner L.R., Rush G.E. 2004. Ex-situ and in-situ mineral carbonation as a means to sequester carbon dioxide. In: Proceedings of Twenty-First Annual International Pittsburgh Coal Conference; Sep 13−17; Osaka, Japan. Pittsburgh, PA: Pittsburgh Coal Conference (PCC).

Goff, F., Lackner, K.S., 1998. Carbon dioxide sequestering using ultramafic rocks. Environ. Geosci. 5 (3), 89−101.

Grace, P.R., Antle, J., Aggarwal, P.K., Ogle, S., Paustian, K., Basso, B., 2012. Soil carbon sequestration and associated economic costs for farming systems of the Indo-Gangetic Plain: a meta-analysis. Agric, Ecosyst. Environ. 146, 137−146.

Grigore, R., Capat, C., Hazi, A., Hazi, G., 2016. Eco-efficiency indicators used for the environmental performance evaluation of a thermal power plant. Environ. Eng. Manag. J. 15 (1), 143−149.

Gupta, A.K., Sinha, S., 2008. Decontamination and/or revegetation of fly ash dykes through naturally growing plants. J. Hazard. Mater. 153, 1078−1087.

Haynes, R.J., 2009. Reclamation and revegetation of fly ash disposal sites: challenges and research needs. J. Environ. Manage. 90, 43−53.

Huijgen, W.J.J., Comans, R.N.J., 2003. Carbon Dioxide Sequestration by Mineralcarbonation. Energy Research Centre of The Netherlands, Petten, NL.

Iqbal, J., Siegrist, J.A., Nelson, J.A., McCulley, R.L., 2012. Fungal endophyte infection increases carbon sequestration potential of southeastern USA tall fescue stands. Soil Biol. Biochem. 44, 81−92.

Israelsson, P.H., Chow, A.C., Adams, E.E., 2009. An updated assessment of the acute impacts of ocean carbon sequestration by direct injection. Energy Procedia 1, 4929−4936.

Jala, S., Goyal, D., 2006. Fly ash as a soil ameliorant for improving crop production − a review. Bioresour. Technol 97, 1136−1147.

Kaur, B., Gupta, S.R., Singh, G., 2002. Carbon storage and nitrogen cycling in silvopastoral systems on a sodic in northwestern India. Agrofor. Syst. 54, 21−29.

Krishnamoorthy, R., 2000. Ash utilisation in India—prospect and problems. Barrier and utilisation option for large volume application of fly ash in India. In: Hajela, V. (Ed.), Proceedings of theWorkshop on USAID/India Greenhouse Gas Pollution Prevention Project. p. 63−7.

Kuimei, Q., Liping, W., Ningning, Y., 2012. Effects of AMF on soil enzyme activity and carbon sequestration capacity in reclaimed mine soil. Int. J. Min. Sci. Technol. In Press).

Kumari, A., Lal, B., Pakade, Y.B., Chand, P., 2011. Assessment of Bioaccumulation of Heavy Metal by Pteris Vittata L. Growing in the Vicinity of Fly Ash. Int. J. Phytoremediat. 13 (8), 779−787.

Lackner, K.S., 2002. Carbonate chemistry for sequestering fossil carbon. Annu. Rev. Energ. Environ. 27, 193−232.

Lackner, K.S., Wendt, C.H., Butt, D.P., Joyce, E.L., Sharp, D.H., 1995. Carbon dioxide disposal in carbonate minerals. Energy 20 (11), 1153−1170.

Lal, R., 2003. Global potential of soil carbon sequestration to mitigate the greenhouse effect. Crit. Rev. Plant. Sci. 22, 151−184.

Lim, S.S., Choi, W.J., 2014. Changes in microbial biomass $CH_4$ and $CO_2$ emissions, andsoil carbon content by fly ash co-applied with organic inputs with contrasting-substrate quality under changing water regimes. Soil. Biol. Biochem. 68, 494−502.

Lim, S.S., Choi, W.J., Lee, K.S., Ro, H.M., 2012. Reduction in $CO_2$ emission from nor-maland saline soils amended with coal fly ash. J. Soil. Sediment. 12, 1299−1308.

Lim, S.-S., Choia, W.-J., Chang, S.-X., Arshad, M.A., Yoon, K.-S., Kim, H.-Y., 2017. Soil carbon changes in paddy fields amended with fly ash. Agric. Ecosyst. Environ. 245, 11−21.

Love, A., Tandon, R., Banerjee, B.D., Babu, C.R., 2009. Comparative study on elemen-tal composition and DNA damage in leaves of a weedy plant species, *Cassia occidentalis*, growing wild on weathered fly ash and soil. Ecotoxicology 18, 91−801.

Maiti, S.K., Nandhini, S., 2006. Bioavailability of metals in fly ash and their bioaccumula-tion in naturally occurring vegetation: a pilot scale study. Environ. Monit. Assess. 116, 263−273.

Maiti, S.K., Jaiswal, S., 2008. Bioaccumulation and translocation of metals in the natural vegetation growing on fly ash lagoons: a field study from Santaldih thermal power plant, West Bengal, India. Env. Monit. Assess. 136, 355−370.

Maiti S.K., Singh G., Srivastava S.B., 2005. Study of the possibility of utilizing fly ash for back filling and reclamation of opencast mines: plot and pot scale experiments with Chandrapura FA. In: International Congress on FA, TIFAC, December 4−7, 2005, New Delhi.

Maroto-Valer, M.M., Kuchta, M.E., Zhang, Y., Andrésen, J.M., Fauth, D.J., 2005. Activation of magnesium rich minerals as carbonation feedstock materials for $CO_2$ sequestration. Fuel Process. Technol. 86, 1627−1645.

Matamala, R., et al., 2003. Impact of fine root turnover on forest NPP and soil C seques-tration potential. Nature 302, 1385−1387.

McBride, A.C., West, T.O., 2005. Estimating net $CO_2$ emissions from agricultural lime applied to soils in the US. In: Proceedings of the Fall meeting 2005. American Geophysical Union (Abstract #B41B-0191).

Montes-Hernandez, G., Pérez-López, R., Renard, F., Nieto, J.M., Charlet, L., 2009. Mineral sequestration of $CO_2$ by aqueous carbonation of coal combustion fly-ash. J. Hazard. Mater. 2009 (161), 1347−1354.

Naik, T.R., Tyson, S.S., 2000. Environmental benefits from the use of coal combustion products (CCP). In: Verma, C.V.J., Rao, S.V., Kumar, V., Krishnamoorthy, R. (Eds.), Proceedings of the Second International Conference on Fly Ash Disposal, Utilisation. pp. 40−43.

Palumbo, A.V., McCarthy, J.F., Amonettec, J.E., Fisher, L.S., Wullschleger, S.D., Daniels, W.L., 2004. Prospects for enhancing carbon sequestration and reclamation of degraded lands with fossil-fuel combustion by-products. Adv. Environ. Res. 8, 425−438.

Pan, S.-Y., Chang, E.E., Chiang, P.-C., 2012. $CO_2$ capture by accelerated carbonation of alkaline wastes: a review on its principles and applications. Aerosol Air Qual. Res. 2012 (12), 770−791.

Pandey, V.C., 2012a. Invasive species based efficient green technology for phytoremediation of fly ash deposits. J. Geochem. Explor. Available from: https://doi.org/10.1016/j.gexplo.2012.05.008.

Pandey, V.C., 2012b. Phytoremediation of heavy metals from fly ash pond by *Azolla caroliniana*. Ecotoxicol. Environ. Saf. 82, 8−12.

Pandey, V.C., Singh, N., 2010. Impact of fly ash incorporation in soil systems. Agric. Ecosyst. Environ. 136, 16−27.

Pandey, V.C., Singh, B., 2012. Rehabilitation of coal fly ash basins: current need to use ecological engineering. Ecol. Eng. Available from: https://doi.org/10.1016/j.ecoleng.2012.08.037.

Pandey, V.C., Mishra, T., 2018. Assessment of *Ziziphus mauritiana* grown on fly ash dumps: Prospects for phytoremediation but concerns with the use of edible fruit. Int. J. Phytoremediat. 20 (12), 1250−1256.

Pandey, V.C., Abhilash, P.C., Singh, N., 2009. The Indian perspective of utilizing fly ash in phytoremediation, phytomanagement and biomass production. J. Environ. Manag. 90, 2943−2958.

Pandey, V.C., Singh, K., Singh, R.P., Singh, B., 2012. Naturally growing *Saccharum munja* on the fly ash lagoons: a potential ecological engineer for the revegetation and stabilization. Ecol. Eng. 40, 95−99.

Pandey, V.C., Singh, N., Singh, R.P., Singh, D.P., 2014. Rhizoremediation potential of spontaneously grown Typha latifolia on fly ash basins: Study from the field. Ecol. Eng. 71, 722−727.

Pandey, V.C., Bajpai, O., Pandey, D.N., Singh, N., 2015. Saccharum spontaneum: an underutilized tall grass for revegetation and restoration programs. Genet. Resour. Crop. Ev. 62, 443−450.

Pandey, V.C., Sahu, N., Behera, S.K., Singh, N., 2016. Carbon sequestration in fly ash dumps: comparative assessment of three plant association. Ecol. Eng. 95, 198−205.

Parajuli R., Chang S.J., 2012. Carbon sequestration and uneven-aged management of loblolly pine stands in the Southern USA: a joint optimization approach. For. Policy Econ. 22, 65−71.

Parr, J.F., Sullivan, L.A., 2005. Soil carbon sequestration in phytoliths. Soil Biol. Biochem. 37, 117−124.

Pausstian, K., Cole, V., Sauerbeck, D., Sampson, N., 1998. $CO_2$ by Agriculture: an overview. Clim. Change 40, 135−162.

Ram, L.C., Jha, S.K., Tripathi, R.C., Masto, R.E., Selvi, V.A., 2008. Remediation of fly ash landfills through plantation. Remediation 18, 71−90.

Rau, N., Mishra, V., Sharma, M., Das, M.K., Ahaluwalia, K., Sharma, R.S., 2009. Evaluation of functional diversity in rhizobacterial taxa of a wild grass (*Saccharum ravennae*) colonizing abandoned fly ash dumps in Delhi urban ecosystem. Soil Biol. Biochem. 41, 813−821.

Reichle, D., Houghton, J., Kane, B., Ekmann, J., Benson, S., Clarke, J., et al., 1999. Carbon sequestration: State of the science. U.S. Department of Energy, Office of Science, and Office of Fossil Energy, Washington, D.C.

Renforth, P., Mayes, W.M., Jarvis, A.P., Burke, I.T., Manning, D.A.C., Gruiz, K., 2012. Contaminant mobility and carbon sequestration downstream of the Ajka (Hungary) red mud spill: the effects of gypsum dosing. Sci. Total. Environ. 421−422, 253−259.

Robertson, G.P., Paul, E.A., Harwood, R.R., 2000. Greenhouse gases in intensive agriculture: contributions of individual gases to the radiative forcing of the atmosphere. Science 289, 1922−1925. Available from: https://doi.org/10.1126/science.289.5486.1922.

Rovira, P., Vallejo, V.R., 2002. Labile and recalcitrant pools of carbon and nitrogen in organic matter decomposing at different depths in soil: an acid hydrolysis approach. Geoderma 107, 109−141.

Saleska, S.R., Miller, S.D., Matross, D.M., Goulden, M.L., Wofsy, S.C., et al., 2003. Carbon in Amazon forests: unexpected seasonal fluxes and disturbance-induced losses. Nature 302, 1554−1557.

Sanna, A., Dri, M., Hall, M.R., Maroto-Valer, M., 2012. Waste materials for carbon capture and storage by mineralisation (CCSM)-a UK perspective. Appl. Energ. 99, 545–554.

Selvam, A., Mahadevan, A., 2002. Distribution of mycorrhizas in an abandoned fly ash pond and mined sites of Neyveli Lignite Corporation, Tamil Nadu, India. Basic Appl. Ecol. 3, 277–284.

Shi, X.Z., Wang, H.J., Yu, D.S., Weindorf, D.C., Cheng, X.F., Pan, X.Z., et al., 2009. Potential for soil carbon sequestration of eroded areas in subtropical china. Soil. Til. Res. 105, 322–327.

Shrestha, R.K., Lal, R., 2006. Ecosystem carbon budgeting and soil carbon sequestration in reclaimed mine soil. Environ. Int. 32, 781–796.

Silveira, M.L., Comerford, N.B., Reddy, K.R., Prenger, J., DeBusk, W.F., 2010. Influence of military land uses on soil carbon dynamics in forest ecosystems of Georgia, USA. Ecol. Indic. 10, 905–909.

Singh, B.K., Milard, P., Whitely, A.S., Murrell, J.C., 2004. Unravelling rhizosphere-microbial interactions: opportunities and limitations. Trends Microbiol. 12, 386–393.

Sinsabaugh, R.L., Lauber, C.L., Weintraub, M.N., Ahmed, B., Allison, S.D., Crenshaw, C., et al., 2008. Stoichiometry of soil enzyme activity at global scale. Ecol. Lett. 11 (11), 1252–1264. Available from: https://doi.org/10.1111/j.1461-0248.2008.01245.x.

Soong, Y., Fauth, D.L., Howard, B.H., Jones, J.R., Harrison, D.K., Goodman, A.L., et al., 2006. $CO_2$ sequestration with brine and fly ashes. Energy Convers. Manag. 47, 1676–1685.

Sun, Y., Parikh, V., Zhang, L., 2012. Sequestration of carbon dioxide by indirect mineralization using Victorian brown coal fly ash. J. Hazard. Mater. 209–210, 458–466.

Thistle, D., Sedlacek, L., Carman, K.R., Fleeger, J.W., Brewer, P.G., Barry, J.P., 2006. Simulated sequestration of Industrial carbon dioxide at a deep-sea site: effects on species of harpacticoid copepods. J. Exp. Mar. Biol. Ecol. 330, 151–158.

Tietenberg, T., 2003. Environmental and Natural Resource Economics, sixth ed. Addison Wesley, USA, pp. 404–412.

Tiwari, S., Kumari, B., Singh, S.N., 2008. Evaluation of metal mobility/immobility in fly ash induced by bacterial strains isolated from the rhizospheric zone of *Typha latifolia* growing on fly ash dumps. Bioresour. Technol. 99, 1305–1310.

Uliasz-Boche_nczyk, A., Mokrzycki, E., Piotrowski, Z., Pomykala, R., 2009. Estimation of $CO_2$ sequestration potential via mineral carbonation in fly ash from lignite combustion in Poland. Energy Procedia 1, 4873–4879.

Vieira, F.C.B., Bayer, C., Zanatta, J.A., Dieckow, J., Mielniczuk, J., He, Z.L., 2007. Carbon management index based on physical fractionation of soil organic matter in an Acrisol under long-term no-till cropping systems. Soil Till. Res. 96, 195–204.

Wang W., Wei X., Liao W., Blanco J.A., Liu Y., Liu S., et al., 2012. Evaluation of the effects of forest management strategies on carbon sequestration in evergreen broad-leaved (*Phoebe bournei*) plantation forests using FORECAST ecosystem model. Forest Ecology and Management, Accepted.

West, T.O., McBride, A.C., 2005. The contribution of agricultural lime to carbon dioxide emissions in the United States: dissolution, transport, and net emissions. Agric. Ecosys. Environ. 108 (2), 145–154.

Zehetner, F., 2010. Does organic sequestration in volcanic soils offset volcanic $CO_2$ emissions? Quat. Sci. Rev. 29, 1313–1316.

Zevenhoven, R., Fagerlund, J., 2010. Fixation of carbon dioxide into inorganic carbonates: the natural and artificial "weathering of silicates". In: Aresta, M. (Ed.), Carbon dioxide as chemical feedstock. Wiley- VCH, Weinheim, Germany, pp. 353–379.

Zhu, Y.-G., Miller, R.C., 2003. Carbon cycling by arbuscularmycorhizal fungi in soil plant systems. Trends Plant. Sci. 8, 407–409.

# Further reading

Basile-Doelsch, I., Amundson, R., Stone, W.E.E., Borschneck, D., Bottero, J.Y., Moustier, S., et al., 2007. Mineral control of carbon pools in a volcanic soil horizon. Geoderma 137, 477−489.

Kalra Y.P., Maynard D.G., 1991. Methods manual for forest soil and plant analysis. Northern Forestry Centre, Northwest Region, Forestry Canada, Edmonton, AB. Information Report NOR-X-319.

Manrique, L.A., Jones, C.A., 1991. Bulk density of soil in relation to soil physical and chemical properties. Soil Sci. Soc. Am. J. 55, 476−481.

Ogawa, T., Nakanishi, S., Shidahara, T., Okumura, T., Hayashi, E., 2011. Saline-aquifer $CO_2$ sequestration in Japan-methodology of storage capacity assessment. Int. J. Greenh. Gas. Control. vol. 5, 318−326.

Pandey, V.C., Singh, K., 2011. Is *Vigna radiata* suitable for the revegetation of fly ash landfills? Ecol. Eng. 37, 2105−2106.

Pandey, V.C., Singh, J.S., Singh, R.P., Singh, N., Yunus, M., 2011. Arsenic hazards in coal fly ash and its fate in Indian scenario. Resour. Conserv. Recy. 55, 819−835.

Singh, K., Singh, B., Singh, R.R., 2012. Changes in physico-chemical, microbial and enzymatic activities during restoration of degraded sodic land: ecological suitability of mixed forest over monoculture plantation. Catena 96, 57−67.

Tabatabai, M.A., 1994. Soil enzymes. In: Weaver, R.W., Angle, S., Bottomley, P., Bezdicek, D., Smith, S., Tabatabai, A., et al.,Methods of Soil Analysis Part 2. Microbiological and Biochemical Properties. SSSA Book Series, pp. 755−833.

Vance, E.D., Brookes, P.C., Jenkinson, D.S., 1987. An extraction method for measuring soil microbial biomass. Soil Biol. Biochem. 19, 703−707.

Wang, Y., Hsieh, Y.P., 2002. Uncertainties and novel prospects in the study of the soil carbon dynamics. Chemosphere 49, 791−804.

World carbon dioxide emissions from the consumption of coal. Available from: Washington, DC: Energy Information Administration, International Energy Annual, Department of Energy <http://www.eia.doe.gov/iea/coal.html > ; 2006.

# CHAPTER 9

# Fly ash ecosystem services

## Contents

## 9.1 Introduction

The ever-increasing numbers and areas of fly ash (FA) dumpsites is a major concern all over the world in meeting the demand of coal-based energy. Therefore the continuous increase in the chain of FA dumping sites should be termed as FA catena. It has been recognized as environmental nuisance. There are many countries worldwide that have high rates of FA generation such as China, the United States, India, Europe, South Africa, Australia, Japan, Italy, and Greece. The worldwide FA generation is projected about 600 million tons annually, and its disposal would require approximately 3235 km$^2$ of land by 2015 (Pandey and Singh, 2012). Besides, FA also contains more than one dozen heavy metals that leach out in the environment after FA dumping and causes water, air, and soil pollution (Pandey et al., 2009, 2011). The primary method of preventing the pollution of FA catena is technical surface

*Phytomanagement of Fly Ash*
ISBN: 978-0-12-818544-5
DOI: https://doi.org/10.1016/B978-0-12-818544-5.00009-2
**257**

stabilization, but it is very expensive. In comparison with this, biological stabilization of FA catena mainly consists of planting trees after a previous application of an amendment in the form of fertile layer (Jusaitis and Pillman, 1997; Pandey et al., 2009; Pandey and Singh, 2010). The introduced flora on FA catena is also an important criterion, which initiates the pedogenesis and the process of ecological succession (Pandey and Singh, 2012). In many countries, such as North America, England, Poland, India, and Serbia, afforestation has been introduced to test the adaptability of trees and shrubs at pilot scale level (Scanlon and Duggan, 1979; Juwarkar and Jambhulkar, 2008; Ram et al., 2008; Pietrzykowski et al., 2010; Kostić et al., 2012), and is planned in the future in large areas. This is a very challenging task due to the harsh conditions of FA catena for plant establishment and growth, including mainly high pH and high variability of electrical conductivity, high susceptibility to compaction, poor air and water ratio, absence of nitrogen and available phosphorus, and sometimes high content of meta(loid)s (Pillman and Jusaitis, 1997; Haynes, 2009; Pandey et al., 2009).

In addition, coal-based thermal power stations should be committed to environmental management to conserve the heritage and resources of that nation. It is with this commitment in vision that all thermal power stations have to set up a rehabilitation division for the ash disposal sites. Rehabilitation involves the covering of the ash dumps with soil or organic wastes (press mud, sewage sludge, municipal waste, farmyard manure, etc.) whatever available near the ash dumps and the planting of grasses and trees. These abandoned areas are rehabilitated to the extent that it becomes a habitat for a diversity of flora, microbes, and animal and bird species. Sincere efforts are required to rehabilitate FA catena by developing a desired designed ecosystem into which target tree species could be planted to obtain maximum benefits of afforestation. Thus the rehabilitated FA ecosystem displays various goods and services, which can be classified on the basis of the Millennium Ecosystem Assessment (MEA) (2005) in the following four main categories: (1) provisioning services; (2) regulating services; (3) supporting services; and (4) cultural services. The concept of ecosystem services is increasingly gaining vast attraction and being used to evaluate combined social-ecological systems in order to enhance eco-sustainability and human well-being. The outcomes of such analyses can update land-use managers, planners, and decision and policy makers by providing ecological and socioeconomic knowledge (MEA, 2005; TEEB, 2010; Müller and Burkhard, 2012). The global significance of this concept is evidenced by the creation of the Intergovernmental Panel on Biodiversity and Ecosystem Services (Díaz et al., 2015). The FA ecosystem is increasing and is very similar to desert-like ecosystem, and our knowledge of FA ecosystem is very little and partial. There are still significant knowledge gaps concerning the occurrence and functioning of FA ecosystems and their important role in different biogeochemical cycles. This chapter aims to understand how floral diversity on FA catena contributes to the provision of ecosystem services that benefit the local livelihoods, sustainability, and productive and resilient ecosystems.

## 9.2 Fly ash ecosystem services

The FA ecosystem services are given in Fig. 9.1, which will be described in detail in the next subsections.

### 9.2.1 Provisioning services

Provisioning services of rehabilitated FA catena are the products used by human beings, which are obtained directly from FA ecosystems. In this context, the important goods provided by ecosystem services include in particular food, fodder, fiber, timber fuelwood, chemicals, and compounds (e.g., latex and gums), and genetic resources are likely to become important; of these, only food and fodder should be considered to use after toxicological risk assessment, and many "ecosystem service" assessments do include them in analysis. Afforestation on FA catena may potentially support livelihood improvements near the residents of coal-fired power plants

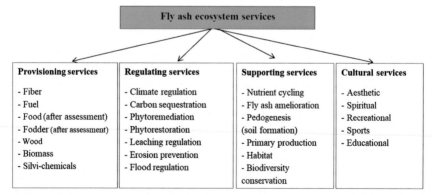

**Figure 9.1** FA ecosystem services on naturally or manually rehabilitated ash dumpsites.

through the exploitation of provisioning services of FA ecosystem. FA ecosystems sustain the livelihood of many local communities (Pandey et al., 2012). But, tribal dependence on FA-related jobs for their livelihood is also reported in Korba, India (Shrivastava, 2007). Their work is tree planting on FA disposal sites and their maintenance. Tribal groups are overrepresented in planting on FA deposits and their maintenance. Traditionally, it is well known that people of the scheduled tribes depend mostly on the forest products for their livelihood. The examples of provisioning services of FA ecosystems are given in Table 9.1. In addition, Fig. 9.2 also shows the provisioning services of FA ecosystems—naturally colonized plants and/or man-made plantations on FA dumpsites offer a wide range of provisioning services.

## 9.2.2 Regulating services

Regulating services are the main goal of the rehabilitated FA catena through afforestation to mitigate pollutions such as leaching of heavy metals, soil erosion, suspended particulate matter, and excess atmospheric carbon dioxide via carbon sequestration and storage by the biomass of herbs, shrubs, and trees. The examples of regulating services of FA ecosystems are phytoremediation, phytostabilization, rhizoremediation, carbon sequestration, and ecological restoration (Pandey, 2012a,b; Pandey, 2013; Pandey et al., 2012, 2014, 2015a,b, 2016a,b; Pietrzykowski et al., 2015; Gajić et al., 2016, 2019), as provided in Table 9.1. Capture of carbon dioxide emitted from coal-fired power plants and long-term storage of carbon dioxide in the substrate-vegetation system of FA catena will play a vital role for developing sustainable environment. Singh and Singh (2012) proposed that the reforestation or restoration of abandoned degraded land may significantly contribute to the recovery of methanotrophic activity in the soil and thereby the soil—methane sink potential. Thus afforestation on FA catena will serve as a potential sink for the buildup of both atmospheric pollutants (carbon dioxide and methane) due to coal-fired power plants. Besides the above regulating services of rehabilitated FA catena, it will be supportive in decreasing the atmospheric temperature nearby thermal power stations. The rehabilitated FA catena will also reduce noise pollution by the absorption of sound waves by vegetation. Fig. 9.3 shows the regulating services of FA ecosystems—dense green cover development on FA dumps through spontaneous colonized plants or planted species provides regulatory services.

**Table 9.1** Ecosystem services of fly ash ecosystems: the classification of services follows the MEA (2005).

| Ecosystem service categories | Products and services | Flora/fauna | References |
|---|---|---|---|
| *Provisioning services* | | | |
| Fiber | Roof thatch, fencing, paper, and craft (i.e., ropes, hand fans, baskets, brooms, and mats) | *S. munja* and *S. spontaneum* | Pandey et al. (2012, 2015a) |
| Energy | Biomass, biodiesel, and biofuel | *Miscanthus giganteus, Ricinus communis, Panicum virgatum,* and *Jatropha curcas* | Lixandru et al. (2013), Pandey (2013), Dzantor et al. (2015), Jamil et al. (2009), Pandey (2017), and Roy et al. (2018) |
| Food | Pollen and fruit | *Aster novae-angliae, Z. mauritiana,* and *Py. communis* | De Joung and Morse (1977), Pandey and Mishra (2016), and Gajić et al. (2019) |
| Fodder | Whole parts except roots and leaves | *C. dactylon* and *D. glomerata* | Maiti et al. (2016) and Gajić (2014) |
| Ceremonial objects | Hindu marriage | *S. munja* and *C. dactylon* | Pandey et al. (2012) |
| *Regulating services* | | | |
| Check air pollution | Dust pollution remediation | *C. martinii, V. zizanioides, C. flexuosus,* and *C. winterianus* | Verma et al. (2014) |
| Carbon sequestration | Carbon in the soil−plant system of fly ash deposits | *S. spontaneum, P. juliflora,* and *T. latifolia* | Pandey et al. (2016a,b) |
| Rhizoremediation | Fly ash pond remediation | *T. latifolia* | Pandey et al. (2014) |
| Phytostabilization | Stabilization of heavy metals through roots | *S. munja, Alnus viridis, F. rubra, C. epigejos, D. glomerata,* and *O. biennis* | Pandey et al. (2012), Pietrzykowski et al. (2015), Gajić (2014), and Gajić et al. (2016) |

(*Continued*)

**Table 9.1** (Continued)

| Ecosystem service categories | Products and services | Flora/fauna | References |
|---|---|---|---|
| Phytoremediation | Heavy metal remediation | *A. caroliniana, I. carnea,* and *R. communis* | Pandey (2012a,b) and Pandey (2013) |
| Ecological restoration | Improvement of organic carbon, nitrogen, phosphorus, and microbial biomass carbon in naturally vegetated fly ash substrate | *S. spontaneum, T. latifolia, C. dactylon, S. munja, P. juliflora, I. carnea, A. nelotica, C. epigejos, Medicago sativa, D. glomerata, L. corniculatus, O. biennis, P. alba, Tamarix tetandra,* and *F. rubra* | Pandey et al. (2015a,b) and Gajić et al. (2016, 2019) |

**Supporting services**

| Ecosystem service categories | Products and services | Flora/fauna | References |
|---|---|---|---|
| Pedogenesis | Fly ash amelioration (substrate quality enhancement—increase in organic carbon, nitrogen, and phosphorus in fly ash substrate) | *R. pseudoacacia, Fraxinus* spp., *Erigeron canadensis, D. glomerata, M. sativa, F. rubra, C. epigejos, O. biennis, T. tetandra, P. alba, Amorpha fruticosa, Pinus sylvestris,* and *Alnus* spp. | Uzarowicz et al. (2017, 2018), Kostić et al. (2018), Woch et al. (2018), and Pietrzykowski et al. (2018a,b) |
| Nutrient cycling | In fly ash substrate | *Alnus* spp. | Świątek et al. (2019) |
| Photosynthesis | Chlorophyll fluorescence and photosynthetic efficiency | *Zizyphus mauritiana, R. pseudoacacia, T. tetandra, P. alba, A. fruticosa, F. rubra, C. epigejos, D. glomerata,* and *O. biennis* | Prajapati and Meravi (2015), Mitrović et al. (2012), Mitrović et al. (2008), Gajić et al. (2013, 2016), and Gajić (2014) |
| Primary production | Tree growth | *P. sylvestris* and *Alnus* spp. | Pietrzykowski et al. (2010, 2018a,b) |
| Habitat | Nestling | Collembola, *Quiscalus quiscala, S. hispidus,* and *O. palustris* | Shaw (2003), Bryan et al. (2012), and Peles and Barrett (1997) |
| Biodiversity conservation | Conservation | Vanishing insects (bees and wasps) | Tropek et al. (2013, 2014) |

***Cultural services***

| | | | |
|---|---|---|---|
| Aesthetic | Aesthetically pleasant landscape | *R. communis* | Pandey (2013) and Pandey et al. (2012) |
| Recreational | Public park | | See Fig. 9.4 |
| Educational | Scientific knowledge—effect of revegetation on fly ash properties and fast green capping on fly ash basins | *D. glomerata, M. sativa, F. rubra, E. canadensis, C. epigejos, O. biennis, T. tetrandra, P. alba, A. fruticosa,* and *S. spontaneum* | Kostić et al. (2018), Pandey and Singh (2014), and Gajić et al. (2019) |

**Figure 9.2** Provisioning services of FA ecosystems—naturally colonized plants and/or man-made plantations on FA dumpsites offer a wide range of provisioning services (Pandey et al., 2012, 2015a; Pandey and Mishra, 2016). (A) Fiber, roof thatching, fencing, crafts (viz., rope, mats, and hats), and energy—*S. spontaneum*. (B) Local women are harvesting the leaves of *S. munja*. (C) The leaf of *S. munja* offers numerous provisioning services. (D) *T. latifolia*—naturally colonized on FA dumpsites, Unchahar Thermal Power Station, Raebareli, India. (E) Energy via firewood—*P. juliflora* plantation on FA dumps, Panki Thermal Power Station (PTPS), Kanpur, India. (F) Local women are

## 9.2.3 Supporting services

Supporting services are those functions that are necessary for the production of all other FA ecosystem services. The changes in supporting services have relatively direct impacts on people. The examples of supporting services of FA ecosystems are pedogenesis, photosynthesis, primary production, nutrient cycling, water cycling, resilience, and habitat (Table 9.1). Parks can be developed at abandoned FA disposal sites for local people. Many countries have started to transform abandoned FA dumpsites into productive FA ecosystems that provide habitats for biotic communities to establish the biogeochemical cycle. Costanza et al. (1997) defined the ecosystem service "nutrient cycling" as the storage and recycling of nutrients by living organisms within ecosystems. These cycles play an important role in the development of FA ecosystems as biogeochemical cycles. In this context, flora development on FA catena contributes to the increase of the organic matter content in the proving substrate through the inputs of litter and fine roots (Pandey and Singh, 2012), which enhances the activities of relevant microbes to biogeochemical cycles. Afforestation with any species on degraded lands favors soil reclamation, if it could be used rationally (Shukla and Misra, 1993). The gradual increase in the detritus mass through litterfall, root decay, and grass turnover changes the soil properties by a number of ways: lowering the C/N ratio (N is generally released below 20; making the soil lighter for better root growth and anchorage; increasing water permeability, moisture retention, soil aeration, and nutrient availability; providing a favorable soil temperature for greater microbial population and diversity; acting as a buffer medium to release the nutrients slowly). A study was done on vegetation (birch and willow) and soil development in FA deposits of southern England (Shaw, 1992). Recently, pedogenesis on FA deposits through vegetation development was studied in detail by several researchers (Uzarowicz et al., 2018; Kostić et al., 2018; Woch et al., 2018; Pietrzykowski et al., 2018a,b). This new habitat (FA ecosystem) gives rise to potential plant species and microbes with amazing adaptations to harsh environmental conditions of FA (Rau et al., 2009; Pandey et al., 2012). Several naturally growing plant species,

cutting dry stems of *P. juliflora* for their firewood needs. Food—*Z. mauritiana* (G) and *P. dulce* (H) fruit tree plantations on FA dumpsites, PTPS (Caution: the fruit tree plantation on FA dumps needs to check toxicological risk assessment for the edibility of fruits to avoid any health hazard to the local people). *Courtesy V.C. Pandey.*

**Figure 9.3** Regulating services of FA ecosystems—dense green cover development on FA dumps through naturally colonized plants (A–G) or man-made plantations (H) provide regulatory services. Phytostabilization—*C. dactylon* (A) and *S. munja* (C) have been noted in the stabilization of heavy metals and dust particles of FA landfills (Maiti et al., 2016, Pandey et al., 2012). Carbon sequestration—*S. spontaneum* (B) and

bacteria, and fungi have been reported from the FA ecosystem worldwide. Rehabilitated FA catena or naturally restored ash dumpsites as FA ecosystems will also offer wildlife shelter. A number of birds made their nest on the grown or planted trees at the habitat of FA ecosystem (Bryan et al., 2012). In the United States, it is reported that successional vegetation dominated with panic grass *(Panicum hemitomon)* and broomsedge *(Andropogon virginicus)* on FA basins that offer habitats for small mammals such as *Sigmodon hispidus* (cotton rat) and *Oryzomys palustris* (rice rat) (Peles and Barrett, 1997). In the United Kingdom, few abandoned FA disposal areas are considered as locally valuable conservation areas for the dense birch/willow woodland with glades of orchids (Greenwood and Gemmill, 1978; Shaw, 1994). Fig. 9.4 shows the supporting services of FA ecosystems—spontaneous colonized plant species and/or planted tree species on FA dumpsites offer a wide range of supporting services.

## 9.2.4 Cultural services

Cultural services of the rehabilitated FA catena are the often nonmaterial (intangible) benefits that people obtain from FA ecosystems through aesthetic landscape, nature, cultural heritage, spiritual and religious enrichment, cognitive development, scientific knowledge, sense of place, reflection experiences, recreation, and ecotourism. FA dumpsites in urban or peri-urban areas are a unique chance to develop spaces for recreational, sport, and educational uses. If FA dumpsites are dry, then there are many possibilities to transform them for cultural services, for example, bike racing, mountain biking, car racing, horse or foot trails, soccer fields, golf courses, tennis courts, climbing school, botanical gardens, car parking place, public park, political meeting place, auditoriums for outdoor activities and wildlife habitats, and leisure (Havrlant and Krtička, 2014), whereas in rural areas, abandoned FA dumpsites also have a substantial recreational potential. In this order, reconversion operations may involve

*P. juliflora* (D) are potentially suitable for the sequestering atmospheric $CO_2$ in the fresh FA deposited sites (Pandey et al., 2016a). Phytoremediation—*A. caroliniana* (E), *T. latifolia* (F), and *Eichhornia crassipes* (G) are grown naturally on FA pond, Unchahar Thermal Power Station, and have been reported for the remediation of heavy metals (Pandey, 2012a; Pandey et al., 2014; Pandey, 2016). Ecological restoration—*S. spontaneum* (B) is a promising perennial grass and valuable genetic resource for ecological restoration of FA dumps (Pandey et al., 2015a), and the mixed plantation (H) on FA dumpsites of PTPS offers a wide range of regulatory services. *Courtesy V.C. Pandey.*

**Figure 9.4** Supporting services of FA ecosystems—naturally colonized plants and/or man-made plantations on FA dumpsites offer a wide range of supporting services (Shaw, 2003; Bryan et al., 2012; Tropek et al., 2013, 2014; Kostić et al., 2018). (A) Pedogenesis—the decomposition of leaf litter at the rehabilitated FA ecosystem improves substrate quality (organic carbon, nitrogen, and phosphorus) of FA dumps, PTPS. (B) The decay of fine roots of *S. spontaneum* also helps in the development of the carbon and nutrient pools in FA dumps (Pandey et al., 2015a), Unchahar Thermal

the development of picnic park, biodiesel park, and aromatic grass garden, for example. Adaptations to recreational forests, horse riding tracks, and scenic overlooks are other feasible possibilities of use (Favas et al., 2018). Some cultural services of FA ecosystems are presented in Table 9.1. In addition, Fig. 9.5 depicts the cultural services of FA ecosystems—the rehabilitated FA dumpsites either naturally or man-made offer a number of cultural services such as aesthetically pleasant landscape, recreational, educational, etc.

## 9.3 Strategies for enhancing fly ash ecosystem services

*Significance of plant—microbe interactions*—The importance of plant—microbe interactions in the revitalization of FA ecosystems is increasingly gaining vast attraction. It is an efficient and low biotechnology for FA ecosystem revitalization and has been suggested for a possible implementation to restore abandoned FA dumpsites into productive landscapes in order to provide ecosystem goods and services. As FA dumpsites have been well documented in the form of man-made stressed and hazardous sites on the earth, rehabilitation of FA dumps and their transformation into biologically productive sites are matters of prime concern over the world. This could be facilitated by using ecologically important functional groups of rhizomicrobes that have the following positive traits such as heavy metal tolerance, stimulating the plant growth, developing tolerance in plants against biotic and abiotic stress, and assisting the plants for the revitalization of FA dumpsites (Selvam and Mahadevan, 2002; Juwarkar and Jambhulkar, 2008; Kumar et al., 2008, 2009; Pandey et al., 2009; Rau et al., 2009; Babu and Reddy, 2011). In addition, such plant—microbe interactions can modify the physicochemical and biochemical properties of the substrate of FA ecosystems. Plant secretes phytosiderophores that help in the sequestration of metallic micronutrients from the soil. Root exudates also have secondary metabolites that help in the plant—microbe communication. However, these interactions are interestingly complex

Power Station. (C) Habitat—the hanging retort-shaped elaborately woven nest of the baya weaver at the FA ecosystem. (D) Baya weaver (*Ploceus philippinus*) sitting on twig of *P. juliflora* grown in the FA ecosystem. (E) Koel (*Eudynamys scolopaceus*) at the FA ecosystem, PTPS. (F) Peahen (*Pavo cristatus*) at the FA ecosystem, PTPS. (G) Cows (*Bos taurus*) resting on the rehabilitated FA dumps, PTPS. (H) Naturally vegetated FA dumps act as habitats for the buffalo (*Bos bubalis*). *Courtesy V.C. Pandey.*

**Figure 9.5** Cultural services of FA ecosystems—the rehabilitated FA dumpsites either naturally or man-made offer a number of cultural services such as aesthetically pleasant landscape, recreational, educational, etc. Public park—the development of a public park on the passive FA deposited site of the Renusagar Thermal Power Station, Renukoot, Sonbhadra, Uttar Pradesh, India. The main road to reach the park (A), grass lawns (B), aromatic garden (C), track for walkers (D), children play ground (E), washroom and drinking water for public (F), sitting bench for public (G), and park located at Hindalco Industries' Renusagar Power Plant (H). *Courtesy V.C. Pandey.*

and dynamic and quite difficult to decode, as it takes places at various interfaces (i.e., rhizosphere, phyllosphere, and endosphere). Hence, a deep understanding of the intertwined processes taking place at various interfaces is vital for unraveling the role of the entire player for the FA ecosystem well-being. Thus it is utmost important to understand the key processes of the plant—microbe interactions in order to assess the contribution of the plant-associated microbes to enhance FA ecosystem services. The ecologically important functional groups of rhizomicrobes may be used for enhancing the potential of FA ecosystems, but they have to be the following traits: (1) improving the nutrient supply (releasing phytohormones for better growth of root system; producing siderophores for acquisition of iron and other trace metals); (2) augmenting the nutrient pool (nitrogen fixation; solubilization of phosphate and other mineral complexes); (3) disrupting the transmission of stress signal to the plants (decrease in stress ethylene due to the activity of ACC-deaminase enzyme); and (4) controlling the deleterious organisms (producing hydrogen cyanide, siderophores, and antibiotics).

In this direction, a very limited research work is focused. The restoration practices on FA deposits have been unsuccessful in declining high mortality of plantlets due to the absence of nitrogen and available phosphorus, high soluble salt concentrations of trace elements, low microbial activity, and the presence of compacted and cement layers (Pandey et al., 2009; Pandey and Singh, 2012). The lack of mycorrhizal fungi may be partially answerable for the revegetation problems (Sylvia and Williams, 1992). Mycorrhizal fungi, a ubiquitous group of rhizomicrobes, form a symbiotic association with the roots of most terrestrial plants (Smith and Read, 2008), which enhance the nutrient supply to the host plants and alleviate biotic and abiotic stresses. Arbuscular mycorrhizal fungi (AMF) are ecologically significant, as they help in restoration of polluted lands by enhancing plant nutrition and soil fertility (Chen et al., 2001). Its application may increase the restoration rate and phytoremediation efficiency, and speeds up these processes (Vosatka, 2001). It is well proved that the application of adapted AMF can be more effective in bioremediation and restoration programs than the nonadapted AMF (Vivas et al., 2005). Selvam and Mahadevan (2002) reported the distribution of AMF naturally associated with the roots of plants in a passive FA dumpsite. The incorporation of farmyard manure and AMF to the FA dump improved the growth of plant, the physical properties of FA, and the decrease of toxic metals (Juwarkar and Jambhulkar, 2008). Many investigators have stated the impacts of certain strains of AMF on the plant growth, nutrient uptake, and aggregation of FA (Enkhtuya et al., 2005; Wu et al., 2009). Mycorrhiza potentially influences the growth of plants on two anthropogenic substrates, namely FA dumps and coal mine spoils. Native grasses that naturally colonize FA ecosystems seem to represent the significant agents of mycorrhizal distribution, which can facilitate the establishment of mycorrhizal association of transplanted tree seedlings during the revegetation programs (Enkhtuya et al., 2005). Babu and Reddy (2011) studied the AMF diversity in the rhizosphere of plants grown on FA dumps. They noticed that its diversity was variable among the different plant species. *Glomus rosea* was the most dominant fungus from the studied rhizosphere. *Lantana camara* was only one plant species that is associated to all recognized AMF species excluding *Glomus magnicaule* compared with other plants. The FA-adapted AMF inoculation increased the growth (84.9%), chlorophyll (54%), and total P content (44.3%) of *Eucalyptus tereticornis* seedlings grown on FA dumps compared with noninoculated seedlings and reduced Al, Fe, Zn, and Cu accumulations. Therefore these results encouraged the rehabilitation programs on the application of FA-adapted AMF strains to aid the establishment of plants at the passive FA dumpsites through using inoculated tree seedlings.

Kumar et al. (2008) isolated the metal-tolerant plant growth promoting bacteria (PGPR) NBRI K28 *Enterobacter* sp. from FA-contaminated soils. The strain NBRI K28 and its siderophore overproducing mutant NBRI K28 SD1 were found capable of stimulating plant biomass and enhance the phytoextraction of heavy metals from FA by the metal accumulating plant *Brassica juncea*. Simultaneous production of siderophores, indole acetic acid, and phosphate solubilization exposed it as a potential plant growth promoter. The strain also showed 1-aminocyclopropane-1-carboxylic acid (ACC)-deaminase activity. The *Enterobacter aerogenes* strain NBRI K24 and *Rahnella aquatilis* strain NBRI K3 were also identified as potential plant growth promoters as well as protect the plant against the inhibitory effect of $Cr^{2+}$ and $Ni^{2+}$ (Kumar et al., 2009). Therefore these above strains and their mutants can be exploited as effective plant growth promoters in the development of a sustainable FA ecosystem that will be able to provide multiple ecosystem services. Rau et al. (2009) reported a taxonomically and functionally diverse group of rhizobacteria from the rhizosphere of *Saccharum ravennae* grown on FA dumpsites. Sixty five morphologically distinct rhizobacteria having dominant nature were purified, identified, and functionally characterized, which belonged to 18 genera and 38 species. *Bacillus* spp. and *Paenibacillus* spp. were dominant. Gram-positive bacteria were dominating at dumpsites. Some isolated rhizobacteria were found to have multiple-metal tolerance and the ability to produce siderophore. These are: *Bacillus* sp. IPSr80, *Brochothrix campestris* BPSr3, *Enterococcus casseliflavus* BPSr32, *Microbacterium barkeri* IPSr74, *Pseudomonas aeruginosa* BPSr43, and *Serratia marcescens* IPSr90 and IPSr82. Maximum rhizobacteria were able to grow on a nitrogen-deficient medium. *S. marcescens* IPSr90 was reported as the only rhizobacterium that exhibited ACC-deaminase activity. Proportion of phosphate-solubilizing bacteria was high. Some observations (i.e., a significant improvement in the seedling establishment, plant weight, and shoot length in the *S. ravennae* grass) revealed the importance of inoculated rhizobacteria in colonization and the spread of *S. ravennae* at FA conditions. This FA-tolerant ecologically diverse functional group of rhizobacteria (especially good plant growth promoting traits) may be potentially and commercially useful for the development of inoculation technologies to transform barren FA dumps into ecologically and socioeconomically productive habitats. These are: *Arthrobacter ureafaciens* BPSr55, *E. casseliflavus* BPSr32, *Paenibacillus larvae* BPSr106, *Paenibacillus azotofixans* BPSr107, *P. aeruginosa* BPSr43, and *S. marcescens* IPSr82 and IPSr90.

*Importance of plant selection criteria*—Plant species selection for the revegetation/rehabilitation of FA dumpsites is another important criterion and should be considered with the following traits such as perennial nature, fast growth with high biomass production, unpalatable, FA tolerant, regeneration potential, phytoremediation potential, ecological and socioeconomic values (Pandey and Singh, 2011, Pandey et al., 2012, 2016a,b), as well as fast decomposition rate of leaf litter and fine root. The physicochemical characteristics of FA substrate will mostly determine which plant species can be used to create a sustainable FA ecosystem. Therefore the selection of plant species for FA dumps' rehabilitation is site specific. Because of large differences in species response to the same substrate, a variety of plant species should be sampled when assessing possible trace element deficiencies or toxicities in plants growing on abandoned ash basins. Several plant species have been suggested for effective revegetation of abandoned FA catena worldwide. The keystone species for fast revegetation of FA catena may be divided in two categories. The first category is naturally colonized plant species, which includes especially grasses but sometimes trees also, and the second one is introduced ecologically and socioeconomically tree species on FA catena.

In the first category, a number of naturally colonized plant species on FA deposits have been noted and suggested for potential revegetation and phytoremediation of new FA catena (Pavlović et al., 2004; Mitrović et al., 2008; Pandey et al., 2012, 2015a, 2016a; Pandey, 2015; Kumari et al., 2016; Gajić et al., 2019). In this context, grasses that are noted as promising species due to being the "nurse species," having economic importance and perennial nature, would be a reasonable strategy to convert the FA dumpsites into biologically and sociologically significant habitats. In India, some highly potential plant species that naturally colonized on FA deposits are *Saccharum spontaneum*, *Saccharum bengalense* (syn. *Saccharum munja*), *Cynodon dactylon*, *S. ravennae*, *Typha latifolia*, *Azolla caroliniana*, *Ipomoea carnea*, *Ipomoea aquatica*, *Sonchus asper*, *Eragrostis pilosa*, *Tridex procumbens*, *Dactyloctenium aegyptium*, *Desmodium triflorum*, *Prosopis juliflora*, *Acacia nelotica*, etc. (Pandey et al., 2012, 2015b; Pandey, 2012a,b; Maiti and Jaiswal, 2008; Rau et al., 2009). In England, Shaw (1996) recorded several herbaceous plant species on the FA dumpsite: *Achillea millefolium*, *Agrostis stolonifera*, *Arrenatherrum elatius*, *Atriplex prostrate*, *Chenopodium rubrum*, *Cirsium arvense*, *Dactylis glomerata*, *Epilobium hirsitum*, *Festuca rubra*, *Lotus corniculatus*, *Lycopus europaeus*, *Melilotus officinalis*, *Oenothera biennis*, *Plantago lanceolata*, *Reseda lutea*, *Rorippa sylvestris*, *Rumex obtusifolia*, *Senecio vulgaris*, *Sonchus oleraceus*, *Suaeda maritime*, *Dactylorhiza* sp., *Trifolium repens*, *Tussilago farfara*, *Vicia sativa*, etc. In Serbia, some perennial plant species having the capacity to grow on the FA deposits are: *Calamagrostis epigejos*, *Chenopodium album*, *Crepis setosa*, *Cichorium intybus*, *C. arvense*, *Carduus accanthoides*, *Descurainia sophia*, *D. glomerata*, *Daucus carota*, *Epilobium hirsutum*, *Euphorbia cyparissias*, *Eupathorium cannabinum*, *Echium vulgare*, *Hypericum perforatum*, *Linaria vulgaris*, *Medicago lupulina*, *Melissa officinalis*, *Papaver rhoes*, *P. lanceolata*, *Pyrus communis*, *Populus alba*, *Rosa canina*, *Silene vulgaris*, *Sinapis arvense*, *T. farfara*, *Verbascum phlomoides*, etc. (Gajić et al., 2019). These naturally colonized plant species act as pioneers by beginning the revegetation process on FA dumps due to having a high ability to tolerate the toxicity and/or deficiency of some important nutrients in FA as well as drought, high temperatures, and radiation (Kostić et al., 2012; Pandey, 2012a,b; Gajić et al., 2013, 2016). These naturally colonized plant species can be propagated easily on new FA dumpsites for rehabilitation programs by their seeds or vegetative organs (rhizomes, stolons, bulbs, tubers, etc.) with low-cost inputs and limited maintenance. The rhizomatous plants can tolerate seasonally adverse conditions, may recover from disturbance because of the accumulation of reserve substances (carbohydrate), and serve as a link between the above-ground shoots (*S. spontaneum, S. bengalense* (syn. *S. munja*), *S. ravennae, C. epigejos, Phragmites communis, F. rubra, T. farfara, C. arvense, Heliatnhus tuberosus,* and *Epilobium* sp.). Over time, sown, seeded, and naturally colonized plant species on FA landfills bind FA particles with a dense root system, form a dense green cover that provides erosion control, improve the physicochemical properties of FA substrate, and retain moisture and nutritional substances that will be later used by introduced ecologically and socioeconomically tree species for multiple ecosystem services (Pandey and Singh, 2012). In addition, some aromatic grasses such as *Cymbopogon martini* (palmarosa grass), *Vetiveria zizanioides* (vetiver grass), *Cymbopogon flexuosus* (lemon grass), and *Cymbopogon winterianus* (Java citronella grass) are stress tolerant, perennial nature, and unpalatable by livestock. The main product of aromatic grasses is essential oil that is free from heavy metal risk (Lal et al., 2013). These promising aromatic grasses have the ability for the revegetation/phytoremediation of FA dumps along with a wide range of ecosystem services (Verma et al., 2014) and are recommended for in-situ remediation of FA catena.

In the second category, several ecologically and socioeconomically tree species have been introduced successfully for the plantation on FA catena and recommended for rehabilitation programs of new FA catena (Carlson and Adriano,1991; Juwarkar and Jambhulkar, 2008; Ram et al., 2008). Struggles are ongoing in this direction, and numerous tree species are identified to grow a vital forest cover on FA basins all over the world, though the choice of tree species should be based on the basis of economic significance and the wood demand of forest-based industries. However, it may aid to some amount for catering of various industrial purposes like biodiesel, fuelwood, pulp paper, timber, and plywood. Some FA ecosystems have been created on FA basins worldwide by potential tree species having economic values. In India, pulp-paper tree species (*Dendrocalamus strictus, Eucalyptus* sp., and *Leucaena leucocephala*), firewood tree species with nitrogen-fixing capacity (*Acacia auriculiformis, Acacia nilotica, Albizia lebbek, Cassia siamea, Casuarina equisetifolia, Dalbergia sissoo,* and *P. juliflora*), plywood/firewood tree species (*Eucalyptus hybrid, Eucalyptus globulus, Melia azedarach, Populus deltoids, Syzigium cumini,* and *Tamarindus indica*), timber tree species (*D. sissoo, Shorea robusta,* and *Tectona grandis*), and plywood species (*Bombex ceiba, E. tereticornis,* and *Populus euphratica*) have been well planted on the barren substrate of FA catena (Juwarkar and Jambhulkar, 2008; Ram et al., 2008; Pandey et al., 2009). In the United States, the FA ecosystem has been created fruitfully on acidic and alkaline FA basins with *Liquidambar styraciflua* (Sweetgum) and *Platanus occidentalis* (Sycamore), respectively, by Carlson and Adriano (1991). *L. styraciflua* and *P. occidentalis* are the important timber trees recognized as leading furniture woods for making of barrels, boxes, cabinetwork, and veneer. In Serbia, a number of shrubs and trees that have been planted on FA deposits are: *Ailanthus glandulosa, Elaegnus angustifolia, Gleditchia triacanthos, P. alba, Robinia pseudoacacia,* and *Tamarix* sp. (Gajić et al., 2019). The regeneration potential of plant species is also an utmost important feature during the selection of plants on which rehabilitation potential, ecosystem services, and ecosystem sustainability depend. In this regard, Pandey and Mishra (2016) assessed the status of regeneration potential of planted tree species, such as *Ailanthus excelsa, Albizia lebbeck, Albizia procera, D. sissoo, E. tereticornis, Holoptelea integrifolia, L. leucocephala, Pithecellobium dulce, Pongamia pinnata, P. juliflora,* and *Ziziphus mauritiana,* at the FA ecosystem. Three plant species namely *P. juliflora, P. dulce,* and *P. pinnata* are noted as dominant species based on an importance value index in the FA ecosystem. The regeneration study concludes that the above three tree species (68.91%) have a good regeneration capacity, which can be used for the revegetation of new FA basins as well as enhancement of FA ecosystem services.

## 9.4 Biogeochemical services

Studies on substrate-vegetation system in FA basin functions as nutrient cyclings in inhospitable environments are very limited. The FA ecosystem is not able to become so productive and diverse without microbial and mycorrhizal associations in rhizosphere. Compared with other terrestrial ecosystems, generally nitrogen is absent in FA catena, which may be fixed by using leguminous plant species during the rehabilitation programs that have the nitrogen–fixing ability. FA catena appears to act as sinks for carbon dioxide through covering of vegetation, which will help in reducing environmental pollution and global warming related to climate change. In this context, rehabilitated FA catena will also help to sequester carbon dioxide released from thermal power plants. Rehabilitated FA ecosystem can stabilize heavy metals, and sequester carbon dioxide released by thermal power plants, thus providing multiple services. There are several structural components of the vegetation of FA catena, namely biomass, life form (herb, shrub, tree, plant height, and stratification), diameter, age, species richness, species diversity, distribution, and plant canopy.

These are very important for ensuring the proper functioning of the abiotic components (e.g., nutrient dynamics, productivity, the water cycle, soil formation, soil structure, and decomposition) and the biotic components (bacterial, mycorrhizal, etc.) of the FA ecosystem. Some biogeochemical processes are vital to the functioning of FA ecosystems, and their services to the society are facilitated by microbes. These are carbon sequestration and nutrient mobilization, especially nitrogen and phosphorus. Soil microbes provide an important ecosystem functioning by mediating biogeochemical cycles. Thus the research on the FA ecosystem will help to improve our understanding on the linkage between microbial communities and biogeochemical cycles.

## 9.5 Science lessons from the fly ash ecosystems

The key lessons for ecological engineers, practitioners, and soil scientists are given here for creating a sustainable FA ecosystem. Further research needs are suggested below for developing a more stable and self-sustainable FA ecosystem on the dumped ashes worldwide.

- There is a current need for the quantitative information on the scope and importance of FA ecosystem services across the world. An urgent collaborative research should be done, linking a range of disciplines, including ecological engineering, forestry, and economics, to jointly model the provision and value of FA ecosystem services. This information is needed to recognize the importance of FA ecosystem services across the world to protect the natural ecosystem from the toxicant of FA.
- Rehabilitated FA catena to provide ecosystem services needs to be sited and designed according to the climatic condition and local flora of that particular region. In ecological engineering context, the FA ecosystem services can be enhanced with proper design and maintenance by human interventions. Such a designed FA ecosystem may be better to provide eco-services than an unmanaged FA ecosystem. So, it is an urgent need to develop a designer FA ecosystem toward sustainability.
- Naturally grown flora on FA catena should be screened across the world for better revegetation of abandoned FA catena of a particular region.
- Three types of flora namely natural, introduced, and mix (natural + introduced) may be used during the rehabilitation programs of FA

catena. It should be analyzed that which type of flora have more ecological suitability over the above-mentioned flora for revegetating FA catena. Better management is needed in how to adapt FA basin design to optimize the performance for better ecosystem services.

- Type of plant community also plays an important role in the restoration programs (Pandey et al., 2011). Therefore the role of annuals versus perennials should be considered in FA catena rehabilitation programs.
- The main aim of FA catena rehabilitation is to establish a self-sustainable ecosystem that needs minimum human interventions in order to provide sufficient goods and services (Pandey and Singh, 2012; Pandey et al., 2012).
- Comparing the rehabilitated FA catena with the reference sites gives an indication of improvement with respect to physical–chemical, microbial, and enzymatic activities in comparison with the reference site plantation, and it can be supposed that the rehabilitation FA catena will be sustainable in the long term or not.

## 9.6 Policy lessons for the fly ash catena

The key lessons for policy makers of the environmental management related to FA pollution are necessary in the current scenario because of the proliferation of the coal-based thermal power plants worldwide.

- FA disposal is occupying thousands of hectares of land worldwide, and this approach of FA disposal will be continued in the future. Regular and periodic monitoring of the dumped FA area should be done at national and international levels. Dumped FA areas are the potential hazards for the local people and ecosystems. These hazards affect the livelihoods of residents of dumped FA areas. The dumped FA sites represent a very poor ecosystem service in comparison with near the sites.
- There are several policy implications for the encouragement of FA utilization in India. But, policy related to FA catena rehabilitation should be considered for involved agencies to FA generation. It should also be considered worldwide for the rehabilitation of FA dumping sites and mitigation of FA pollution and its hazards.
- One ideal policy related to FA catena rehabilitation may be identified at world level that can be implemented to aid in decision-making about the restoration and phytomanagement of the FA lagoon.

- Promotion of research on FA catena rehabilitation is a current need. Low-cost biotechnology approaches should be discovered for the FA catena revegetation as a stable and self-sustainable methodology.
- Governmental organization, nongovernmental organization, voluntary local community participations should be involved in the rehabilitation programs of FA catena.

## 9.7 Valuation and policy approaches to promote fly ash ecosystem services

FA catena across the world is a source of environmental pollution to the residents of local livelihoods near the coal-based thermal power plants, despite the various agreements and policies related to FA utilization at national and international levels. This suggests that the importance of the FA catena rehabilitation as well as goods and services provided by FA ecosystem is not adequately recognized. Hence, the policies and agreements related to rehabilitation of FA catena in order to protect the environment are also inadequate at the world level. The lack of understanding of the economic value of the goods and services provided by FA ecosystems is not well recognized. They are often not taken into account in decision-making and can generate marketable goods and services. Some research papers present the benefits of FA catena rehabilitation, heavy metal stabilization/remediation through spontaneously grown plants on FA dumping sites, biodiversity conservation, trapping of FA particles through nearby plantation of thermal power stations, etc.

## 9.8 Eco-toxicological risk assessment

FA ecosystem provides a number of services, namely provisioning, regulating, supporting, and cultural, which needs an urgent attention for eco-toxicological risk assessment to avoid any health hazard to the wildlife and nearby society who depends on this system. If fruit plants are used in FA dumps' rehabilitation, then risk assessment is very essential due to the presence of toxicants mainly heavy metals, which may be accumulated in plant parts and may ultimately enter into the food chain via fruits. An assessment of the edible fruit tree *Z. mauritiana* grown on FA dumps was done to evaluate its suitability for phytoremediation but concerns with the use of edible fruit. The results showed that edible fruit trees are not safe and should not be considered for the restoration of FA dumps (Pandey and Mishra, 2016).

It may be avoided by using unpalatable grasses (viz., aromatic grasses) and trees for such programs. The regular monitoring of FA ecosystem should also be done before recommending it for public domain to avoid metal toxicity to local residents (Pandey and Mishra, 2016). De Joung and Morse (1977) assessed that selenium in pollen gathered by bees foraging on FA−grown aster plants was significantly higher compared with soil-grown aster plants. An assessment revealed that FA basins act as an attractive nuisance to birds, and parental provisioning exposes nestlings to harmful trace elements (Bryan et al., 2012). First time, in-situ investigations were done to know the effects of FA toxins on the species of small mammals in FA ecosystems. No evidence of genetic damage was detected as a result of exposure to FA toxicants in both mammals, although enhanced metal concentrations were noted in the tissues of both species collected from FA basin compared with the reference site (Peles and Barrett, 1997). These small mammals are a significant link for the transfer of toxic metals to higher trophic levels (Laurinolli and Bendell-Young, 1996). Thus these results have important suggestions about potential bioaccumulation of metals in the food chain (Brewer and Barrett, 1995). Energy or timber plantation could be a better alternative than fruit tree plantation for the restoration of FA dumps. Chakraborty and Mukherjee (2011) revealed that there was no damage in nuclear DNA in *V. zizanioides* (vetiver) grown on FA, which recommends the long-term survival of this aromatic grass on the FA dumpsite. Therefore vetiver is an unpalatable, aromatic, and socioeconomic grass, suitable for the rehabilitation programs of waste dumpsite uses all over the world.

## 9.9 Designer fly ash ecosystem—the sustainability of the fly ash ecosystem

The rapidly increasing area of FA ecosystems over the world has raised expectations and concerns as to their impact on ecosystem services. The appropriate design of plantations may improve the delivery of FA ecosystem's services such as biodiversity enhancement, wood availability for furniture, firewood for energy, fiber for craft, stabilization of heavy metals and FA dust particles, restoration, and carbon sequestration.

Here, first time, we define the designer FA ecosystems that are intentionally created through the inclusion of selected plant species combinations toward their sustainability and potential delivery of ecosystem services. Designer FA ecosystems should be developed or created de novo on the passive FA deposited sites. The terrestrial site of FA dumps poses ecological and socioeconomic threats; luxurious growth of perennial diversity rich plantations on the terrestrial site of FA dumps may support the livelihood of local people. In addition, the FA pond (temporary aquatic site of FA dumps) should be also managed for biomass production especially biofuel proposes and the lessening of heavy metals' leaching. Designer FA ecosystems will offer a wide range of ecosystem services, but it urgently needs regulatory services (especially phytostabilization of fine ash particles and phytoremediation of heavy metals to avoid leaching). Biodiversity boosts the sustainability of ecosystem functioning and makes ecosystems more robust against abiotic disturbances and exotic invasions (Fargione and Tilman, 2005). Hence, as far as possible, the designer FA ecosystems should be rich in biodiversity with importance on ecologically and socioeconomically suitable bioproducts, for example, timber, essential oil, bioenergy crops, fibers, etc. The designer FA ecosystems should be tolerant to persistent animal and human disturbance and will provide a package of ecosystem services mainly FA dumps' restoration, phytoremediation, and bioenergy production, which are being delivered by historical biodiversity in well-maintained ecosystems. Naturally grown native species pool on FA deposits is a valuable genetic resource for new FA dumps' restoration. Some appropriate plant species, to make desired biodiversity, from naturally grown FA deposits should be selected on the basis of dominance, most tolerant, perennial nature, fast-growing nature with high biomass productivity, unpalatable, regeneration potential, phytoremediation prospective, and ecological and socioeconomic values. Desired biodiversity can be created in the designer FA ecosystems through managing FA dumps' abiotic factors, such as high pH, compacting nature, poor aeration, heavy metal's presence, absence of nitrogen, and unavailable phosphorus, by using suitable wastes (press mud, organic wastes, etc.) as well as FA-tolerant PGPR and AMF for assisting preferred plant establishment, instead of trusting only on the native species pool. Moreover, habitat heterogeneity, disturbance, and the sequence of species colonization also affect the community association (Fukami and Morin 2003). Introduction of invasive plants, which also offer significant services, can be used in the later stages of advancement of designer FA ecosystems. Nonnative species incorporated in the community after the native community that has been established through restoration practices were found to be significantly less invasive (Martin and Wilsey, 2012). "Understanding the roles of these factors requires exploration of numerous combinations of biotic and abiotic factors and monitoring ecosystem function and human well-being until we obtain the desired diversity with the desired services" (Awasthi et al., 2016). Government can support this notion through "pay for ecosystem services" schemes (Naeem et al., 2015) to encourage the involvement of local people for sustainable management and utilization of ecosystem services toward confirming nature conservation. This will further enhance the climate change adaptability and ecological resilience of the whole landscape (Hobbs et al., 2014).

Fast-growing plantations based on native flora share many ecosystem services, but the extent to which this takes place depends significantly on what type of land use a plantation replaces and how plantations are established and managed. Industrial plantations are targeted mainly for demanding wood production; the links between business goals and ecosystem services are fashioned by the markets, by financing, by matters of image, and by regulatory and legal frameworks. A firm's participation in the management of ecosystem services is driven, for instance, by the identification of new business risks and opportunities, by the expectation of new markets, and by the opportunities to influence government policies, strengthen existing approaches to environmental management, and improve stakeholder relationships (Hanson et al., 2008).

In final, using promising strategies such as low maintenance and minimum inputs, reduction in time frame of rehabilitation process, low risk to animal including human being, rich biodiversity, and multiple benefits are urgently required to implement in rehabilitation programs of FA dumps to make designer FA ecosystems. The strategy we proposed here for the designer FA ecosystem is indeed a safe and sustainable approach and must be in the mind of practitioners (Table 9.2). Both researchers and practitioners are required to play their role for producing relevant knowledge and using

**Table 9.2** Strategies for the sustainability of designer fly ash ecosystems.

| S. no. | Strategies | Putting into practice |
|--------|-----------|----------------------|
| 1. | Low maintenance and minimum inputs | Perennial native flora with high regeneration potential |
| 2. | Reduction in time frame of rehabilitation process | Fast-growing plant species |
| 3. | Low risk to animal including human being | Unpalatable plant species |
| 4. | Rich biodiversity | Herbs, shrubs, and trees |
| 5. | Multiple benefits | Ecologically and socioeconomically valuable species |

that knowledge to make the difference on the ground. Designer FA ecosystems will also open up potential opportunities for local people.

## 9.10 Conclusion and prospects

Ecosystem services refer to various benefits that ecosystems provide to people that include food, fodder, fiber, clean water, nature, wildlife, and protection against anthropogenic pollution or natural disasters such as flooding as well as aesthetic landscape development. Scientific information related to goods and services derived from the FA ecosystem can be used to update decision-making for sustainable development of FA ecosystems. There are some examples of FA rehabilitation worldwide, where ample information can lead to better and adaptive FA management.

Designer FA ecosystem can be made in an accurate scientific approach during the development of a green cover by incorporating ecologically and socioeconomically valuable plant species at later stages of restoration, which not only reduces air pollution, dust problem, and heavy metal leaching, but also sequesters carbon dioxide, attenuates noise pollution, lessens atmospheric temperature, provides habitats for wildlife, and enhances aesthetic appearance, as well as offers a wide range of phytoproducts to local societies. Coal-based thermal power generation is the main culprit in global warming due to contributing about one-quarter of total emissions of carbon dioxide. Therefore the ever-increasing numbers and areas of FA deposits need an urgent attention toward its phytomanagement, from commitment to action, which will cut in the carbon emissions of power station through sequestering carbon in the ash-vegetation system of FA disposal sites.

## References

Awasthi, A., Singh, K., O'Grady, A., Courtney, R., Kalra, A., Singh, R.P., et al., 2016. Designer ecosystems: a solution for the conservation-exploitation dilemma. Ecol. Eng. 93, 73–75.

Babu, A.G., Reddy, M.S., 2011. Diversity of arbuscular mycorrhizal fungi associated with plants growing in fly ash pond and their potential role in ecological restoration. Curr. Microbiol. 63, 273–280. Available from: https://doi.org/10.1007/s00284-011-9974-5.

Brewer, S.R., Barrett, G.W., 1995. Heavy metal concentrations in earthworms following long-term nutrient enrichment. Bull. Environ. Contam. Toxicol. 54, 120–127.

Bryan, A.L., Hopkins, W.A., Parikh, J.H., Jackson, B.P., Unrine, J.M., 2012. Coal fly ash basins as an attractive nuisance to birds: parental provisioning exposes nestlings to harmful trace elements. Environ. Pollut. 161, 170–177.

Carlson, C.L., Adriano, D.C., 1991. Growth and elemental content of two tree species growing on abandoned coal fly ash basins. J. Environ. Qual. 20 (3), 581–587.

Chakraborty, R., Mukherjee, A., 2011. Technical note: vetiver can grow on coal fly ash without DNA damage. Int. J. Phytoremediat. 13 (2), 206–214.

Chen, B.D., Christie, P., Li, X.L., 2001. A modified glass bead compartment cultivation system for studies on nutrient uptake by arbuscular mycorrhiza. Chemosphere 42, 185–192.

Costanza, R., dArge, R., deGroot, R., Farber, S., Grasso, M., Hannon, B., et al., 1997. The value of the world's ecosystem services and natural capital. Nature 387, 253–260.

De Joung, D., Morse, R.A., 1977. Selenium in pollen gathered by bees foraging on fly ash–grown plants. Bull. Environ. Contam. Toxicol. 18 (4), 442–444. Available from: https://doi.org/10.1007/BF01683714.

Díaz, S., Demissew, S., Carabias, J., Joly, C., Lonsdale, M., Ash, N., et al., 2015. The IPBES conceptual frame-work—connecting nature and people. Curr. Opin. Environ. Sustain. 14, 1–16.

Dzantor, E.K., Adeleke, E., Kankarla, V., Ogunmayowa, O., Hui, D., 2015. Using coal fly ash in agriculture: combination of fly ash and poultry litter as soil amendments for bioenergy feedstock production. CCGP 7, 33–39.

Enkhtuya, B., Poschl, M., Vosatka, M., 2005. Native grass facilitates mycorrhizal colonization and P uptake of tree seedlings in two anthropogenic substrates. Water Air Soil Pollut. 166, 217–236.

Fargione, J., Tilman, D., 2005. Diversity decreases invasion via both sampling and complementarity effects. Ecol. Lett. 8, 604–611.

Favas, P.J.C., Martino, L.E., Prasad, M.N.V., 2018. Abandoned mine land reclamation—Challenges and opportunities (Holistic Approach). In: Prasad, M.N.V., Favas, P.J.C., Maiti, S.K. (Eds.), Bio-Geotechnologies for Mine Site Rehabilitation. Elsevier, Netherlands, pp. 3–31. Available from: http://dx.doi.org/10.1016/B978-0-12-812986-9.00001-4.

Fukami, T., Morin, P.J., 2003. Productivity–biodiversity relationships depend on the history of community assembly. Nature 424, 423–426.

Gajić, G., 2014. Ecophysiological adaptations of selected species of herbaceous plants at the fly ash landfill of the thermal power plant 'Nikola Tesla-A' in Obrenovac (Ph.D. thesis). Faculty of Biology, University of Belgrade, Belgrade, 406 p (in Serbian).

Gajić, G., Pavlović, P., Kostić, O., Jarić, S., Djurdjević, L., Pavlović, D., et al., 2013. Ecophysiological and biochemical traits of three herbaceous plants growing of the disposed coal combustion fly ash of different weathering stage. Arch. Biol. Sci. 65 (1), 1651–1667.

Gajić, G., Djurdjević, L., Kostić, O., Jarić, S., Mitrović, M., Stevanović, B., et al., 2016. Assessment of the phytoremediation potential and an adaptive response of Festuca rubra L. sown on fly ash deposits: native grass has a pivotal role in ecorestoration management. Ecol. Eng. 93, 250–261.

Gajić, G., Mitrović, M., Pavlović, P., 2019. Ecorestoration of fly ash deposits by native plant species at thermal power stations in Serbia. In: Pandey, V.C., Bauddh, K. (Eds.), Phytomanagement of Polluted Sites. Elsevier, ISBN: 978-0-12-813912-7pp. 113–177. Available from: https://doi.org/10.1016/C2017-0-00586-4.

Greenwood, E.F., Gemmill, R.P., 1978. Derelict industrial land as a habitat for rare plants in S. Lancs (v.c. 59) and W. Lancs (v.c. 60). Watsonia 12, 33–40.

Hanson, C., Finisdore, J., Ranganathan, J., Iceland, C., 2008. The Corporate Ecosystem Services Review. Available at Version 1.0. World Resources Institute, World Business Council for Sustainable Development, Meridian Institute. <http://www.wri.org/publication/corporate-ecosystem-services-review> (retrieved 9.02.10).

Havrlant, J., Krtička, L., 2014. Reclamation of devastated landscape in the Karviná region (Czech Republic). Environ. Socio. Econ. Stud. 2 (4), 1–12.

Haynes, R.J., 2009. Reclamation and revegetation of fly ash disposal sites: challenges and research needs. J. Environ. Manag. 90, 43–53.

Hobbs, R.J., Higgs, E., Hall, C.M., Bridgewater, P., Chapin III, F.S., Ellis, E.C., et al., 2014. Managing the whole landscape: historical, hybrid, and novel ecosystems. Front. Ecol. Environ. 12, 557–564.

Jamil, S., Abhilash, P.C., Singh, N., Sharma, P.N., 2009. *Jatropha curcas*: a potential crop for phytoremediation of coal fly ash. J. Hazard. Mater. 172, 269–275.

Jusaitis, M., Pillman, A., 1997. Revegetation of waste fly ash lagoons. I. Plant selection and surface amelioration. Waste Manage. Res. 15 (3), 307–321.

Juwarkar, A.A., Jambhulkar, H.P., 2008. Restoration of fly ash dump through biological interventions. Environ. Monit. Assess. 139, 355–365.

Kostić, O., Mitrović, M., Knezević, M., Jaric, S., Gajić, G., Djurdjević, L., et al., 2012. The potential of four woody species for the revegetation of fly ash deposits from the 'Nikola Tesla-A' thermoelectric plant (Obrenovac, Serbia). Arch. Biol. Sci. Belgr. 64, 145–158.

Kostić, O., Jariæ, S., Gajić, G., Pavlović, D., Pavlović, M., Mitrović, M., et al., 2018. Pedological properties and ecological implications of substrates derived 3 and 11 years after the revegetation of lignite fly ash disposal sites in Serbia. Catena 163, 78–88.

Kumar, K.V., Singh, N., Behl, H.M., Srivastava, S., 2008. Influence of plant growth promoting bacteria and its mutant on heavy metal toxicity in *Brassica juncea* grown in fly ash amended soil. Chemosphere. 72 (4), 678–683. Available from: https://doi.org/10.1016/j.chemosphere.2008.03.025.

Kumar, K.V., Srivastava, Singh, S.N., Behl, H.M., 2009. Role of metal resistant plant growth promoting bacteria in ameliorating fly ash to the growth of *Brassica juncea*. J. Hazard. Mater. 170, 51–57.

Kumari, A., Lal, B., Rai, U.N., 2016. Assessment of native plant species for phytoremediation of heavy metals growing in the vicinity of NTPC sites, Kahalgaon, India. Int. J. Phytoremediat. 18, 592–597.

Lal, K., Yadava, R.K., Kaurb, R., Bundelaa, D.S., Khana, M.I., Chaudharya, M., et al., 2013. Productivity, essential oil yield, and heavy metal accumulation in lemon grass (*Cymbopogon flexuosus*) under varied wastewater–groundwater irrigation Regimes. Ind. Crop. Prod. 45, 270–278.

Laurinolli, M., Bendell-Young, L.I., 1996. Copper, zinc, and cadmium concentrations in *Peromyscus maniculatus* sampled near an abandoned copper mine. Arch. Environ. Contam. Toxicol. 30, 481–486.

Lixandru, B., Dragomir, N., Morariu, F., Popa, M., Coman, A., Savescu, N., et al., 2013. Researches regarding the adaptation process of the species *Miscanthus giganteus* under the conditions of fly ash deposit from Utvin, Timis County. Sci. Pap. Anim. Sci. Biotechnol. 46 (1), 199–203.

Maiti, S.K., Jaiswal, S., 2008. Bioaccumulation and translocation of metals in the natural vegetation growing on fly ash lagoons: a field study from Santaldih thermal power plant, West Bengal, India. Environ. Monit. Assess. 136, 355–370.

Maiti, S.K., Kumar, A., Ahirwal, J., Das, R., 2016. Comparative study on bioaccumulation and translocation of metals in bermuda grass (*Cynodon dactylon*) naturally growing on fly ash lagoon and topsoil. Appl. Ecol. Environ. Res. 14 (1), 1–12. Available from: https://doi.org/10.15666/aeer/1401_001012.

Martin, L.M., Wilsey, B.J., 2012. Assembly history alters alpha and beta diversity, exotic–native proportions and functioning of restored prairie plant communities. J. Appl. Ecol. 49, 1436–1445.

Millennium Ecosystem Assessment (MEA), 2005. Ecosystems and Human Well-Being: Current State and Trends, vol. 1. Island Press, Washington, DC.

Mitrović, M., Jaric, S., Kostić, O., Gajić, G., Karadzić, B., Lola Djurdjević, L., et al., 2012. Photosynthetic efficiency of fourwoody species growing on fly ash deposits of a Serbian 'Nicola Tesla-A'Thermoelectric Plant. Pol. J. Environ. Stud. 21 (5), 1339–1347.

Mitrović, M., Pavlović, P., Lakusić, D., Stevanović, B., Djurdjevic, L., Kostić, O., et al., 2008. The potential of *Festuca rubra* and *Calamagrostis epigejos* for the revegetation on fly ash deposits. Sci. Total Environ. 72, 1090—1101.

Müller, F., Burkhard, B., 2012. The indicator side of ecosystem services. Ecosyst. Serv. 1, 26—30.

Naeem, S., Ingram, J.C., Varga, A., Agardy, T., Barten, P., Bennett, G., et al., 2015. Get the science right when paying for nature's services. Science 347, 1206—1207.

Pandey, V.C., 2012a. Phytoremediation of heavy metals from fly ash pond by *Azolla caroliniana*. Ecotoxicol. Environ. Saf. 82, 8—12.

Pandey, V.C., 2012b. Invasive species based efficient green technology for phytoremediation of fly ash deposits. J. Geochem. Explor. 123, 13—18.

Pandey, V.C., 2013. Suitability of *Ricinus communis* L. cultivation for phytoremediation of fly ash disposal sites. Ecol. Eng. 57, 336—341.

Pandey, V.C., 2015. Assisted phytoremediation of fly ash dumps through naturally colonized plants. Ecol. Eng. 82, 1—5.

Pandey, V.C., 2016. Phytoremediation efficiency of *Eichhornia crassipes* in fly ash pond. Int. J. Phytoremediat. 18 (5), 450—452.

Pandey, V.C., 2017. Managing waste dumpsites through energy plantations. In: Bauddh, K., Singh, B., Korstad, J. (Eds.), Phytoremediation Potential of Bioenergy Plants. Springer, Singapore, pp. 371—386. Available from: https://doi.org/10.1007/978-981-10-3084-0-15.

Pandey, V.C., Singh, K., 2011. Is *Vigna radiata* suitable for the revegetation of fly ash landfills? Ecol. Eng. 37 (12), 2105—2106.

Pandey, V.C., Singh, N., 2010. Impact of fly ash incorporation in soil systems. Agric. Ecosyst. Environ. 136, 16—27.

Pandey, V.C., Singh, B., 2012. Rehabilitation of coal fly ash basins: current need to use ecological engineering. Ecol. Eng. 49, 190—192. Available from: https://doi.org/10.1016/j.ecoleng.2012.08.037.

Pandey, V.C., Singh, N., 2014. Fast green capping on coal fly ash basins through ecological engineering. Ecol. Eng. 73, 671—675.

Pandey, Mishra, 2016. Assessment of *Ziziphus mauritiana* grown on fly ash dumps: prospects for phytoremediation but concerns with the use of edible fruit. Int. J. Phytoremediat. Available from: https://doi.org/10.1080/15226514.2016.1267703.

Pandey, V.C., Abhilash, P.C., Singh, N., 2009. The Indian perspective of utilizing fly ash in phytoremediation, phytomanagement and biomass production. J. Environ. Manag. 90, 2943—2958.

Pandey, V.C., Singh, J.S., Singh, R.P., Singh, N., Yunus, M., 2011. Arsenic hazards in coal fly ash and its fate in Indian scenario. Resour. Conserv. Recy. 55, 819—835.

Pandey, V.C., Singh, K., Singh, R.P., Singh, B., 2012. Naturally growing *Saccharum munja* on the fly ash lagoons: a potential ecological engineer for the revegetation and stabilization. Ecol. Eng. 40, 95—99.

Pandey, V.C., Singh, N., Singh, R.P., Singh, D.P., 2014. Rhizoremediation potential of spontaneously grown *Typha latifolia* on fly ash basins: study from the field. Ecol. Eng. 71, 722—727.

Pandey, V.C., Bajpai, O., Pandey, D.N., Singh, N., 2015a. *Saccharum spontaneum*: an underutilized tall grass for revegetation and restoration programs. Genet. Resour. Crop. Evol. Available from: https://doi.org/10.1007/s10722-014-0208-0.

Pandey, V.C., Prakash, P., Bajpai, O., Kumar, A., Singh, N., 2015b. Phytodiversity on fly ash deposits: evaluation of naturally colonized species for sustainable phytorestoration. Environ. Sci. Pollut. Res. 22, 2776—2787.

Pandey, V.C., Sahu, N., Behera, S.K., Singh, N., 2016a. Carbon sequestration in fly ash dumps: comparative assessment of three plant association. Ecol. Eng. 95, 198—205.

Pandey, V.C., Bajpai, O., Singh, N., 2016b. Plant regeneration potential in fly ash ecosystem. Urban. For. Urban Green. 15, 40−44.

Pavlović, P., Mitrović, M., Djurdjevic, L., 2004. An ecophysiological study of plants growing on the fly ash deposits from the "Nikola Tesla-A" thermal power station in Serbia. Environ. Manage. 33, 654−663.

Peles, J.D., Barrett, G.W., 1997. Assessment of metal uptake and genetic damage in small mammals inhabiting a fly ash basin. Bull. Environ. Contam. Toxicol. 59, 279−284.

Pietrzykowski, M., Krzaklewski, W., Gaik, G., 2010. Assessment of forest growth with plantings dominated by Scots pine (*Pinus sylvestris* L.) on experimental plots on a fly ash disposal site at the Bełchatów power plant. Environ. Eng. 137 (17), 64−74 (In Polish).

Pietrzykowski, M., Krzaklewski, W., Woś, B., 2015. Preliminary assessment of growth and survival of green alder (*Alnus viridis*), a potential biological stabilizer on fly ash disposal sites. J. For. Res. 26 (1), 131−136. Available from: https://doi.org/10.1007/s11676-015-0016-1.

Pietrzykowski, M., Woś, B., Pająk, M., Wanic, T., Krzaklewski, W., Chodak, M., 2018a. The impact of alders (*Alnus* spp.) on the physico-chemical properties of technosols on a lignite combustion waste disposal site. Ecol. Eng. 120, 180−186.

Pietrzykowski, M., Woś, B., Pająk, M., Wanic, T., Krzaklewski, W., Chodak, M., 2018b. Reclamation of a lignite combustion waste disposal site with alders (*Alnus sp.*): assessment of tree growth and nutrient status within 10 years of the experiment. Environ. Sci. Pollut. Res. Available from: https://doi.org/10.1007/s11356-018-1892-7.

Pillman, A., Jusaitis, M., 1997. Revegetation of waste fly ash lagoons II. Seedling transplants and plant nutrition. Waste Manage. Res. 15 (4), 359−370.

Prajapati, S.K., Meravi, N., 2015. Effect of fly ash on the photosynthetic parameters of *Zizyphus Mauritiana*. Adv. For. Lett. 4. Available from: https://doi.org/10.14355/afl.2015.04.001.

Ram, L.C., Jha, S.K., Tripathi, R.C., Masto, R.E., Selvi, V.A., 2008. Remediation of fly ash basins through plantation. Remediation 18, 71−90.

Rau, N., Mishra, V., Sharma, M., Das, M.K., Ahaluwalia, M., Sharma, R.S., 2009. Evaluation of functional diversity in rhizobacterial taxa of a wild grass (*Saccharum ravennae*) colonizing abandoned fly ash dumps in Delhi urban Ecosystem. Soil Biol. Biochem. 41, 813−821.

Roy, M., Roychowdhury, R., Mukherjee, P., 2018. Remediation of fly ash dumpsites through bioenergy crop plantation and generation: a review. Pedosphere 28 (4), 561−580. Available from: https://doi.org/10.1016/S1002-0160(18)60033-5.

Scanlon, D.H., Duggan, J.C., 1979. Growth and element uptake of woody plant on fly ash. Environ. Sci. Technol. 13 (3), 311−315.

Selvam, A., Mahadevan, A., 2002. Distribution of mycorrhizas in an abandoned fly ash pond and mined sites of Neyveli Lignite Corporation, Tamil Nadu, India. Basic. Appl. Ecol. 3, 277−284.

Shaw, P.J.A., 1992. A preliminary study of successional changes in vegetation and soil development on unamended fly ash (PFA) in southern England. J. Appl. Ecol. 29, 728−736.

Shaw, P.J.A., 1994. Orchid woods and floating islands: the ecology of fly ash. Br. Wildl. 5, 149−157.

Shaw, P.J.A., 1996. Role of seedbank substrates in the revegetation of fly ash and gypsum in the United Kingdom. Restor. Ecol. 4 (1), 61−70.

Shaw, P., 2003. Collembola of pulverised fuel ash sites in east London. Eur. J. Soil Biol. 39, 1−8.

Shrivastava, S., 2007. Tribal dependence on fly ash in Korba. J. Ecol. Anthropol. 11, 69−73.

Shukla, A.K., Misra, P.N., 1993. Improvement of sodic soil under tree cover. Indian For. 119, 43−52.

Singh, J.S., Singh, D.P., 2012. Reforestation: a potential approach to mitigate excess atmospheric CH4 build-up. Ecol. Manag. Restor. 13 (3), 245–248.

Smith, S.E., Read, D.J., 2008. Mycorrhizal Symbiosis, third ed. Academic Press, London.

Świątek, B., Woś, B., Chodak, M., Maiti, S.K., Jozefowska, A., Pietrzykowski, M., 2019. Fine root biomass and the associated C and nutrient pool under the alder (*Alnus* spp.) plantings on reclaimed technosols. Geoderma 337, 1021–1027.

Sylvia, D.M., Williams, S.E., 1992. Vesicular-arbuscular mycorrhizae and environmental stresses. In: Bethlenfalvay, G.J., Linderman, R.G. (Eds.), Mycorrhizae in Sustainable Agriculture. ASA, Madison, WI, pp. 101–124.

TEEB, 2010. The economics of ecosystems and biodiversity: mainstreaming the economics of nature: a synthesis of the approach, conclusions and recommendations of TEEB. <www.teebweb.org/our-publications/all-publications/>.

Tropek, R., Cerna, I., Straka, J., Cizek, O., Konvicka, M., 2013. Is coal combustion the last chance for vanishing insects of inland drift sand dunes in Europe? Biol. Conserv. 162, 60–64.

Tropek, R., Cerna, I., Straka, J., Kadlec, T., Pech, P., Tichanek, F., et al., 2014. Restoration management of fly ash deposits crucially influence their conservation potential for terrestrial arthropods. Ecol. Eng. 73, 45–52.

Uzarowicz, è., Zagórski, Z., Mendak, E., Bartmiński, P., Szara, E., et al., 2017. Technogenic soils (Technosols) developed from fly ash and bottom ash from thermal power stations combusting bituminous coal and lignite. Part I. Properties, classification, and indicators of early pedogenesis. Catena 157, 75–89. Available from: https://doi.org/10.1016/j.catena.2017.05.010.

Uzarowicz, L., Kwasowski, W., Śpiewak, O., Świtoniak, M., 2018. Indicators of pedogenesis of technosols developed in an ash settling pond at the Belchatów thermal power station (central Poland). Soil Sci. Annual. 69 (1), 49–59. Available from: https://doi.org/10.2478/ssa-2018-0006.

Verma, S.K., Singh, K., Gupta, A.K., Pandey, V.C., Trevedi, P., Verma, S.K., et al., 2014. Aromatic grasses for phytomanagement of coal fly ash hazards. Ecol. Eng. 73, 425–428.

Vivas, A., Barea, J.M., Azcón, R., 2005. Interactive effect of *Brevibacillus brevis* and *Glomus mosseae*, both isolated from Cd contaminated soil, on plant growth, physiological mycorrhizal fungal characteristics and soil enzymatic activities in Cd polluted soil. Environ. Pollut. 134, 257–266.

Vosatka, M., 2001. A future role for the use of arbuscular mycorrhizal fungi in soil remediation: a chance for small-medium enterprises? Minerva Biotechnol. 13, 69–72.

Woch, M.W., Radwańska, M., Stanek, M., èopata, B., Stefanowicz, A.M., 2018. Relationships between waste physicochemical properties, microbial activity and vegetation at coal ash and sludge disposal sites. Sci. Total Environ. 642, 264–275.

Wu, F.Y., Bi, Y.L., Wong, M.H., 2009. Dual inoculation with an arbuscular mycorrhizal fungus and rhizobium to facilitate the growth of Alfalfa on coal mine substrates. J. Plant. Nutr. 32, 755–771.

# Further reading

Krzaklewski, W., Pietrzykowski, M., Woś, B., 2012. Survival and growth of alders (*Alnus glutinosa* (L.) Gaertn and *Alnus incana* (L.) Moench) on fly ash technosols at different substrate improvement. Ecol. Eng. 49, 35–40.

Pandey, V.C., Pandey, D.N., Singh, N., 2015c. Sustainable phytoremediation based on naturally colonizing and economically valuable plants. J. Clean. Prod. 86, 37–39.

# CHAPTER 10

# An appraisal on phytomanagement of fly ash with economic returns

## Contents

## 10.1 Introduction

Fly ash (FA), a by-product of coal combustion, has been recognized as a hazardous material over the world, dumped at open land due to low utilization. Two methods specifically dry and wet are used for disposing FA near the abandoned site of thermal power plants. In the dry method, FA is disposed in landfills or lagoons or basins. While in the wet method, FA is mixed with water to form slurry for dumping in ash pond through pipes. FA disposal in slurry form needs $1040 \, mm^3$ water per year, which is an additional burden on water assets. Both the methods finally lead to pollution near the FA dumping sites that not only destroys surrounding

*Phytomanagement of Fly Ash*
ISBN: 978-0-12-818544-5
DOI: https://doi.org/10.1016/B978-0-12-818544-5.00010-9
**289**

air but also pollutes adjacent land and water ecosystems through leaching and surface runoff. Residents nearby the coal-based thermal power stations have been recognized to suffer from heart disease, genetic disorders, respiratory disorders, and cancer (USEPA, 2007). Furthermore, Bryan et al. (2012) reported the effects of FA on birds nesting and detected adverse results due to the accumulation of Se, As, Cd, and Sr in their offspring. In fact, the unused FA dumped into poorly designed ash ponds pollutes the nearby environment. In India, it has been forecasted that 300 million tons per annum and 600 million tons per annum of FA will be generated by the year 2021−22 and 2031−32, respectively. In this context, total land occupied by ash ponds would be about 82,200 ha by the year 2020 (CEA, 2014). This expanding trend of FA production is going to be a most important concern across the nations. Therefore potential efforts are necessary to restrict its invasion and FA should be treat as a treasure rather than waste mainly thinking the presence of essential micro- and macronutrients in FA (Bilski et al., 2011). The stabilization of abandoned FA dumps is generally achieved through different know-hows, such as watering FA dumps by sprinkler method, soil layering on FA dumps, and green covering on FA dumps through vegetation.

Therefore the FA management is a most important concern, which requires primarily four aspects: (1) a safe method of FA disposal; (2) land requirement; (3) control air−soil−water pollution; and (4) heavy metals risk by food chain. The last two constraints can be managed through the choice of plant species and techniques for vegetation establishment. Phytoremediation is an eco-friendly and cost-effective approach as it would stabilize FA basins, oversight wind, and water erosion as well as invite gradual FA restoration. A number of researchers over the world have been worked on phytoremediation of FA dumpsites, but few researchers are working on the phytomanagement of FA to generate additional benefits instead of remediation. Phytomanagement of FA delivers a number of direct and indirect benefits, which depends on the FA characteristics, followed by the type of amendment and its application rate as well as the selection of plant species. Native grass species is an effective choice at initial stages. On later stages, we can introduce ecologically and socioeconomically valuable plant species. They must be perennial and unpalatable, have high biomass producing capacity, regeneration potential, and extensive root system, and must be tolerant to harsh conditions. There are some benefits of vegetation development on FA deposits, which include binding of ash particles, reduction in leaching of ash solution, stabilization/accumulation of metals, substrate quality improvement, carbon sequestration in ash–plant system, aesthetically pleasing landscape, and habitat for wildlife (Pandey and Singh, 2012). Likewise, planned rehabilitation programs also provide phytoproducts that will enhance the socioeconomic status of local residents or villagers. It will be better if we create a park on abandoned FA deposits toward the development of tourism. In this regard, a park has been developed on old FA deposit site of Renusagar thermal power station, which is located near Renukoot in the Sonbhadra district of Uttar Pradesh. It is operated by the Hindalco Industries. Currently, there is a pressing need to develop designer FA ecosystem toward the sustainability of nature.

Green cover development on abandoned FA deposits is a challenging task. In this regard, the insight of field studies provides the accurate knowledge of phytodiversity of bare FA deposits, which is the most vital tool for a right choice of plant species toward the phytomanagement of FA (Pandey et al., 2015c). In general, grasses and legumes are used for this purpose. Grasses are also known as nurse plants, which initiate succession on hostile conditions as well as restore the properties and functions of that site. As we know, FA deposit contains basically all the elements present in soil except organic carbon and nitrogen (Pandey and Singh, 2010). If we choose legumes, then the nitrogen deficiency problem of FA deposits can be solved, because they would add nitrogen in the ash layer. Thus the combination of grass—legume is the best choice, as they can easily colonize the ash deposits and develop a green cover in short term. In the primary succession of FA deposits, some plants first colonize the barren substrate of FA deposits. On a new FA deposit, after the settling of ash slurry, seeds blown by the wind may reserve on the ash substrate. On favorable condition, they can germinate. Often these first colonizing plants are grass species that do not only add a layer of organic matter above the ash substrate but also do reproduce quickly. This would fabricate the ash layer for the establishment of secondary colonizers followed by bioenergy-based trees species (Pandey and Singh, 2012; Maiti and Maiti, 2015). Now, a special attention is necessary to establish a range of valuable crops for green-cover development on FA dumpsite with economic returns, to stop air—soil—water pollution, through phytomanagement.

## 10.2 Phytomanagement: linking bioeconomy with phytoremediation

In general, the phytoremediation is well known for their low operational costs, but the process of phytoremediation is often time-consuming, which increases operational costs (Pandey and Bajpai, 2019a). By adding economically valuable crops with fast growing nature in phytoremediation programs, we can eliminate both the limitations (operational cost and time). Final operational cost will be reduced by obtaining bioeconomy as phytoproducts that were produced from contaminated areas during phytoremediation. The above overall approach is termed the concept of phytomanagement (Pandey and Souza-Alonso, 2019b). Therefore the selection of plant species remains the most important criterion to determine the success of phytomanagement of polluted sites or FA dumpsites (Pandey and Souza-Alonso, 2019b). The number of suitable plant species for phytomanagement will increase with more research on biodiversity exploration. The following factors must be considered in selected plants for the success of phytomanagement of polluted sites toward ecological and socioeconomic sustainability.

- The plants must have high biomass, fast growing capacity, extensive root system, and woody and native nature including tolerance to local conditions such as changing climate, enriched levels of trace elements, highly alkaline substrate, metal-contaminated site, and pests.
- Economically valuable plant species such as bioenergy crops, essential oil bearing aromatic crops, and valuable timber tree species.
- They have high metal accumulation capacity in the shoots than roots or leaves.
- They able to secrete chelating agents actively.
- They should be responsive toward agriculture practices, propagation, and management.

## 10.3 Commercial crops in the phytomanagement of fly ash

The presence of essential macro- and micronutrients, high moisture content, and high porosity with low water retention capacity has been the significant property of FA that favors plant growth upon carbon and nitrogen supplementation. But the presence of heavy metals (Cd, Cr, Pb, As, Hg, Mn, B, Co, Mo, Se, Sr, etc.) in FA makes it unsafe for agronomy. Though the plantation of bioenergy and aromatic crops on FA deposits is an ideal way for the phytomanagement of FA deposits, these crops are not directly connected to food chain. The major domain of FA utilization is in biomass production, which covers forestry, floriculture, and agriculture (only evidence-based crop like rice). Moreover, numerous economically important crops, such as pulp and paper crops, biodiesel crops, fiber crops, aromatic crops, firewood, timber wood, plywood crops, and floriculture crops, have been planted on FA deposits with the help of organic amendments and microbial inoculation (Pandey et al., 2009b).

Some important issues are also responsible for mishandling of FA deposits, which involves lack of awareness and regulations. In this regard, there is a pressing need toward the remediation and management of FA deposits as it will save precious topsoil and reduce land and water requirement. Phytomanagement is a holistic and evergreen strategy that will provide nature sustainability in the nearby surroundings of thermal power stations. There are some limitations on FA deposits that inhibit plant establishment and growth. These are high pH, deficiency of organic carbon, nitrogen, and phosphorous, high soluble salts, high concentration of toxic elements, lack of microbial activity, and self-cementing property due to pozzolanic activity. FA deposits form cemented or compacted layers that reduce aeration, water infiltration, and root penetration (Haynes, 2009; Pandey and Singh, 2010). Its presence on FA deposits is a major problem, which hinders vegetation development. The incorporation of organic material into ash deposits can support to break up these layers.

An integrated approach would be needed for initial green cover development on FA deposits followed by successive high biomass producing vegetation (Haynes, 2009). The most FA tolerant plant species have been identified, which belong to the families of Brassicaceae, Chenopodiaceae, Leguminosae, and Poaceae. Phytostabilization is an important task to check and minimize wind and water erosion of FA deposits. The woody native plant species are more appropriate for the phytomanagement of FA than other species, because these are well adapted that tolerate the hostile conditions. They are long living with high biomass producing capacity that can accumulate high metals from FA deposits and store them for a long term. Yet, litter fall should be monitored time to time so that bioaccumulated toxic elements are not reintroduced into the ash substrate.

The cultivation of biofuel crops on abandoned FA deposits is a superb example of the phytomanagement of FA (Pandey et al., 2016a). This review discusses all the features of biofuel crops regarding its establishment on FA dumps. Biofuel is known as carbon neutral fuel, because the $CO_2$ released during coal combustion is recaptured by the growth of the plant biomass during photosynthesis. Abandoned FA dumpsites can be utilized for the cultivation of biofuel crops after proper amendments like vermicompost, farmyard manure, and organic manure. Most of the biomass generating plants especially grasses can grow aggressively on the waste dumpsites such as FA deposits with minimum nutritional needs and offer good economic return. There is a potential option as green capping on FA dumpsites for growing bioenergy crops with multiple benefits.

Different case studies of tree plantation on FA deposits have been done with a number of different species such as Salix (Willow), Populus (Poplar), bamboo, aromatic crops, etc. Willow and bushy Salix species, fast growing species, have been recognized as the most suitable plant species for biomass coppice. Furthermore, *Salix viminalis* is one of the most extensively used species. *Acacia auriculiformis* and *Leucaena leucocephala* have been proved to have a high survival and tolerance on infertile, arid, and metal-enriched FA deposits (Cheung et al., 2000). In addition, legume plant *Cassia surattensis* and symbiotic $N_2$-fixing bacteria can enhance the nitrogen content of barren FA landfills (Vajpayee et al., 2000). The plants belonging to the family Leguminosae grow well on FA amended soils without appearance of any damage symptoms (Singh et al., 1997; Pandey et al., 2009a). There are two ways (such as natural restoration and plantation) for the utilization of FA deposits through phytomanagement with economic returns. Natural restoration will take more time than plantation. Fig. 10.1 shows the excellent tree species *Prosopis juliflora* grown naturally on FA deposits. Fig. 10.2 shows most dominant plant species growing naturally on FA deposits and their economic value among society and local residents. Likewise, Fig.10.3 depicts key plant species of man-made plantation on FA deposits at Panki Thermal Power Station, Kanpur, Uttar Pradesh, India, and their economic significance among society and local residents.

**Figure 10.1** Revegetation of abandoned FA deposits by naturally growing *P. juliflora* (Sw.) DC. *Courtesy V.C. Pandey.*

**Figure 10.2** Naturally grown plant species on FA deposits at Feroze Gandhi Unchahar Thermal Power station, Raebareli, Uttar Pradesh, India, and their economic value among society and local residents. (A) *P. juliflora* (Sw.) DC. on FA deposits—a potential firewood, (B) *S. munja* Roxb.—an economic perennial grass, (C) *S. sponta-neum* L.—a multiple-use and multifunctional species, (D) *T. latifolia* L.—having a very wide range of traditional and modern uses, (E) Mat like view of *Azolla caroliniana* (Mosquito fern), spread across the FA pond, (F) a closer view of *Azolla caroliniana*—a natural nitrogen source, and (G) and (H) *E. crassipes* (Mart.) Solms—a potential of bioenergy. *Courtesy V.C. Pandey.*

**Figure 10.3** Planted species on FA deposits at Panki Thermal Power Station, Kanpur, Uttar Pradesh, India, and their economic value among society and local residents. (A) Well established *P. juliflora* (Sw.) DC. on FA deposits—a potential firewood, (B) a local woman is cutting dry part of *P. juliflora* tree for cooking her food, (C) and (D) *Pithecellobium dulce* (Roxb.) Benth.—normally used as firewood for brick kilns, (E) *P. pinnata* (L.) Pierre—an important biodiesel plant, (F) Eucalyptus species—mostly uses in pole and construction, (G) and (H) *L. leucocephala* (Lam.) de Wit.—an important tree for Pulp and paper industry. *Courtesy V.C. Pandey.*

## 10.3.1 Energy crops

Coupling bioenergy production with phytoremediation programs has been suggested as a synergistic bonding to resolve the sustainability issues and efficiently solve the problems of ever-increasing polluted sites and biofuel demands across the nations (Pandey et al., 2016a). Pandey (2017) also suggested evidence-based phytomanagement of FA deposits through energy plantation as a holistic approach toward ecological and socioeconomic development. Vegetation cover development on FA deposits through the cultivation of bioenergy crops is a novel approach for the multiple benefits such as control of ash and water erosion, addition of organic matter, improvement of the ash structure toward soil development, enhancement of the microbial activity, bioaccumulation of the toxic elements, and binding of the ash particles. Several researches over the world confirmed the phytoremediation potential of some bioenergy crops with respect to FA deposits. In addition, FA has been suggested to be used as soil amendment for growing bioenergy crops. Therefore the bioenergy crops have ability to grow successfully on FA deposits or in FA amended soil. These are *Ricinus communis* (Singh et al., 2010; Pandey, 2013); *Salix* spp. (Hrynkiewicz et al., 2009), *Populus* spp. (Pavlović et al., 2004; Čermák, 2008; Lombard et al., 2011; Kostic et al., 2012; Mitrovic et al., 2012), *Miscanthus* × *giganteus* (Técher et al., 2012), *Panicum virgatum* (Gomes, 2012; Dzantor et al., 2013, 2015; Awoyemi and Dzantor, 2017a, b,c; Awoyemi and Adeleke, 2017a), *Pongamia pinnata* (Juwarkar and Jambhulkar, 2008; Jambhulkar and Juwarkar, 2009; Kapur et al., 2002), *Jatropha curcas* (Bagchi, 2013; Jamil et al., 2009; Mohan, 2011; Pandey et al., 2012a; Chaudhary and Ghosh, 2013; Das et al., 2013; Raj and Mohan, 2016), *Saccharum munja* (Pandey et al., 2012b), *Saccharum spontaneum* (Pandey et al., 2015b), *Water hyacinth* (*Eichhornia crassipes*) (Pandey, 2016) and *Prosopis juliflora* (Pandey et al., 2015c). Recently, it was observed that *J. curcas* has better survivability in FA amended mine spoil instead of mine spoil (Vurayai et al., 2017). Therefore FA tolerant energy crops are appropriate candidates for long-term FA field remediation with economic returns. This approach could be of specific significance in those countries with increasingly less arable land due to the fast development of urbanization and industrialization.

## 10.3.2 Aromatic crops

Aromatic grasses are unpalatable and perennial, are of fast growing nature with multiple harvests, are essential oil bearing crops, and have tolerance against stress conditions such as drought, heavy metal toxicity, and pH variability (Pandey and Singh, 2015; Verma et al., 2014). The cultivation of aromatic crops on polluted sites has been suggested as a holistic approach to avoid the entry of contaminants into the food chain (Pandey and Singh, 2015; Verma et al., 2014). The reason behind this approach is essential oil that is extracted through hydrodistillation method and free from the risk of toxic heavy metals (Khajanchi et al., 2013; Zheljazkov et al., 2006). Some important aromatic grasses, that is, *Vetiveria zizanioides, Cymbopogon flexuosus, Cymbopogon martini*, and *Cymbopogon winterianus*, are famous in the world due to their valued essential oil, which are used in perfumery, pharmaceuticals, cosmetics, and aromatherapy as well as toiletry products. Its suitability for FA vegetation can be supposed by the adaptability of wild grasses *S. munja and S. spontaneum* that grows well on FA deposits. Both aromatic and wild grasses belong to Poaceae family. Grasses are well known as nurse species, which improve the fertility of the substrate for the upcoming plant species. In general, it is considered that phytoremediation is a time-consuming process, but it mostly depends on the choice of plants. Aromatic grasses are fast growing in nature, which can reduce "remediation period" that is one of the most important constraints of phytoremediation programs. *V. zizanioides* quickly flourishes on abandoned FA deposits and its roots grow up to 19.68-ft (6-m) deep and produce high biomass (Lavania and Lavania, 2009). It is suggested by researchers that aromatic grasses have ability toward phytomanagement of FA deposits and to get several direct and indirect benefits (Pandey and Singh, 2015; Verma et al., 2014). Recently, in a pot scale study, it was observed that lower rate of amendment application (2%—5% farmyard manure and 5%—10% topsoil on weight basis) in FA improves the growth and biomass of aromatic grass *Cymbopogon citratus* (Maiti and Prasad, 2017). Recently, it is also confirmed by several researchers that *C. citratus* has ability to grow in FA amended soil along with metal polluted sites (Israila et al., 2015; Gautam et al., 2017; Panda et al., 2018). All these studies of *C. citratus* belong to physiology, bioaccumulator, growth performance, and antioxidant response.

### 10.3.3 Floriculture crops

Huge FA disposal covers thousands of hectares of land and creates several environmental problems and health risks such as silicosis, asthma, and other breathing problems due to fine ash particles (Pandey et al., 2009b;

Pandey and Singh, 2012). In addition, the leaching of toxic metals from FA deposits is also a major problem, which poses water and land pollution. In this regard, the cultivation of floriculture crops on FA deposits is a suitable and viable approach toward economic returns along with the utilization of FA deposits. Floriculture is known as an ancient occupation to generate remunerative self-employment among farmers. Now, floriculture activity is also popularizing across the developed countries. Floriculture crop-based revegetation of FA deposits will also provide direct and indirect benefits such as bioaesthetic environment, carbon sequestration, reduction in leaching of heavy metals, binding of fine ash particles, green cover development, and improvement in ash substrate and wildlife habitat (Pandey et al., 2009b). As we know that FA deposits cannot support vegetation establishment and growth but when amended with organic manure and mycorrhizae, it can support flora on FA dumps without using soil (Adholeya et al., 1997). The mycorrhizal fungi are helpful for absorbing nutrients from FA to the plants for their survival and establishment on FA deposits. Adholeya et al. (1997) showed that 5% farmyard manure and inoculation of mycorrhiza in FA improve physicochemical properties such as pH, EC, P, total N, organic C, etc. and biological properties that resulted in the establishment of plants on FA deposits. FA disposal site near Badarpur has been converted into lavish green with flower plants, that is, *Calendula* spp., *Dianthus* spp., *Helianthus annuus*, *C. citratus*, *Polianthes tuberosa*, *Gladiolus* spp., and *Lilium* spp. (Adholeya et al., 1997). The successful blooming of bougainvillea on FA landfills has been demonstrated by amending it with organic and biological resources such as sewage sludge, biofertilizers, humic acid, etc. (Ram et al., 2005). Thus it is proved that floriculture is possible on FA deposits but it is still an unexploited resource. Such inputs in vast area of FA dumps have been covered with luxurious flowering plants (Ram et al., 2005, Sharma and Adholeya, 2005). Table 10.1 shows floriculture on FA deposits or in FA amended soils.

Furthermore, FA mixed with sewage sludge and lime can be applied to barren soils for increasing plant height, biomass, and flower production of *Aster* sp. (Rethman et al., 1999). FA amended with commercial peat moss has also been used for the cultivation of Chrysanthemums (Chou et al., 2005). One potential use in floriculture is to mix FA into container substrates for the production of ornamental plants. The FA is easily accessible in most of the countries, and its fusion with container substrates could be a potential and safe opportunity of beneficial use in floriculture. It has been

**Table 10.1** Floriculture on FA amended soils or FA deposits.

| Substrate | Amendments | Floriculture crops grown | Effects on substrate | Response on plant growth | References |
|---|---|---|---|---|---|
| FA overburdens, India (pH:7.41. EC:1.13, P:11.6, total N%:0.014, and organic C %:0.60) | FA + 5% farmyard manure + mycorrhiza innoculation | *Calendula* sp., *Dianthus* sp., Sunflower, *C. citratus, Polianthes tuberose, Gladiolus* sp., and *Lilium* sp. | FA properties improved pH: 6.60, EC:1.51, total N %:0.31, and organic C %:1.84 | Good growth and flower production | Adholeya et al. (1998) |
| Low fertility soil, South Africa (20% silt and 80% sand) P(5ppm), Ca(270 ppm), Mg(125 ppm), and K(20 ppm) | 95% soil + 5%SLASH (27% sewage sludge + 64% FA + 9% lime on dry matter basis) | *Aster* sp. (Asters) | Increase in pH, P content:61 ppm, Ca: 1845 ppm, Mg:145 ppm, and improved in soil texture | Significant increase in plant height and biomass, and some increase in flower production | Rethman et al. (1999) |
| Commercial peat moss/FA mixed media | Alkaline FA at 25%–100% | *Chrysanthemums* | Increase in alkalinity and surface tension predicted | Plant height and quality decreased with increase in FA | Chou et al. (2005) |
| Pond ash, India | Sewage sludge, biofertilizer like Rhizobium and Phosphobacterium cultures, humic acid, etc. | *Bougainvillea* sp. | Decrease in pH and increase in organic carbon and available NPK | Increase in microbial activity and population | Ram et al. (2005) |

| | | | | | |
|---|---|---|---|---|---|
| Soil | Fly ash dose at 5%, 10%, 15%, and 20% | *Tagetes erecta* (Marigold) | Physicochemical parameters increased with increase in fly ash amendment except N, P, and K | Up to 10% fly ash dose in soil increases the growth with respect to pigment content | Pradhan et al. (2015) |
| Soil | FA at 20%, 40%, 60%, 80% and 100% (v/v) | *H. annuus* L. | | In seeds, except Fe, Pb, Mn, Zn and other heavy metals (Cu, Cr, Cd, Co, and Ni) remained untraced up to 40% of the FA, above that their quantity slightly increased, but the values are very much under the permissible limits | Siddiqui et al. (2004) |
| Fly ash deposits | Pit size (60 × 60 × 60 cm), filled with various amendments, such as soil + cow manure (2:1 ratio), 25 g of di-ammonium phosphate as chemical fertilizers, 10 g of rhizobium and phosphobacterium as biofertilizers, and 5 g of humic acid | Champa (*Michelia champaca*), China Rose (*Hibiscus rosa-sinensis*), and Bougainvillea (*Bougainvillea glabra*) | Suitably reclaimed for social forestry purposes | Floriculture species can be grown on fly ash deposits. The microbial activity of fly ash deposits could be successfully improved through the plantation of *B. glabra* | Ram et al. (2008) |

established that FA can be substitutes to commercial dolomites used as amendments to soilless substrates for the production of ornamental plants. Beneficial use of FAs as a container substrate material has a new market opportunity (Chen and Li, 2006). It is well proved that application of FA in floriculture or agriculture depends on species to species and its doses (Pandey et al., 2009b; Ram and Masto, 2014). There is a pressing need to determine proper dose of FA for a specific FA—floriculture crop combination. However, the most important aspect of FA utilization is the risk of food contamination with heavy metal enriched FA via biomagnification through food chain. It can be avoided if we choose floriculture crops rather than growing food crops on FA contaminated sites or FA deposits.

## 10.3.4 Fiber-yielding crops

Fiber-yielding crops are widely grown at the world level due to their high economic value and multiple uses. Fiber crops are well recognized for their some important features such as fast growth, large biomass, and strong adaptability. The fibers are mostly used in clothing, adsorbents, building materials, as well as papermaking (Pandey and Gupta, 2003; Ververis et al., 2007). A special attention is given to fiber crops as a potential species for heavy metal polluted soils due to its tolerance (Ho et al., 2008; Yang et al., 2010). A number of field- and lab-based investigations revealed a certain degree of metal tolerance in Hemp (*Cannabis sativa*) (Citterio et al., 2003; Arru et al., 2004) and Ramie (*Boehmeria nivea*) (Yang et al., 2010) as well as FA tolerance in *Crotalaria juncea* (Maiti, 2017) and *Gossypium hirsutum* (Stevens and Dunn, 2004), without phytotoxicity symptoms. The ameliorative effects of lime, dolomite, FA, and compost were tested on kenaf crop (*Hibiscus cannabinus*), which was grown on the farmland soil polluted with mine tailing (Huang, 2009). The impact of dolomite and FA was better on kenaf crop (*H. cannabinus*), which was grown on heavy metal polluted soils, compared with limestone and organic fertilizer (Yang et al., 2013). Some other important fiber-yielding crops such as *Corchorus capsularis* L. and *Linum usitatissimum* should also be examined for the application of FA as soil ameliorants for growth and high biomass production.

## 10.3.5 Trees in paper industry

Raw materials of paper industries come mostly from plant biomass. These trees are planted and grown specifically on managed timberlands to make paper. They are just like agricultural crop. These timberlands can be managed to produce trees forever for papermaking as trees are a renewable resource. In this regard, ever-increasing numbers and areas of FA deposits can be used to produce tree biomass for paper-making. Numerous trees have been grown successfully on FA dumpsites (Pandey et al., 2009b). In addition, several researchers have suggested the use of FA as a fertilizer supplement for plantation on the degraded lands (Malewar et al., 1998; Cheung et al., 2000). The following tree species such as *L. leucocephala*, *Bambusa* sp., *Populus euphratica*, and *Eucalyptus* sp. have been successfully demonstrated to grow on FA dumpsites. *L. leucocephala* is a fast growing leguminous tree species in most tropical areas globally. Its wood biomass is suitable for the manufacturing of paper and packaging material (Pandey and Kumar, 2013). *L. leucocephala* has been effectively cultivated on FA dumps due to high tolerance nature (Cheung et al., 2000; Kapur et al., 2002). Nutrient enriched FA has been examined to increase the yield of leguminous trees (Rai et al., 2004; Tripathi et al., 2000). It is also stated that 10% FA amended sand is a suitable rooting media for vegetative propagation of *L. leucocephala* (Pandey and Kumar, 2013). Bamboo is the fastest growing plant on the earth. It is able to grow in diverse conditions owing to a unique rhizome-root system. In India, a novel research project was funded by the International Development Research Centre of Canada to develop bamboo and FA-based technologies to rehabilitate barren land. The result was that it can be successfully cultivated to restore barren land by the use of soil + sludge + FA combination, because it can produce the maximum biomass per unit area and time (Anonymous, 1997). *Dendrocalamus strictus* is the most common bamboo species that is widely distributed in dry regions of India. *D. strictus* has been successfully planted through demonstration on FA dumps. It is noticed that its plantation has an efficient restoration potential and positive rehabilitation effect on FA dumps (Juwarkar and Jambhulkar, 2008; Jambhulkar and Juwarkar, 2009; Babu and Reddy, 2011; Das et al., 2013). It is able to give higher biomass within a short time. In addition, Coal-based thermal power stations have grown bamboo on their FA dumps with the support of National Thermal Power Corporation (NTPC). Therefore the utilization of FA dumps for cultivating high biomass producing bamboo plant is a viable and efficient approach toward nature sustainability. *P. euphratica* can be established and grown on FA dumps (http://www.kuenvbiotech.org/casestudy/.htm). *Eucalyptus* sp. has been planted well on FA deposits because of it has adaptability to a wide range of edaphoclimatic conditions (Juwarkar and Jambhulkar, 2008; Adholeya et al., 1998; Malewar et al., 1998; Zevenbergen et al., 2000; Kapur et al., 2002). *Eucalyptus tereticornis* was also used for the reclamation of FA dumpsites using amendments such as farmyard manure and mycorrhizal biofertilizer (Das et al., 2013). *Eucalyptus* species has capability to develop arbuscular mycorrhizal (AM) fungi association to perform ectomycorrhizal symbiosis. Therefore the use of FA-adapted AM fungal strains will be more effective, for better tolerance and high biomass production, than applying nonadapted strains on FA dumps. It is reported that the inoculation of AM fungi supports the growth of plant, nutrient uptake, and FA aggregation (Enkhtuya et al., 2005; Wu et al., 2009).

## 10.3.6 Firewood crops

Mostly shrub and tree species comes in the category of firewood crops for energy production. It is concluded that variations in biomass yield and ecological tolerance are more significant in the selection of species for energy crops than the differences in their heating values (Neenan and Steinbeck, 1979). Therefore the choice of energy crop species for the revegetation purpose should be mainly based on the biomass yields and ecological tolerance for a specific site such as FA deposits. $N_2$-fixing firewood tree (*Acacia* sp., *A. auriculiformis, Albizia lebbeck, Dalbergia sissoo,* and *Cassia siamea*) and nonfixing firewood tree (*Melia azadirachta, Populus deltoides, Tamarindus indica, Eucalyptus globulus, Eucalyptus hybrid, Syzygium cumini,* and *Morus alba*) have been reported to grow well on FA overburdens or ameliorated FA dumps (Adholeya et al., 1998; Malewar et al., 1998; Cheung et al., 2000; Singh et al., 2000; Tripathi et al., 2000; Selvam and Mahadevan, 2002; Kumar et al., 2002; Kapur et al., 2002; Das et al., 2013). The growth of pine (*Casuarina cunninghamiana* Miq.) doubled with the incorporation of FA in sandy soil (Riekerk, 1984). In addition, *Pongamia glabra* and *Parkinsonia aculeata* have been cultivated on FA dumps (Kapur et al., 2002). Wood Pole-based tree (*A. auriculiformis, E. globulus, Casuarina equisetifolia*) have been grown directly on FA dumps (Zevenbergen et al., 2000). Numerous studies have revealed that *P. juliflora* has the ability to grow well on FA deposits (Rai et al., 2004; Kapur et al., 2002). *P. juliflora* is also reported as an excellent naturally growing species on abandoned FA deposits (Pandey et al., 2015b). *P. juliflora* is the most promising firewood species as a short rotation crop for the revegetation of FA deposits due to their fast growth rate, high tolerance, and high biomass productivity with good energy value.

### 10.3.7 Valuable timber tree

Though valuable timber trees do not yield significant wood on FA dumps, yet certain plant species like *D. sissoo, Tectona grandis, and Shorea robusta* were developed successfully on FA deposits (Juwarkar and Jambhulkar, 2008). *D. sissoo* has also been shown to grow well on barren FA landfills (Ram et al., 2008; Das et al., 2013). *Platanus occidentalis* and *Liquidambar styraciflua* are noted as valuable timber trees for a variety of products including furniture, cabinets, barrels, boxes, veneer, crates, and butcher blocks. *P. occidentalis* and *L. styraciflua* trees have been demonstrated well on abandoned alkaline FA lagoon and acidic FA lagoon, respectively (Carlson and Adriano, 1991). Integrated biomycophytoremediation technology can increase the wood productivity on FA deposits along with other benefits such as aesthetic appearance, habitat for wildlife, carbon sequestration, reduction in dust pollution along with noise, as well as substrate quality improvement toward soil development.

### 10.3.8 Trees in plywood industry

*Bombax ceiba, P. euphratica, and E. tereticornis* have been grown well on FA dumpsites (http://www.kuenvbiotech.org/casestudy/.htm; Das et al., 2013). These trees are being most frequently grown in agroforestry system due to their fast growth rate and good returns. Nowadays, agroforestry is a novel trend that assures good economic returns with no fear of crop failure within shorter time among the farming community with multiple benefits. Therefore growing trees for plywood industry in agroforestry is become more profitable than cereal crops. Thus the utilization of ever-increasing areas of FA dumps all over the world for biomass production for the plywood industries is an innovative approach that will reduce pressure on natural forest along with reduction in FA dust pollution. The wood of these trees are used mainly in plywood industry and also paper, matches, packaging, etc.

### 10.3.9 Paddy crops

Nutrient enriched FA has been tested all over the world and found beneficial to the modification of the physicochemical and biological properties of degraded soil (Pandey and Singh, 2010), which elevates micro- and macronutrients and porosity, optimizes soil pH, improves soil texture (by lowering water holding capacity, hydraulic conductivity, and bulk density), and reduces its water holding capacity (Pandey and Singh, 2010;

Ram and Masto, 2014). The next major area for FA utilization apart from construction is in biomass production with respect to crop growth and yield, which includes agroforestry, floriculture, and agriculture. In agriculture, FA has been broadly used as soil ameliorant to accelerate growth and yield of crops such as *Cicer arietinum* (Pandey et al., 2010), *Cajanus cajan* (Pandey et al., 2009a), *Oryza sativa* (Lee et al., 2006; Singh and Pandey, 2013), *Phaseolus vulgaris, Medicago sativa, Zea mays* (Wong and Wong, 1989), Lettuce (Lau and Wong, 2001), *Brassica parachinensis* and *Brassica chinensis* (Wong and Wong, 1990), *Brassica campestris* (Jayasinghe and Tokashiki, 2012), *Brassica* spp. (Kim et al., 1997), and many more (Pandey et al., 2009b). Moreover, alkaline FA has also demonstrated to be helpful for neutralizing acidic soils (Taylor and Schuman, 1988), thus facilitating the utilization of degraded lands toward agriculture.

Paddy is a single crop on which maximum researchers worked all over the world for the utilization of nutrient enriched FA for its growth and yield. In most of the studies, FA is reported as beneficial soil ameliorant for increasing the paddy yield as well as improving the paddy field. But the presence of heavy metals is the negative part of FA, which limits its agricultural use. Though previous studies reveal that lower dose of FA can bring success and proves it as an effective soil improver but in long term, it may produce harmful effects through food chain (Singh et al., 2008). In the United States, less than 4% of produced FA (68 million tons) is used in agriculture sector owing to the fear of heavy metal contamination (Dzantor et al., 2013). The problem of heavy metal could be avoided by the integrated use of FA with organic waste or organic fertilizer such as farmyard manure, press mud, green manure, cow dung manure, vermicompost, etc., which stabilize heavy metals in the soil system as well as improve physicochemical properties of paddy soil and paddy yield. Therefore paddy is only one crop of agriculture systems, which can be used safely with FA application.

## 10.4 Pros and cons of phytoproducts produced from fly ash and their SWOT analysis

The most important advantage of phytoproduct timber produced from FA dumps can be used to make furniture and household accessories. Thus there is no danger for accumulated toxic metals in timber and can be preserved for a long time more than 100 years (Pandey and Singh, 2012).

In addition, heavy metal accumulated biomass can be used as bioenergy through using the process of compaction (pelletization), pyrolysis, and incineration (Sas-Nowosielska et al., 2004). The thermal degradation (combustion, gasification) of contaminated biomass into manageable volumes of metal-enriched ash is a viable and promising disposal option, because the energy production affects the rate of return, and the toxic elements are enriched in a small amount of solid remains, depending on that the ash content in the plant may vary between 1% and 2% (Kovacs and Szemmelveisz, 2017). Likewise, such heavy metal accumulated woody biomass produced from FA dumps can be used as bioenergy through thermal degradation process. Phytomining is a technique for metal recovery from plant biomass. This skill is still in its infancy stage and may be used a potential way for mineral industry. In this method, a metal-hyperaccumulating plant species is used to extract metal from the contaminated sites and accumulate it in plant biomass. The burning of harvested biomass produces a bio-ore. Phytomining of some elements such as Ni, gold, and thallium has been carried out successfully (Sheoran et al., 2009). Biosorption is one more technique that can be used for metal recovery from the biomass during the composting process (Volesky and Holan, 1995; Wang and Chen, 2006). Successful heavy metal removal from FA from the combustion of municipal solid waste was studied by some researchers (Ferreira et al., 2005, Liao et al., 2014). In this direction, Keller et al. (2005) also achieved evaporative separation of heavy metal from the leaves of the Cd and Zn hyperaccumulator *Thlaspi caerulescens* and *S. viminalis* that were grown on contaminated site. The temperature was raised from 250°C to 9500°C using a laboratory scale reactor, and the evaporation/volatilization of heavy metals was examined by a thermodesorption spectrometer. They noticed that gasification under reducing condition was a better method for increase in volatilization cum recovery of the selected heavy metals than incineration under oxidizing conditions (Keller et al., 2005).

The main disadvantage of phytoremediation is the concern of toxic metal entering into the food chain from contaminated plant parts to human through grazers. A number of plant species have been recommended for phytoremediation programs; however, they are palatable by animals, livestock, or grazers. Therefore this type of polluted sites should be banned/fenced or made unreachable to livestock or grazers. In the case of fencing of polluted sites, it is a very costly method and mostly unacceptable from economic point of view. This problem can be avoided by using unpalatable plant species in phytoremediation programs (Pandey, 2015). In this regard, a number of unpalatable plant species such as *V. zizanioides* (Verma et al., 2014), *Ipomoea carnea* (Pandey, 2012), *Jatropha curcus* (Jamil et al., 2009; Pandey et al., 2012a), *R. communis* (Pandey, 2013), and *Typha latifolia* (Pandey et al., 2014) have been suggested for phytoremediation of heavy metals enriched FA deposits or FA contaminated soil. As we know very well that heavy metals are nondegradable, one very common question arises from the scientific society about the fate of heavy metals in accumulated plant biomass. Harvested biomass from such contaminated sites can be used for bioenergy generation through thermal degradation (Sas-Nowosielska et al., 2004). Biomass crops offer a carbon neutral fuel to society and would be helpful to mitigate the global warming by reducing the carbon emissions. Following the Short Rotation Intensive Culture system into the FA dumpsites could further reduce global carbon emissions from fossil fuels by 20% (Dixon et al., 1994). This would also lessen the sulfur and nitrogen emissions that are liable for acid rain by around 1% and 0.5% $\pm$ 1.5%, respectively (Dixon et al., 1994). These reductions may be helpful to meet the various air quality standards to power producers. However, all these initiatives, sustainability point of view, will need subsidies and technology transfer to make a successful venture. Decreasing the cost of biomass fuel production, as compared with presently used fossil fuels, is significant to power company and consumer acceptance (Abrahamson et al., 1998). The main hidden drawback is the lack of awareness on the benefits of FA usage among producers and end users, as well as vital instructions from governments are often absent in many nations. Even though giving importance to increase coal-based power production, authorities neglect how to use the massive ash positively. Forthcoming instructions must include innovative policies so that every coal-based thermal power station must utilize ever-increasing FA dumping sites toward bioenergy production through using of perennial grasses or fast growing trees. The utilization of FA amended soil toward agriculture must be examined on bench scale and field scale. It is urgently needed to inspire the use of FA deposits through reclamation for bioenergy production. For this program, thermal power stations, practitioners, farmers, scientists, stakeholders, and other possible users must be involved to promote generation of livelihood by framing instructions and guidelines. Table 10.2 shows a SWOT analysis of the phytomanagement of FA with economic returns.

**Table 10.2** SWOT analysis of phytomanagement of FA with economic returns.

| Strength | References |
| --- | --- |
| Nutrient enriched fly ash is a potential source of soil amender to enhance crop production in barren soil system. | Pandey et al. (2009a), Pandey and Singh (2010), Pandey (2017), Sheoran et al. (2014), and Haynes (2009) |
| Fly ash may improve the physicochemical and biological properties of the degraded soils. It can be used as a liming agent. | Pandey and Singh (2010) and Ram and Masto (2014) |
| Integrated phytoremediation would be very effective in planting of economically valuable plants especially biofuel crops and reducing level of toxic metals from fly ash deposits. | Pandey et al. (2016a), Pandey (2017), Dzantor et al. (2013), and Pandey (2013) |
| Application of fly ash on the growth and yield of agricultural crops towards its safe utilization. | Pandey et al. (2009a), Singh et al. (2011), and Singh and Pandey (2013) |
| Reduction in $CO_2$ emission of thermal power plants through sequestering carbon in ash vegetation system of fly ash deposits. | Pandey et al. (2016a,b,c) and Pandey and Singh (2010) |
| Aesthetically pleasant landscape, continuous biomass yield and its utilization, bioaccumulation of toxic metals, wildlife habitat, and evergreen cover development against climate change issue would be very much acceptable and would be much admired by all. | Pandey et al. (2015a,b,c, 2016a,b,c) |
| Heavy metal loaded biomass can be used for bioenergy generation through using gasification, thermochemical conversion, and pyrolysis method. The produced heavy metal loaded biomass may be from waste dumpsites (fly ash deposits) or polluted sites. | Zhang et al. (2010), Verhulst et al. (1996), Shao et al. (2012), and Pandey (2017) |
| **Weakness** | **References** |
| Presence of toxicants especially heavy metals in fly ash is main weakness. Therefore it is needed to reduce toxic metal load of fly ash. | Inthasan et al. (2002) and Pandey and Singh (2010) |

*(Continued)*

**Table 10.2** (Continued)

| Weakness | References |
| --- | --- |
| Self-cementing property of fly ash is another major problem, which hinders vegetation development on fly ash deposits due to its pozzolanic activity that reduce aeration, water infiltration, and root penetration. The mixture of organic material into ash deposits can support to break up cemented or compacted layers. | Haynes (2009) and Pandey and Singh (2010) |
| Absence of organic carbon and nitrogen in fly ash is also a problem but leguminous crop-based plantation can add both the nutrient in fly ash deposits. | Pandey and Singh (2010) |
| Fast growing perennial grasses would be vital to reduce the heavy metal load of fly ash deposits. | Pandey et al. (2012, 2015) |
| Accumulated heavy metals in the biomass may hamper its combustion. Therefore it is necessary to eliminate heavy metals from biomass before combustion. | Kovacs and Szemmelveisz (2017) and Sas-Nowosielska et al. (2004) |
| Awareness about biomass production in fly ash amended soil is lacking among farmers and practitioners. | Pandey et al. (2009b) and Pandey and Singh (2012) |
| Limited practical knowledge as a phytoremediation action in fly ash deposits; some case studies are accessible. | Haynes (2009), Pandey et al. (2009b), and Pandey and Singh (2012) |

| Opportunities | References |
| --- | --- |
| Being nutrient rich, fly ash has vast potential for improving the soil system and food productivity towards socioenvironmental sustainability. | Pandey and Singh (2010) and Ram and Masto (2014) |
| Several opportunities exist for improving fly ash deposits by using different type of organic amendments such as vermicompost, cow dung, press mud, farmyard manure, sewage sludge, etc. | Rautaray et al. (2003), Sheoran et al. (2014), and Ram and Masto (2014) |
| Phytoremediation along with biological amendments would greatly enhance substrate quality of fly ash deposits. | Pandey et al. (2009b) and Pandey (2012) |

(*Continued*)

**Table 10.2** (Continued)

| Opportunities | References |
|---|---|
| Mycorrhizoremediation and bioremediation of fly ash would greatly remediate the fly ash deposits. | Channabasava et al. (2014) |
| A number of economically valuable crops can be grown on fly ash deposits toward economic returns. | Verma et al. (2014), Pandey (2015), and Pandey and Singh (2012) |
| Incineration (or coincineration) has been recognized as the most feasible, potential, economically acceptable, and environmentally sound option for the management of the heavy metal-enriched biomass produced from phytoremediation. | Keller et al. (2005), Haynes (2009), and Sas-Nowosielska et al. (2004) |
| Aromatic crops or grasses have been suggested as a promising option for phytomanagement of FA deposits with economic returns as well as it could be cultivated on heavy metal polluted sites, without causing any major risks of metal transfer from soil to essential oil and alterations in essential oil quality. The main reason is the use of steam distillation process to obtain essential oil. Heavy metals persist in the extracted plant residues during the oil extraction process (steam distillation). Finally, essential oil is free from heavy metal risk. | Verma et al. (2014), Pandey and Singh (2015), Zheljazkov and Nielsen (1996a,b), and Zheljazkov et al. (2006) |
| The most recent biomass combustion and cofiring system has been attained that disposes the biomass readily and produces heat/electricity as biomass to energy with least emission of harmful gases. The volatile toxic metals can be captured from the flue gas, while the nonvolatile toxic metals stay in the ashes. Therefore the produced biomass from phytoremediation has potential opportunity as biomass to energy. | James et al. (2012) and Singh and Shukla (2014) |
| Low-volume ash is generated with an efficient boiler system that can be returned back to dumping site again to top up back the nutrients lost during the growth of plants. | Kovacs and Szemmelveisz (2017) and Keller et al. (2005) |

| Threats | References |
| --- | --- |
| FA deposit has high risk of containing various toxic metals and other contaminants, which may cause serious damage to wild animals, human, and environment. | Pandey et al. (2011), Haynes (2009), and Ram and Masto (2014) |
| Biomagnification of heavy metals through food chain in cattle or human being is a risk if appropriate attention is not taken. It can be avoided if we select unpalatable plant species for phytoremediation program. | Gall et al. (2015) and Pandey (2015) |
| A study suggested that additional care (regular monitoring) should be undertaken to avoid metal toxicity to local residents, if edible fruit trees are considered for the afforestation programs on FA dumps via phytoremediation. Energy or timber plant-based remediation of FA dumps is a viable and promising approach than using fruit tree plants. | Pandey and Mishra (2018) |
| The threat of heavy metal accumulation in the edible crop during the application of fly ash can be minimized or reduced by using organic amendments. | Ram and Masto (2014) and Pandey and Singh (2010) |
| The pH and texture of FA dump as well as the presence of pollutants in FA dump and its concentrations must be within the limits of plant tolerance. In highly hostile FA deposits, produced from very low-grade coal, native microbes may be inhibited to assist pollutant degradation or support plant growth. | Koptsik (2014) |
| Ash dumps reduce heat transfer and may also result in severe corrosion at high temperatures. | Kovacs and Szemmelveisz (2017) |
| Disposal of ash produced from incineration of heavy metal loaded biomass and produced from fly ash deposits during phytoremediation will be threat again. | Košnář et al. (2016) and Freire et al. (2015) |

Source: Modified from Roy, M., Roychowdhary, R., Mukherjee, P., 2018. Remediation of fly ash dumpsites through bioenergy crop plantation and generation: A review. Pedosphere, 28 (4), 561−580.

## 10.5 Conclusion and future prospects

In conclusion, this chapter highlights on the phytomanagement of FA with ecological and socioeconomic sustainability issues, which has both opportunities and constraints. The ecological sustainability can be achieved by the choosing fast growing native species for the revegetation of FA deposits. In this regard, fast growing perennial native grasses is a best choice of plant species for initiating green cover on hostile condition of FA dumps. Being tolerant in nature, grasses have ability to grow on FA deposits without using topsoil and manure. However, the time and cost of phytoremediation of FA deposits could be reduced by using minimum topsoil and less manure. In addition, mixed plantation of grasses and legumes improves the ash substrate, stabilizes toxic elements, prevents erosion, checks leaching of toxic elements, increases litter turnover as organic matter, and produces biomass as a source of income, and it will offer revenue in future for local people. The physicochemical and biological characteristics of FA deposits should be considered before initiating green cover on such site. It will help for selecting the type of ameliorant, its application rate, and type of plant species, mainly perennial grasses, to be used. In this regard, vetiver has been recommended as a potential and promising grass species for revegetation of exposed FA surface with economic returns and can be assisted by using suitable ameliorant along with Jute Geotextile sheet, which provide a quick green cover over exposed FA lessening the dust pollution of ash particles and at the same time reduces the leaching of toxic metals. FA deposits can be rehabilitated easily by the use of organic manure, mycorrhizal inoculation, jute geotextiles, and tolerant socioeconomically plants such as vetiver through ecological engineering without using precious topsoil. The geotextiles and organic manure act as a host for the beneficial microbes to improve ash substrate and support the speedy growth of desired plants to enhance the green cover development toward self-sustaining ecosystems. Mycorrhizal inoculation is also helpful to plants for its nutrient availability.

The socioeconomic sustainability could be accomplished by introducing economically valuable plant species, at later phases of restoration, which generate biomass for paper and plywood industries, biofuel (biodiesel and bioethanol), firewood, timber, and aromatic essential oils, should be practiced efficiently in abandoned FA deposits. It is also dependent on the costs of phytoproduct production from FA deposits, market and consumer demands, and governmental initiatives to introduce market and creating awareness about the prospects of using these phytoproducts. Finally, government should take strong actions to remove the constraints and provide incentives for using phytoproducts. Moreover, a regular monitoring should be taken in the phytomanagement of FA dumps. Especially the analysis of bioavailable toxic metals in FA and their accumulation in the various parts of plants will determine to safe use of the phytoproducts produced from the FA dumpsites or FA amended soil. The ameliorated soil with FA should also be checked time to time for their quality and fertility status, because continuous FA incorporation in soil may create heavy metal toxicity.

# References

Abrahamson, L.P., Robison, D.J., Volk, T.A., White, E.A., Neuhauser, E.F., Benjamin, W.H., et al., 1998. Sustainability and environmental issues associated with willow bioenergy development in New York (U.S.A.). Biomass Bioenerg. 15, 17–22.

Adholeya, A., Sharma, M.P., Bhatia, N.P., Tyagi, C., 1997. Mycorrhizae biofertilzer: a tool for reclamation of wasteland and bioremediation. In: National Symposium on Microbial Technologies for Environmental Management and Resource Recovery. The Energy and Resources Institute, New Delhi, India, pp. 58–63.

Adholeya, A., Bhatia, N.P., Kanwar, S., Kumar, S., 1998. Fly ash source and substrate for growth of sustainable agro-forestry system. In: Proceeding of Regional Workshop Cum Symposium on Fly Ash Disposal and Utilization. Kota Thermal Power Station, RSEB, Kota, Rajasthan, India.

Anonymous, 1997. Healing degraded land [J]. INBAR Magazine 5 (3), 40–45.

Arru, L., Rognoni, S., Baroncini, M., Bonatti, P.M., Perata, P., 2004. Copper localization in *Cannabis sativa* L. grown in a copper-rich solution. Euphytica 140, 33–38.

Awoyemi, O.M., Adeleke, E.O., 2017a. Bioethanol production from switchgrass grown on coal fly ash-amended soil. World J. Agric. Res. 5 (3) 1, 47–155. Available from: https://doi.org/10.12691/wjar-5-3-4.

Awoyemi, O.M., Dzantor, E.K., 2017b. Fate and impacts of priority pollutant metals in coal fly ash-soil-switchgrass plant mesocosms. Coal Combust. Gasif. Prod. 9, 42–51. Available from: https://doi.org/10.4177/CCGP-D-14-00004.1.

Awoyemi, O.M., Dzantor, E.K., 2017c. Toxicity of coal fly ash (CFA) and toxicological response of switchgrass in mycorrhiza-mediated CFA-soil admixtures. Ecotoxicol. Environ. Saf. 144, 438–444.

Babu, A.G., Reddy, M.S., 2011. Dual inoculation of arbuscular mycorrhizal and phosphate solubilizing fungi contributes in sustainable maintenance of plant health in fly ash ponds. Water Air Soil Pollut. 219, 3–10.

Bagchi, S.S., 2013. Study of *Jatropha Curcas* growth in fly ash amended soil. Univers. J. Environ. Res. Technol. 3 (3), 364–374.

Bilski, J., McLean, K., McLean, E., Soumaila, F., Lander, M., 2011. Revegetation of coal ash by selected cereal crops, and trace elements accumulation by plant seedlings. Int. J. Environ. Sci. 1, 1040–1053.

Bryan, A.L., Hopkins, W.A., Parikh, J.H., Jackson, B.P., Unrine, J.M., 2012. Coal fly ash basins as an attractive nuisance to birds: parental provisioning exposes nestlings to harmful trace elements. Environ. Pollut. 161, 170–177.

Carlson, C.L., Adriano, D.C., 1991. Growth and elemental content of two tree species growing on abandoned coal fly ash basins. J. Environ. Qual. 20, 581−587.

Central Electricity Authority (CEA), 2014. Report on Fly Ash Generation at Coal/Lignite Based Thermal Power Stations and Its Utilization in the Country for the Year 2013−14. CEA, New Delhi.

Čermák, P., 2008. Forest reclamation of dumpsites of coal combustion by-products (ccb). J. For. Sci. 54, 2008(6), 273−280.

Channabasava, A., Lakshman, H.C., Muthukumar, T., 2014. Fly ash mycorrhizoremediation through *Paspalum scrobiculatum* L., inoculated with *Rhizophagus fasciculatus*. C. R. Biol. Available from: https://doi.org/10.1016/j.crvi.2014.11.002.

Chaudhary, D.R., Ghosh, A., 2013. Bioaccumulation of nutrient elements from fly ash-amended soil in *Jatropha curcas* L. a biofuel crop. Env. Monit. Assess. 185 (8), 6705−6712. Available from: https://doi.org/10.1007/s10661-013-3058-x.

Chen, J., Li, Y., 2006. Coal fly-ash as an amendment to container substrate for Spathiphyllum production. Bioresour. Technol. 97, 1920−1926.

Cheung, K.C., Wong, J.P.K., Zhang, Z.Q., Wong, J.W.C., Wong, M.H., 2000. Revegetation of lagoon ash using the legume species *Acacia auriculiformis* and *Leucaena leucocephala*. Environ. Pollut. 109, 75−82.

Chou, S.F.J., Chou, M.I.M., Stucki, J.W., Warnock, D., Chemler, J.A., Pepple, M.A., 2005. Plant growth in sandy soil/compost mixture and commercial peat moss both amended with Illinois coal fly ash. In: Proceedings of the World of Coal Ash Conference, Lexington, KE, pp. 1−7.

Citterio, S., Santagostino, A., Fumagalli, P., Prato, N., Ranalli, P., Sgorbati, S., 2003. Heavy metal tolerance and accumulation of Cd, Cr and Ni by *Cannabis sativa* L. Plant Soil 256, 243−252.

Das, M., Agarwal, P., Singh, R., Adholeya, A., 2013. A study of abandoned ash ponds reclaimed through green cover development. Int. J. Phytoremediation 15 (4), 320−329.

Dixon, R.K., Brown, S., Houghton, R.A., Solomon, A.M., Trexler, M.C., Wisniewski, J., 1994. Carbon pools and flux of global forest ecosystems. Science 263, 185−186.

Dzantor, E.K., Pettigrew, H., Adeleke, E., Hui, D., 2013. Use of fly ash as soil amendment for biofuel feedstock production with concomitant disposal of waste accumulations. In: 2013 World of Coal Ash (WOCA) Conference, Lexington, KY, pp. 22−25.

Dzantor, E., Adeleke, E., Kankarla, V., Ogunmayowa, O., Hui, D., 2015. Using coal fly ash in agriculture: combination of fly ash and poultry litter as soil amendments for bioenergy feedstock production. CCGP 7, 33−39. Available from: https://doi.org/10.4177/CCGPD-15-00002.1.

Enkhtuya, B., Poschl, M., Vosatka, M., 2005. Native grass facilitates mycorrhizal colonisation and P uptake of tree seedlings in two anthropogenic substrates. Water Air Soil Pollut. 166, 217−236.

Ferreira, C., Ribeiro, A., Ottosen, L., 2005. Effect of major constituents of MSW fly ash during electrodialytic remediation of heavy metals. Sep. Sci. Technol. 40, 2007−2019.

Freire, M., Lopes, H., Tarelho, L.A., 2015. Critical aspects of biomass ashes utilization in soils: composition, leachability, PAH and PCDD/F. Waste Manag. 46, 304−315.

Gall, J.E., Boyd, R.S., Rajakaruna, N., 2015. Transfer of heavy metals through terrestrial food webs: a review. Environ. Monit. Assess. 187, 201.

Gautam, M., Pandey, D., Agrawal, M., 2017. Phytoremediation of metals using lemongrass (*Cymbopogon citratus* (D.C.) Stapf.) grown under different levels of red mud in soil amended with biowastes. Int. J. Phytoremed. 12 (6), 555−562.

Gomes, H.I., 2012. Phytoremediation for bioenergy: challenges and opportunities. Environ. Technol. Rev. 1 (1), 59−66. Available from: https://doi.org/10.1080/09593330.2012.696715.

Haynes, R.J., 2009. Reclamation and revegetation of fly ash disposal sites-challenges and research needs. J. Env. Manage. 90, 43—53.

Ho, W.M., Ang, L.H., Lee, D.K., 2008. Assessment of Pb uptake, translocation and immobilization in kenaf (*Hibiscus cannabinus* L.) for phytoremediation of sand tailings. J. Environ. Sci. 20, 1341—1347.

Hrynkiewicz, K., Baum, C., Niedojadło, J., Dahm, H., 2009. Promotion of mycorrhiza formation and growth of willows by the bacterial strain Sphingomonas sp. 23L on fly ash. Biol. Fertil. Soil. 45, 385—394. Available from: https://doi.org/10.1007/s00374-008-0346-7.

Huang, S.H., 2009. Study on immobilization of heavy metals polluted soils in Dabao Mountain by amendments and their mechanism. M. S. Thesis, Sun Yat-Sen University.

Inthasan, J., Hirunburana, N., Herrmann, L., Stahr, K., 2002. Effects of fly ash applications on soil properties, nutrient status and environment in Northern Thailand. In: 17th World Congress of Soil Science, Paper no. 249, 14—21 August, Thailand.

Israila, Y.Z., Bola, A.E., Emmanuel, G.C., Ola, I.S., 2015. The effect of application of EDTA on the phytoextraction of heavy metals by *Vetivera zizanioides*, *Cymbopogon citrates* and Helianthus annus. Int. J. Environ. Monit. Anal. 3, 38—43.

Jambhulkar, H.P., Juwarkar, A., 2009. Assessment of bioaccumulation of heavy metals by different plant species grown on fly ash dump. Ecotoxicol. Environ. Saf. 72 (4), 1122—1128.

James, A.K., Thring, R.W., Helle, S., Ghuman, H.S., 2012. Ash management review— applications of biomass bottom ash. Energies 5, 3856—3873.

Jamil, S., Abhilash, P.C., Singh, N., Sharma, P.N., 2009. Jatropha curcas: a potential crop for phytoremediation of coal fly ash. J. Hazard. Mater. 172, 269—275.

Jayasinghe, G.Y., Tokashiki, Y., 2012. Influence of coal fly ash pellet aggregates on the growth and nutrient composition of *Brassica campestris* and physicochemical properties of greysoils in Okinawa. J. Plant Nutr. 35, 453—447.

Juwarkar, A.A., Jambhulkar, H.P., 2008. Restoration of fly ash dump through biological interventions. Environ. Monit. Assess. 139, 355—365.

Kapur, S.L., Shyam, A.K., Soni, R., 2002. Greener management practices ash mound reclamation. TERI Inf. Dig. Energ. Environ. 1 (4), 559—567.

Keller, C., Ludwig, C., Davoli, F., Wochele, J., 2005. Thermal treatment of metal-enriched biomass produced from heavy metal phytoextraction. Environ. Sci. Technol. 39, 3359—3367.

Khajanchi, L., Yadava, R.K., Kaurb, R., Bundelaa, D.S., Khana, M.I., Chaudharya, M., et al., 2013. Productivity essential oil yield, and heavy metal accumulation in lemon grass (*Cymbopogon flexuosus*) under varied wastewater—groundwater irrigation regimes. Ind. Crop. Prod. 45, 270—278.

Kim, B.J., Back, J.H., Kim, Y.S., 1997. Effect of fly ash on the yield of Chinese cabbage and chemical properties of soil. J. Korean Soc. Soil Sci. Fert. 30, 161e167.

Koptsik, G.N., 2014. Problems and prospects concerning the phytoremediation of heavy metal polluted soils: a review. Eurasian Soil. Sci. 47, 923.

Košňář, Z., Mercl, F., Perná, I., Tlustoš, P., 2016. Investigation of polycyclic aromatic hydrocarbon content in fly ash and bottom ash of biomass incineration plants in relation to the operating temperature and unburned carbon content. Sci. Total Environ. 563—564, 53—61.

Kostic, O., Mitrovic, M., Kneževic, M., Jaric, S., Gajic, G., Djurdj, L., et al., 2012. The potential of four woody species for the revegetation of fly ashdeposits from the 'Nikola Tesla-A' thermoelectric plant (Obrenovac Serbia). Arch. Biol. Sci. 64 (1), 145—158.

Kovacs, H., Szemmelveisz, K., 2017. Disposal options for polluted plants grown on heavy metal contaminated brownfield lands-a review. Chemosphere. 166, 8—20.

Kumar, A., Vajpayee, P., Ali, M.B., Tripathi, R.D., Singh, N., Rai, U.N., et al., 2002. Biochemical responses of *Cassia siamea* L. grown on coal combustion residue (fly-ash). B. Environ. Contam. Tox 68, 675—683.

Lau, S.S.S., Wong, J.W.C., 2001. Toxicity evaluation of weathered coal fly ash amended manure compost. Water Air Soil Pollut. 128, 243—254.

Lavania, U.C., Lavania, S., 2009. Sequestration of atmospheric carbon into horizons through deep-rooted grasses-vetiver grass model. Curr. Sci. 97, 618—619.

Lee, H., Ha, H.S., Lee, C.H., Lee, Y.B., Kim, P.J., 2006. Fly-ash effect on improving soil properties and rice productivity in Korean paddy soils. Bioresour. Technol. 97, 1490—1497.

Liao, W.P., Yang, R., Zhou, Z.X., Huang, J.Y., 2014. Electrokinetic stabilization of heavy metals in MSWI fly ash after water washing. Env. Prog. Sustain. Energ. 33, 1235—1241.

Lombard, K., O'Neill, M., Ulery, A., Mexal, J., Blake Onken, B., Forster-Cox, S., et al., 2011. Fly ash and composted biosolids as a source of Fe for hybrid poplar: a greenhouse study. Appl. Environ. Soil Sci. Article ID 475185, 11 pages. Available from: https://doi.org/10.1155/2011/475185.

Maiti, D., 2017. Reclamation of fly ash waste dumps by biological means—a scientific approach. NexGen Technologies for Mining and Fuel Industries (NxGnMiFu 2017) vol. 2, 1011—1022.

Maiti, S.K., Maiti, D., 2015. Ecological restoration of waste dumps by topsoil blanketing, coir-matting and seeding with grass—legume mixture. Ecol. Eng. 77, 74—84. Available from: https://doi.org/10.1016/j.ecoleng.2015.01.003.

Maiti, D., Prasad, B., 2017. Studies on colonisation of fly ash disposal sites using invasive species and aromatic grasses. J. Environ. Eng. Landsc. Manag. 25, 251—263. Available from: https://doi.org/10.3846/16486897.2016.1231114.

Malewar, G.U., Adsul, P.B., Ismail, S., 1998. Effect of different combinations of fly-ash and soil on growth attributes of forest and dryland fruit crops. Indian. J. For. 21 (2), 124—127.

Mitrovic, M., Jaric, S., Kostic, O., Gajic, G., Karadzic, B., Lola Djurdjevic, L., et al., 2012. Photosynthetic efficiency of fourwoody species growing on fly ash deposits of a Serbian 'Nicola Tesla-A' thermoeclectric plant. Pol. J. Environ. Stud. 21 (5), 1339—1347.

Mohan, S., 2011. Growth of biodiesel plant in fly ash: a sustainable approach-response of *Jatropha curcas*, a biodiesel plant in fly ash amended soil with respect to pigment content and photosynthetic rate. Procedia Environ. Sci. 8, 421—425.

Neenan, M., Steinbeck, K., 1979. Calorific values for young sprouts of nine hardwood species. For. Sci. 25, 455—461.

Panda, D., Panda, D., Padhan, B., Biswas, M., 2018. Growth and physiological response of lemongrass (*Cymbopogon citratus* (D.C.) Stapf.) under different levels of fly ash-amended soil. Int. J. Phytoremediation 20 (6), 538—544. Available from: https://doi.org/10.1080/15226514.2017.1393394.

Pandey, V.C., 2012. Invasive species based efficient green technology for phytoremediation of fly ash deposits. J. Geochem. Explor. 123, 13—18.

Pandey, V.C., 2013. Suitability of *Ricinus communis* L. cultivation for phytoremediation of fly ash disposal sites. Ecol. Eng. 57, 336—341.

Pandey, V.C., 2015. Assisted phytoremediation of fly ash dumps through naturally colonized plants. Ecol. Eng. 82, 1—5.

Pandey, V.C., 2016. Phytoremediation efficiency of *Eichhornia crassipes* in fly ash pond. Int. J. phytoremediat. 18 (5), 450—452.

Pandey, V.C., 2017. Managing waste dumpsites through energy plantations. In: Bauddh, K., Singh, B., Korstad, J. (Eds.), Phytoremediation Potential of Bioenergy Plants. Springer, Singapore, pp. 371–386.

Pandey, V.C., Mishra, T., 2018. Assessment of *Ziziphus mauritiana* grown on fly ash dumps: prospects for phytoremediation but concerns with the use of edible fruit. Int. J. Phytoremediat. 20 (12), 1250–1256.

Pandey, V.C., Kumar, A., 2013. *Leucaena leucocephala*: an underutilized plant for pulp and paper production. Genet. Resour. Crop Evol 60, 1165–1171.

Pandey, A., Gupta, R., 2003. Fibre yielding plants of India: genetic resources, perspective for collection and utilization. Nat. Prod. Rad. 2, 194–204.

Pandey, V.C., Singh, J.S., Kumar, A., Tewary, D.D., 2010. Accumulation of heavy metals by chickpea grown in fly ash treated soil: effect on antioxidants. Clean Soil Air Water 38, 1116–1123.

Pandey, V.C., Singh, J.S., Singh, R.P., Singh, N., Yunus, M., 2011. Arsenic hazards in coal fly ash and its fate in Indian scenario. Resour. Conserv. Recy. 55 (9–10), 819–835.

Pandey, V.C., Singh, N., 2010. Impact of fly ash incorporation in soil systems. Agric. Ecosyst. Environ. 136, 16–27.

Pandey, V.C., Singh, B., 2012. Rehabilitation of coal fly ash basins: current need to use ecological engineering. Ecol. Eng. 49, 190–192.

Pandey, V.C., Singh, N., 2015. Aromatic plants versus arsenic hazards in soils. J. Geochem. Explor. 157, 77–80.

Pandey, V.C., Bajpai, O., 2019a. Phytoremediation: from theory toward practice. In: Pandey, V.C., Bauddh, K. (Eds.), Phytomanagement of Polluted Sites. Market Opportunitiesin Sustainable Phytoremediation. Elsevier, Amsterdam, The Netherlands, pp. 1–49.

Pandey, V.C., Souza-Alonso, P., 2019b. Market opportunities in sustainable phytoremediation. In: Pandey, V.C., Bauddh, K. (Eds.), Phytomanagement of Polluted Sites. Market Opportunitiesin Sustainable Phytoremediation. Elsevier, Amsterdam, The Netherlands, pp. 51–82.

Pandey, V.C., Abhilash, P.C., Upadhyay, R.N., Tewari, D.D., 2009a. Application of fly ash on the growth performance and translocation of toxic heavy metals within *Cajanus cajan* L. implication for safe utilization of FA for agricultural production. J. Hazard. Mater. 166, 255–259.

Pandey, V.C., Abhilash, P.C., Singh, N., 2009b. The Indian perspective of utilizing fly ash in phytoremediation, phytomanagement and biomass production. J. Environ. Manage. 90, 2943–2958.

Pandey, V.C., Singh, K., Singh, J.S., Kumar, A., Singh, B., Singh, R.P., 2012a. Jatropha curcas: a potential biofuel plant for sustainable environmental development. Renew. Sustain. Energ. Rev. 16, 2870–2883.

Pandey, V.C., Singh, K., Singh, R.P., Singh, B., 2012b. Naturally growing *Saccharum munja* on the fly ash lagoons: a potential ecological engineer for the revegetationand stabilization. Ecol. Eng. 40, 95–99.

Pandey, V.C., Singh, N., Singh, R.P., Singh, D.P., 2014. Rhizoremediation potential of spontaneously grown *Typha latifolia* on fly ash basins: study from the field. Ecol. Eng. 71, 722–727.

Pandey, V.C., Pandey, D.N., Singh, N., 2015a. Sustainable phytoremediation based on naturally colonizing and economically valuable plants. J. Clean. Prod. 86, 37–39.

Pandey, V.C., Prakash, P., Bajpai, O., Kumar, A., Singh, N., 2015b. Phytodiversity on fly ash deposits: evaluation of naturally colonized species for sustainable phytorestoration. Env. Sci. Pollut. Res. 22, 2776–2787.

Pandey, V.C., Bajpai, O., Pandey, D.N., Singh, N., 2015c. *Saccharum spontaneum*: anunderutilized tall grass for revegetation and restoration programs. Genet. Resour. Crop. Evol. 62 (3), 443–450.

Pandey, V.C., Bajpal, O., Singh, N., 2016a. Energy crops in sustainable phytoremediation. Renew. Sustain. Energ. Rev. 54, 58—73.

Pandey, V.C., Sahu, N., Behera, S.K., Singh, N., 2016b. Carbon sequestration in fly ash dumps: comparative assessment of three plant association. Ecol. Eng. 95, 198—205.

Pandey, S.K., Bhattacharya, T., Chakraborty, S., 2016c. Metal phytoremediation potential of naturally growing plants on fly ash dumpsite of Patratu thermal power station, Jharkhand, India. Int. J. Phytoremediation. 18, 87—93.

Pavlović, P., Mitrović, M., Djurdjević, L., 2004. An ecophysiological study of plants growing on the fly ash deposits from the "Nikola Tesla-A" thermal power station in Serbia. Environ. Manage. 33 (5), 654—663.

Pradhan, A., Dash, A.K., Mohanty, S.S., Das, S., 2015. Potential use of fly ash in floriculture: A case study on the photosynthetic pigments content and vegetative growth of *Tagetes erecta* (MARIGOLD). Eco. Environ. Conser. 21, AS369—AS376.

Rai, U.N., Pandey, K., Sinha, S., Singh, A., Saxena, R., Gupta, D.K., 2004. Revegetating fly-ash landfills with Prosopis juliflora L.: impact of different amendments & rhizobium inoculation. Environ. Int. 30, 293—300.

Raj, S., Mohan, S., 2016. Impact on proline content of Jatropha curcas in fly ash amended soil with respect to heavy metals. Int. J. Pharm. Pharm. Sci. 8 (5), 244—247.

Ram, L.C., Masto, R.E., 2014. Fly ash for soil amelioration: a review on the influence of ash blending with inorganic and organic amendments. Earth Sci. Rev. 128, 52—74.

Ram, L.C., Srivastava, N.K., Singh, G., 2005. Reclamation of wasteland/ob dump and abandoned ash pond through fly ash and biological amendments In: Proc. Fly ash India Natl. Sem. Bus. Meet. on use of fly ash in Agric., New Delhi, India, pp. 37—49.

Ram, L.C., Jha, S.K., Tripathi, R.C., Masto, R.E., Selvi, V.A., 2008. Remediation of fly ash landfills through plantation. Remediation 18, 71—90.

Rautaray, S.K., Ghosh, B.C., Mittra, B.N., 2003. Effect of fly ash, organic wastes and chemical fertilizers on yield, nutrient uptake, heavy metal content and residual fertility in a rice—mustard cropping sequence under acid lateritic soils. Bioresour. Technol. 90, 275—283.

Rethman, N.F.G., Reynolds, K.A., Kruger, R.A., 1999. Crop responses to SlASH (Mixture of sewage sludge, lime and fly ash) as influenced by soil texture, acidity and fertility. In: Proc. International Ash Utilization Symposium, Center for Applied Research, University of Kentucky, Lexington, KY, pp. 1—11.

Riekerk, H., 1984. Coal-ash effects on fuelwood production and runoff water quality. South. J. Appl. For. 8, 99—102.

Roy, M., Roychowdhary, R., Mukherjee, P., 2018. Remediation of fly ash dumpsites through bioenergy crop plantation and generation: A review. Pedosphere 28 (4), 561—580.

Sas-Nowosielska, A., Kucharski, R., Malkowski, E., Pogrzeba, M., Kuperberg, J.M., Kryński, K., 2004. Phytoextraction crop disposal—an unsolved problem. Env. Pollut. 128, 373—379.

Selvam, A., Mahadevan, A., 2002. Distribution of mycorrhizas in an abandoned fly ash pond and mined sites of Neyveli Lignite Corporation, Tamil Nadu, India. Basic. Appl. Ecol. 3, 277—284.

Shao, Y., Wang, J., Preto, F., Zhu, J., Xu, C., 2012. Ash deposition in biomass combustion or co-firing for power/heat generation. Energies 5, 5171—5189.

Sharma, M.P., Adholeya, A., 2005. Demonstration of Mycorrhizal Technology for Reclamation and utilization of fly ash in Agriculture and Plantation Activities In: Proc. Fly ash India Natl. Sem. Bus. Meet. on use of fly ash in Agric., New Delhi, India, pp. 58—68.

Sheoran, H.S., Duhan, B.S., Kumar, A., 2014. Effect of fly ash application on soil properties: a review. J. Agr. Nat. Resour. Manag. 1, 98—103.

Sheoran, V., Sheoran, A.S., Poonia, P., 2009. Phytomining: a review. Miner. Eng. 22, 1007.

Siddiqui, S., Ahmad, A., Hayat, S., 2004. The fly ash influenced the heavy metal status of the soil and the seeds of sunflower—a case study. J. Environ. Biol. 25, 59—63.

Singh, A., Sharma, R.K., Agrawal, S.B., 2008. Effects of fly ash incorporation on heavy metal accumulation, growth and yield responses of Beta vulgaris plants. Bioresour. Technol. 99, 7200—7207.

Singh, J.S., Pandey, V.C., 2013. Fly ash application in nutrient poor agriculture soils: impact on methanotrophs population dynamics and paddy yields. Ecotoxicol. Environ. Saf. 89, 43—51.

Singh, R., Shukla, A., 2014. A review on methods of flue gas cleaning from combustion of biomass. Renew. Sustain. Energ. Rev. 29, 854—864.

Singh, S.N., Kulshreshtha, K., Ahmad, K.J., 1997. Impact of fly ash soil amendment on seed germination, seedling growth and metal composition of Vicia faba L. Ecol. Eng. 9, 203—208.

Singh, V.K., Behal, K.K., Rai, U.N., 2000. Comparative study on the growth of mulberry (Morus alba) plant at different levels of fly ash amended soil. Biol. Mem. 26, 1—5.

Singh, R., Singh, D.P., Kumar, N., Bhargava, S.K., Barman, S.C., 2010. Accumulation and translocation of heavy metals in soil and plants from fly ash contaminated area. J. Environ. Biol. 31, 421—430.

Singh, J.S., Pandey, V.C., Singh, D.P., 2011. Coal fly ash and farmyard manure amendments in dry-land paddy agriculture field: effect on N-dynamics and paddy productivity. Appl. Soil Ecol. 47, 133—140.

Stevens, G., Dunn, D., 2004. Fly ash as a liming material for cotton. J. Environ. Qual. 33 (1), 343—348.

Taylor Jr., E.M., Schuman, G.E., 1988. Fly-ash and lime amendment of acidic coal spoil to aid revegetation. J. Environ. Qual. 17, 120—124.

Técher, D., Laval-Gilly, P., Bennasroune, A., Henry, S., Martinez-Chois, C., D'Innocenzo, M., et al., 2012. An appraisal of Miscanthus x giganteus cultivation for fly ash revegetation and soil Restoration. Ind. Crop. Prod. 36, 427—433.

Tripathi, R.D., Yunus, M., Singh, S.N., Singh, N., Kumar, A., 2000. Final technical report—reclamation of fly-ash landfills through successive plantation, soil amendments and or through integrated biotechnological approach. Directorate of Environment, UP, India.

US EPA, 2007. Human and Ecological Risk Assessment of Coal Combustion Wastes. US EPA (draft).

Vajpayee, P., Rai, U.N., Choudhary, S.K., Tripathi, R.D., Singh, S.N., 2000. Managementof fly-ash landfills with Cassia surattensis Burm: a case study. Bull. Environ. Contam. Toxicol. 65, 675—682.

Verhulst, D., Buekens, A., Spencer, P.J., Eriksson, G., 1996. Thermodynamic behavior of metal chlorides and sulfates under the conditions of incineration furnaces. Env. Sci. Technol. 30, 50—56.

Verma, S.K., Singh, K., Gupta, A.K., Pandey, V.C., Trevedi, P., Verma, S.K., et al., 2014. Aromatic grasses for phytomanagement of coal fly ash hazards. Ecol. Eng. 73, 425—428.

Ververis, C., Georghiou, K., Danielidis, D., Hatzinikolaou, D.G., Santas, P., Santas, R., et al., 2007. Cellulose, hemicelluloses, lignin and ash content of some organic materials and their suitability for use as paper pulp supplements. Bioresour. Technol. 98 (2), 296—301. Available from: https://doi.org/10.1016/j.biortech.2006.01.007.

Volesky, B., Holan, Z.R., 1995. Biosorption of heavy metals. Biotechnol. Prog. 11, 235—250.

Vurayai, R., Nkoane, B., Moseki, B., Chaturvedi, P., 2017. Phytoremediation potential of Jatropha curcas and Pennisetum clandestinum grown in polluted soil with and without coal fly ash: a case of BCL Cu/Ni mine, Selibe-Phikwe, Botswana. J. Biodivers. Environ. Sci. 10 (5), 193−206.

Wang, J.L., Chen, C., 2006. Biosorption of heavy metals by Saccharomyces cerevisiae: a review. Biotechnol. Adv. 24, 427−451.

Wong, M.H., Wong, J.W.C., 1989. Germination and seedling growth of vegetable crops in fly-ash amended soils. Agric. Ecosyst. Environ. 26, 23−35.

Wong, J., Wong, M.H., 1990. Effects of fly ash on yields and elemental composition of two vegetables, *Brassica parachinensis* and *B. chinensis*. Agric. Ecosyst. Environ. 30 (3−4), 251−264.

Wu, F., Bi, Y.L., Wong, M.H., 2009. Dual inoculation with an arbuscular mycorrhizal fungus and rhizobium to facilitate the growth of alfalfa on coal mine substrates. J. Plant Nutr. 32 (5), 755−771. Available from: https://doi.org/10.1080/01904160902787867.

Yang, B., Zhou, M., Shu, W.S., Lan, C.Y., Ye, Z.H., Qiu, R.L., et al., 2010. Constitutional tolerance to heavy metals of a fiber crop, ramie (*Boehmeria nivea*), and its potential usage. Environ. Pollut. 158, 551−558.

Yang, Y.X., Lu, H.L., Zhan, S.S., Deng, T.H., Lin, Q.Q., Wang, S.Z., et al., 2013. Using kenaf (*Hibiscus cannabinus*) to reclaim multi-metal contaminated acidic soil. J. Appl. Ecol. 24 (3), 832−838 (Chinese).

Zevenbergen, C., Bradley, J.P., Shyam, A.K., Jenner, H.A., Platenburg, R.J.P.M., 2000. Sustainable ash pond development in India−a resource for forestry and agriculture. Waste Mater. 1, 533−540.

Zhang, L., Xu, C., Champagne, P., 2010. Overview of recent advances in thermochemical conversion of biomass. Energ. Convers. Manage. 51, 969−982.

Zheljazkov, V.D., Nielsen, N.E., 1996a. Studies on the effect of heavy metals (Cd, Pb, Cu, Mn, Zn and Fe) upon the growth, productivity and quality of lavender (*Lavandula angustifolia* Mill) production. J. Essent. Oil Res. 8, 259−274.

Zheljazkov, V.D., Nielsen, N.E., 1996b. Effect of heavymetals on peppermint and cornmint. Plant Soil 178, 59−66.

Zheljazkov, V.D., Craker, L.E., Xing, B., 2006. Effects of Cd, Pb, and Cu on growth and essential oil contents in dill, peppermint, and basil. Environ. Exp. Bot. 58, 9−16.

## Further reading

Kostić, O., Jarić, S., Gajić, G., Pavlović, D., Pavlović, M., Mitrović, M., et al., 2018. Pedological properties and ecological implications of substrates derived 3 and 11 years after the revegetation of lignite fly ash disposal sites in Serbia. Catena 163, 78−88.

# Index

*Note:* Page numbers followed by "*f*" and "*t*" refer to figures and tables, respectively

Printed in the United States
By Bookmasters